EVOLUTION OF THE VERTEBRATES

Fourth Edition

EVOLUTION OF THE VERTEBRATES

A History of the Backboned
Animals Through Time

Fourth Edition

Edwin H. Colbert

Honorary Curator of Vertebrate Paleontology
The Museum of Northern Arizona
Flagstaff, Arizona

Curator Emeritus
The American Museum of Natural History
New York, New York

Professor Emeritus
Columbia University
New York, New York

Michael Morales

Curator of Geology
The Museum of Northern Arizona
Flagstaff, Arizona

 **WILEY-LISS**

A JOHN WILEY & SONS, INC., PUBLICATION
New York • Chichester • Brisbane • Toronto • Singapore

Address all Inquiries to the Publisher
Wiley-Liss, Inc., 41 East 11th Street, New York, NY 10003

Copyright © 1991 Wiley-Liss, Inc.

The copyright for the drawings prepared by Lois M. Darling,
except for figure 12-4, has been assigned to the estate of the artist.

Printed in the United States of America

Library of Congress Cataloging-in-Publication Data

Colbert, Edwin Harris, 1905–
    Evolution of the vertebrates : a history of the backboned animals
through time / Edwin H. Colbert, Michael Morales. — 4th ed.
        p.   cm.
    Includes bibliographical references and index.
    ISBN 0-471-85074-8
    1. Vertebrates, Fossil. 2. Vertebrates—Evolution. 3. Phylogeny.
I. Morales, Michael. II. Title.
QE841.C68   1990   90-40025
566--dc20   CIP

Cover and book design by Jeff A. Menges

Preface

Ten years after publication of the third edition of *Evolution of the Vertebrates*, a new, revised edition is here presented. This fourth edition of the work is somewhat longer in text than its predecessor and has additional illustrations, as might be expected. Yet in spite of revisions and additions this new edition adheres to the general arrangement and presentation of the previous editions.

As in the case of earlier revisions, it has been necessary to recognize and to deal with numerous advances in our knowledge of vertebrate evolution as known from the fossil record that has developed during the past decade. The more significant additions to our perception of the subject are, we hope, properly treated in this book. But owing to the constraints imposed by available space (if the length of the book is to be kept within practicable limits) some new discoveries, particularly those of a lesser nature, must necessarily be passed over, or at best only briefly mentioned.

Lois Darling, who originally drew the many beautiful reconstructions of extinct vertebrates animals as well as the silhouette diagrams that have lent such distinction to previous editions of *Evolution of the Vertebrates*, very kindly revised some of her illustrations to bring them up to date. Louise Waller has revised some of the earlier illustrations figuring skulls and skeletons, and has drawn a considerable number of new illustrations to enhance the present volume.

At this place we wish to express our thanks to various colleagues and friends for advice in the preparation of this fourth edition of *Evolution of the Vertebrates*. Special appreciation is accorded to Dr. David K. Elliott of Northern Arizona University for his critical review and advice of the text dealing with jawless vertebrates and fishes. In addition we owe a debt of gratitude to the Museum of Northern Arizona for the facilities we enjoy at this institution. Finally we are truly grateful to Wiley-Liss, Publishers, and to the Editors, Dr. Mary M. Conway, Ms. Margery Carazzone, and especially Dr. Brian Crawford and Joseph W. Gill, for their patience and help.

Edwin H. Colbert
Michael Morales
January, 1990
Flagstaff, Arizona

Preface to the Third Edition

During the decade that has elapsed since publication of the Second Edition of *Evolution of the Vertebrates* there has been an inevitable increase of our knowledge concerning the backboned animals, fossil and recent. Much of this increment is due to new paleontological discoveries made throughout the world. Some of it is due to reinterpretations of extant materials—the result of increasingly detailed and sophisticated studies affording new insights into the structure and relationships of the vertebrates. Consequently a new edition of the book is desirable.

The text of this new edition is somewhat longer than that of the second edition, as the text of the second edition was longer than that of its predecessor. New illustrations have been added and some of the old illustrations have been revised.

In spite of revisions and additions, this edition maintains the arrangement and presentation of the first two editions. It is in effect an updating and an augmentation of what has gone before rather than a drastic reorganization of the book.

Included in this edition is the discussion at appropriate places of past continental relationships according to the modern theory of Plate Tectonics and the bearing of such relationships on the evolution and distributions of tetrapods. Much has been learned about this aspect of earth history during the past decade. Of particular paleontological significance has been the discovery of Lower Triassic amphibians and reptiles in Antarctica.

As in the preceding editions an attempt is made at a concise presentation of the subject. For details that it is impossible to include in the present book the reader is referred to other works, especially to *Vertebrate Paleontology* by the late Professor Alfred S. Romer. Additional comprehensive texts are listed in the *References*, at the back of this book.

Many colleagues gave valuable assistance and advice in the preparation of the present edition, as they did for the first two editions. I owe particular thanks to Drs. Robert L. Carroll and Peter Galton for advice concerning early amphibians and reptiles, and to Drs. Robert W. Wilson, Malcolm C. McKenna, and Terry A. Vaughan for help on the mammals. Other text had been previously checked by various authorities acknowledged in the prefaces of the first two editions. The new illustrations for this edition were made by Pamela Lungé and M. Louise Waller under my direction.

Again, as in the case of the second edition, I wish to thank the authorities and personnel of the Museum of Northern Arizona for facilities and help during the progress of the work on this edition. Last but not least, I owe a lasting debt of gratitude to John Wiley and Sons, the publishers, and particularly to Dr. Mary A. Conway, Life Sciences Editor.

Edwin H. Colbert
May, 1980
Flagstaff, Arizona

Preface to the Second Edition

More than thirteen years have passed since this book first appeared, and much has happened in the field of vertebrate paleontology during these years. Therefore, the time has come for a new edition if *Evolution of the Vertebrates* is to be continually useful. An effort has been made to bring the text of *Evolution of the Vertebrates* up to date in accordance with our new knowledge of the field—knowledge that has been augmented by important new discoveries of fossils throughout the world, and by important new interpretations of vertebrate evolution. These interpretations have resulted not only from the discoveries but also from studies of older materials made with the fresh and perceptive eyes of many younger men who have entered this field of scientific endeavor. Thus the text has been revised in many places and augmented in many others.

Of course, the text of the new edition is somewhat longer than the text of the first edition, but additions have been kept to a minimum. It is felt that part of the value of *Evolution of the Vertebrates* is the attempt to make a concise presentation of the subject. For anyone interested in the finer details of vertebrate evolution, based on the fossil record, there are other sources, especially, the excellent text on vertebrate paleontology by Alfred S. Romer, now available in a third edition.

We have also revised and added to the illustrations. A book of this kind can never have enough illustrations and, thanks to the generosity of the publishers, we have been able to add substantially to the number of figures although we have kept in mind the limitations necessary for a book of this scope. As in the first edition, all figures, including the new ones, are original and, in many cases, are based on sources in the extant literature. It has been our good fortune to have Mrs. Lois Darling add eight new restorations to the excellent ones she did for the first edition.

The author has exercised his judgment in accepting new ideas (expressed in the literature of the past decade) that concern the relationships of various vertebrates. Hence, the classification of the vertebrates adopted here is something of a compromise, based on the differing views of authorities in their particular fields. It is hoped that this classification is sufficiently up to date, and also that it is sufficiently conservative to be useful to the general reader and to the student. The same remarks may be applied to the expressions of stratigraphic and correlative relationships for the vertebrate-bearing formations of the earth, as they are here set forth.

I acknowledge the help of many people during the preparation of this revision. I especially express my gratitude to my colleagues, Drs.

Bobb Schaeffer, Richard H. Tedford, and Malcolm C. McKenna for advice concerning fossil fishes and mammals, to Drs. Donn E. Rosen and Walter Bock for advice on teleost fishes and birds, to Dr. A. E. Wood for advice concerning rodents, and to Drs. Alfred S. Romer and Donald Baird for their views on amphibians and reptiles.

Most of this revision was accomplished at the Museum of Northern Arizona, and I thank the personnel of that institution for making its facilities available to me.

Edwin H. Colbert
March, 1969
New York

Preface to the First Edition

This book is intended to be a general textbook on vertebrate paleontology, in which there is set forth an account of the evolution backboned animals as based on the fossil record. It is written in language that tries to avoid as much as possible the use of highly technical terms. It is written for the general student rather than for the specialist, and for the lay reader.

An attempt is made in this book to present a very general review of vertebrate evolution, and to show how animals with backbones developed through more than 400 million years of earth history. Consequently, this is not a book upon the principles of evolution, even though principles are frequently mentioned or discussed in passing. Nor is it a book concerned with the mechanisms of evolution as revealed by genetics. As stated above, this is primarily a survey of the fossil record of backboned animals.

A book such as this must inevitably omit a great deal; it must consider the most important facets of vertebrate history, and leave the details to other more comprehensive books. Therefore no attempt has been made to describe numerous genera in each family of vertebrates, or for that matter to discuss all the families. Rather, there are presented short but comprehensive descriptions of characteristic animals in the various orders of vertebrates, and these serve as examples to illustrate the group under consideration. Additional discussion of the group, whatever its rank, revolves around or develops from the description of the characteristic form. It is hoped that such a presentation may give a fair picture of the vertebrates without the confusion that might result from an attempt to tell the story in detail.

To follow the varied trends of vertebrate evolution through geologic time is not an easy assignment, because so many things were happening at once. Ancient fishes appeared during early Paleozoic times, 400 million years or more ago, and from them the fishes have evolved continuously up to the present day. Amphibians in turn sprang from fishes and have continued to recent times, while similarly reptiles evolved from the amphibians, and birds and mammals from reptiles. So it was that may vertebrates were evolving simultaneously in numerous separate lines through long stretches of geologic time. The author is confronted with the task of describing these evolutionary events, one by one, and of trying, somehow, to relate them to each other in the geologic time scale.

An attempt has been made in this book to maintain a feeling for the passing of geologic time through the story, partly by the order in which the various groups of vertebrates are discussed and partly by the insertion of chapters dealing with the relationships of faunas during certain phases of geologic history. Whether such treatment is successful remains to be seen.

Whatever the merits of the prose or of its arrangement in this book, there can be little doubt as to the excellence of the illustrations, practically all of them new, and drawn specifically for the book. This helps to give a fresh approach to the subject. The illustrations, as will be seen, consist of simple bone drawings, often grouped for purposes of comparison, of restorations and of pictorial phylogenies.

The bone drawings, so far as possible, are intended to illustrate characteristic animals (or their parts), and, like the text, do not pretend to be all inclusive. The restorations and the phylogenies, done with great skill and style by Mrs. Lois Darling, show what many ancient vertebrates looked like when they were alive. Here again, it has been necessary to select certain animals and to leave out many, for which similar restorations would be very desirable. Indeed, I wish there might have been many more illustrations made for this book, but limits had to be imposed and maintained if the volume were to be kept in hand.

Various people have helped in the preparation of the book. My particular thanks go to Mr. John Olstrom for assistance on the proof, and for his help in the preparation of the index.

Finally I wish to express my very deep appreciation to the several people who read the manuscript, or parts of the manuscript, and gave me the benefit of much-needed constructive criticism that has helped to improve the quality and the accuracy of the text. For their critical reviews and suggestions I gratefully acknowledge the help of Dr. Glenn L. Jepsen, who read the entire manuscript, of Dr. Bobb Schaeffer, who checked that portion of the manuscript devoted to the fishes, of Dr. Alfred S. Romer, who checked the section dealing with amphibians, reptiles, and birds, and finally of Dr. George Gaylord Simpson, who checked the section on mammals. No author could ask for a more distinguished and authoritative panel of critics in the field of vertebrate paleontology.

Edwin H. Colbert
New York 1955

Contents

14 Birds

15 Aquatic Reptiles

16 Years of the Dinosaurs

17 Surviving Reptiles

18 Beginning of the Mammals

19 Marsupials

20 Introduction to the Placentals

21 The Diverging Placentals

22   Evolution of the Primates

23   Rodents and Rabbits

24   Cetaceans

25   Creodonts and Carnivores

26   Ancient Hoofed Mammals

27   South American Ungulates

28   Perissodactyls

29 Artiodactyls

30 Elephants and Their Kin

31 The Age of Mammals

Fossil Hunters

Introduction

FOSSILS—THE EVIDENCE IN THE ROCKS

This is the story of vertebrate evolution as revealed by the fossil evidence. It is an account of vertebrate life through millions of years of earth history, as based upon petrified remains that are found in the sediments of the earth's crust. It is the paleontologist's interpretation of a record that begins with the primitive vertebrates of early Paleozoic times and extends, through vast expanses of geologic time, through the varied backboned animals of the great Pleistocene Ice Age, to the present day.

Fossils are the raw materials upon which the paleontologist, the student of ancient life upon the earth, bases his studies. Fossils are the remains or indications of past life. They are generally the hard parts of animals that have been petrified, transformed from shell or bone into stone. But they may be the hard parts as they were originally constituted, preserved intact without benefit of petrification or fossilization. Sometimes the soft parts of animals or plants may be fossilized, although such preservation is not common. Indeed, extinct animals are occasionally preserved completely and without any change. Mammoths and other animals of the last Ice Age, frozen in the ice of the far north, have been preserved in this way.

Fossils are not necessarily direct evidence, in the form of preserved bones or shells, of the organisms they represent. They may be molds in the rocks—imprints left by the animal (or plant) or by some part of the organism. They may be footprints made by an animal. They may be the preserved structures that were built by an animal during its lifetime, such as nests or tubes. Indeed fossils are found in many varied forms, which, among other things, helps to make life interesting for the paleontologist.

The study of fossils is a comparatively new science. To the peoples of the classical civilizations the nature of fossils was hardly realized, and it was not until the days of the Renaissance that a true understanding of fossils was reached by (among others) that great and accomplished Florentine, Leonardo da Vinci. The scientific study of fossils, however, is less than two centuries old, whereas the modern evolutionary interpretation of the fossil record really begins with the work of Charles Darwin, as crystallized in his epochal book *The Origin of Species*. In spite of the relative youth of paleontology as a science, an impressive amount of fossil material has been gathered together and studied by paleontologists all over the world. Consequently it can be said that our knowledge of the history of life as based upon the fossil record is now reasonably complete, and this is as true for the vertebrates as for other groups of organisms.

Of course there are many gaps still to be filled, and there are many new forms still to be discovered. Nevertheless the general aspects of vertebrate history are now known sufficiently well to make a connected and integrated story, and so far as lesser categories of backboned animals are concerned, the story is frequently preserved in the fossil record with considerable detail.

It is the possibility of finding entirely new organisms preserved as fossils that makes life for the paleontologist something like an extended treasure hunt. We never know what may be revealed in the next canyon or on the distant badland slope, and even in the laboratory things appear to surprise and delight the student of ancient life. But whether the fossils are new, or whether they represent animals already well known in the fields of paleontology, the collecting and preparing of fossils for study is more often than not arduous and difficult work. Only the person who has hunted for fossils in the field and worked on them in the laboratory can appreciate their true value in terms of the expenditure of human effort.

It takes much looking to find fossils, especially the fossils of the backboned animals. These generally are not so abundantly preserved as the fossils of sea shells and other invertebrates; consequently their discovery commonly requires much walking and climbing and some careful scrutinizing of rock exposures. Moreover, when they are found these fossils require special techniques of collecting if they are to be brought back to the laboratory with any degree of success. Very commonly it is necessary to expose the fossil carefully and to harden it with liquid preservatives as it is exposed, for although the bones or scales have been turned to stone and thus are hard, they are often at the same time as brittle as fine glass. Because such fossils may break of their own weight it is necessary to jacket them with casts of burlap and plaster of Paris before they can be removed from the ground.

Then in the laboratory the whole process must be reversed. The plaster jackets are removed, and the fossils are carefully and completely cleaned of surrounding rock. Often they must be reinforced by wires or irons or by plaster, filling in the hollow cavities of the bones. Only then are they ready for study.

It is long and involved—often it is tedious—but when one becomes thoroughly interested in the subject, it is fascinating work. No antiquarian, poring over a lovely Greek vase, shows more care or solicitude for his treasure than the vertebrate paleontologist studying the fragile fossil bones of some animal that lived millions of years ago.

In the end a picture of this long extinct animal is reconstructed. Bones are compared with bones of other animals, either fossil or recent. Relationships are established. On the basis of our knowledge of modern animals, the soft parts of the fossil usually can be inferred. Muscles are mentally fastened to the bones, and the general patterns of nerves and blood vessels are indicated. Frequently even the nature of the outer covering of scales or skin is shown by the fossil, so that only such superficial features as color and accessory soft parts are left entirely to the imagination. Such are the methods of the paleontologist, and by their use, compounded many times over through the years, the vast picture puzzle of life in the past has been assembled.

ANIMALS WITH BACKBONES

The phylum Chordata includes those animals having, throughout their life or at some stage in their development, an internal supporting structure, in the form either of a continuous flexible rod known as a notochord or of a series of vertebrae, running dorsally along the midline of the body forming a vertebral column, also known as a backbone or spinal column. The chordates having vertebrae are generally designated as the vertebrates, and they comprise by far the largest division of the chordates. Almost all fossil chordates are backboned animals, or vertebrates, so that virtually all our knowledge of the primitive chordates, in which there is only a rodlike notochord, is derived from a few surviving forms.

The vertebrate normally has the long axis of the body, the backbone, horizontally placed; any deviation from this position represents a specialization. There is a concentration of sense organs at the anterior end of the animal, and these are generally housed in a skull. The internal skeleton of the vertebrate is either cartilaginous or bony, and in the vast majority of forms it is bony. There is an axial skeleton, consisting of the vertebral column, with a skull anteriorly placed, and generally with ribs extending laterally from the vertebrae. There may be median fins associated with the axial skeleton. In most vertebrates, but not all, there is also an appendicular skeleton, consisting of paired fins or limbs for steering and balance, or for propulsion, and these are attached to the body by bony girdles. There is a pectoral girdle in front and a pelvic girdle posteriorly. Moreover the vertebrate is characterized by the position of the spinal nerve chord above the notochord or the vertebral column, and by the location of the circulatory system and the digestive system below the vertebral column. Usually respiration in vertebrates is either by means of gills, in which case oxygen is extracted from the water, or by lungs, in which case oxygen is extracted from the air. The gills are supported by stiffened arches—the branchial arches. In all but the most primitive vertebrates there is an upper and a lower jaw, which are formed by transformation of an anterior pair of branchial arches. The sense organs include paired eyes, and sometimes a median eye, the pineal organ, on top of the head, paired nostrils in all but the most primitive types, and paired ears, which may be for balance and hearing, or for balance alone.

AMPHIOXUS

We shall probably never know what the first chordates were like, because it is unlikely that adequate indications of them are preserved in the fossil record. They must have been small, comparatively simple animals, and it is not likely that they had a hard skeleton, capable of being fossilized and preserved in the sediments of the earth. The first vertebrates appearing in the record of the rocks are characterized by highly developed bony armor, and for this rea-

son it has been argued that bone is very primitive in the history of vertebrate evolution. Yet it seems logical to believe that there might have been a long period of vertebrate evolution preceding the development of bony armor so that the first ancient, bone-encrusted vertebrates in the geologic record in truth may be advanced far beyond the condition of the primitive chordates.

It so happens, however, that there is a modern chordate of such primitive form and organization that it approximates to a considerable degree our conception of the central ancestor for the backboned animals. This is the little sea lancelet known as *Amphioxus* or *Branchiostoma*, of the subphylum Cephalochordata. *Amphioxus*, which lives in the shallow waters along certain coast lines, and spends much of its life buried in the sandy bottom, is a translucent animal of fishlike form and rarely more than two inches in length. There are no verte-

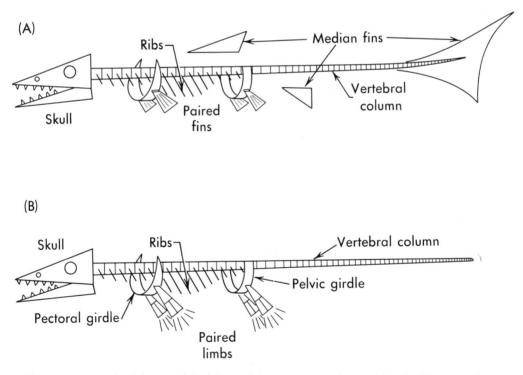

Figure 1-1. Generalized diagram of the skeleton of **(A)** an aquatic vertebrate and **(B)** a land-living vertebrate.

brae in *Amphioxus*, but rather a notochord forming an internal axial support for the animal. This support represents the precursor of the backbone or spinal column. Above the notochord in *Amphioxus* is the nerve chord; below it is a simple digestive tube. There is no real head or brain in this little animal, and no sense organs, except for some pigment spots that seem to be sensitive to light. No matter how primitive and deficient *Amphioxus* is in these respects, it is well supplied with gills, arranged in a long series down either side of the front portion of the body. These extended gills serve to extract oxygen from the water; they also aid the animal in feeding, by functioning as a sort of sieve to strain food from the debris of the ocean floor. Zig-zag bands of muscles, or myotomes, which are characteristic of chordates, run the entire length of the body. Finally, although *Amphioxus* is a capable swimmer, it lacks paired fins of any sort. However, there is a small tail fin.

All in all, *Amphioxus* is a very primitive chordate. The great development of the gill basket shows that this animal may be specialized in some aspects of its anatomy, yet in spite of this specialization the lancelet remains in a general way an approximate structural ancestor for the vertebrates. It is probable that there has been little change in the line of evolution represented by the lancelet since early Paleozoic or even pre-Paleozoic times. In *Amphioxus* we see what our chordate ancestor of six hundred million years ago may have looked like.

PIKAIA

A small, wormlike form called *Pikaia*, from the middle Cambrian Burgess Shale of Canada, probably represents the earliest known chordate. It has the chordate hallmarks of a notochord and zig-zig muscle bands along its body. Distinct head, trunk, and tail regions are present, with a caudal fin wrapped around the posterior end of the tail. Its feeding and breathing organs, however, appear to be more primitive than that of *Amphioxus*. The presence of a primitive chordate so early in the fossil record suggests that the phylum Chordata originated about the same time as many of the invertebrate phyla, at or just before early Cambrian times.

The origin of chordates has been and continues to be a hotly debated topic, with many theories put forth during this century. Based on embryological and biochemical studies of modern forms, it has become clear that the invertebrates to which chordates have the closest relationship are starfish, crinoids, sea urchins, and other groups of the phylum Echinodermata. The exact nature of the relationship, however, is not certain. Chordates and echinoderms may share a common ancestry, as a majority of workers believe. Alternatively, one group of fossil echinoderms, the stylophorans or calcichordates, may themselves be the ancestors of chordates, as a small group of researchers believe. In either case, vertebrates surely are closely related to primitive chordates such as *Amphioxus* and *Pikaia*, and the family tree of chordates as a whole indicates a close relationship with echinoderms.

BONE, CARTILAGE, AND TEETH

Vertebrates possess three main types of body hard parts—bone, cartilage, and teeth—that are not known in any other groups, including primitive chordates. The vast majority of all vertebrate body fossils are the petrified remains of different forms of these three materials; only rarely are the soft parts of vertebrates such as skin and muscles preserved. The distinction between bone, cartilage, and teeth is based on their internal structure or histology.

Bone is produced by the mineralization of a protein matrix by osteocytes, specialized bone-forming cells. The main mineral constituent of bone is a form of calcium called hydroxyapatite, a phosphate rich salt. The earliest specimens of bone, known from ostracoderm dermal armor, are made of three layers: on the upper or external side, small, rounded denticles that resemble little teeth; in the middle, cancellous or spongy bone built like a framework with

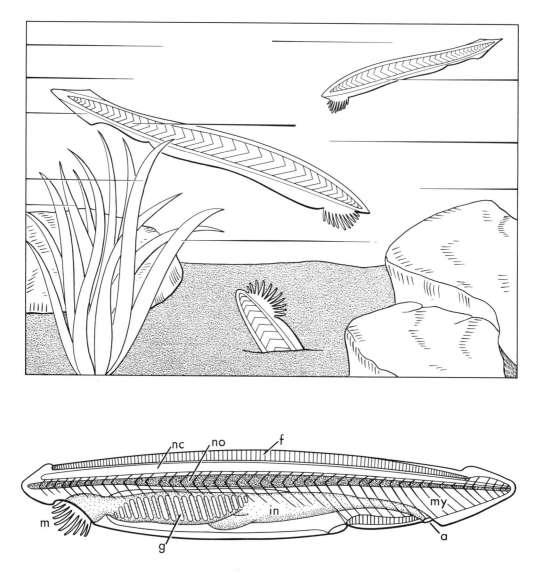

Figure 1-2. The modern lancelet *Amphioxus (Branchiostoma)*, a simple chordate that lives in shallow waters along seacoasts. Abbreviations: a, anus; f, fin ray; g, gills; m, mouth surrounded by tentacles; my, myomeres or segmented muscle bands; nc, dorsal nerve chord; no, notochord.

many openings; and at the bottom, thin layers of bone called lamellae, which lack scattered holes indicating the former positions of osteocytes. Acellular bone of this kind is called aspidin to differentiate it from normal bone, in which the holes for osteocytes are present.

Bone can form by replacement of cartilage, as in limb bones, or by deposition in dermal tissues without the need for a cartilaginous template, as in osteoderms and dermal scales. Normal bone comes in two main structural types, spongy and compact. Spongy bone is associated with the inner parts of bony elements, such as the marrow cavity and growing ends. It is light and full of spaces, which are filled with soft tissue in the living vertebrate.

In most bones of vertebrates, dense compact bone forms the outer surface and takes

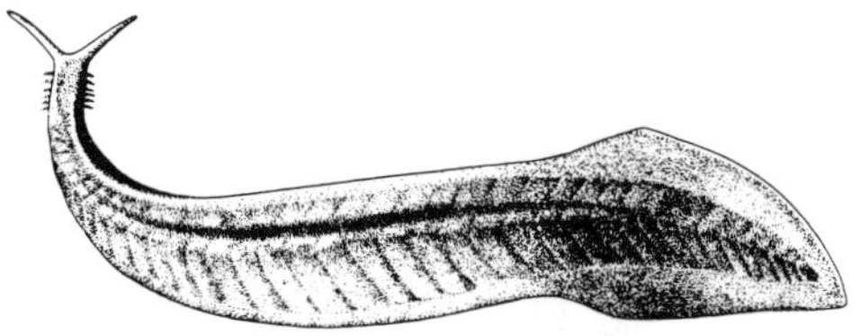

Figure 1-3. The earliest known chordate *Pikaia* from the Middle Cambrian Burgess Shale. Note the zig-zag muscle bands along the body.

up the main forces acting on bones, such as the weight of the animal. Compact bone comes in two varieties, laminar and Haversian, named after the nature of their histology. Laminar compact bone is composed of many thin layers, called laminae, through which a network of blood vessels run. In contrast, Haversian bone is made of a multitude of longitudinal cylinders of bone. Each cylinder or Haversian system consists of a blood vessel that is surrounded by many concentric laminae. Although laminar bone is strong and has a very efficient blood supply, it cannot easily be remodeled during growth. In Haversian bone, however, resorption and redeposition of bone material can be done quite readily. Laminar bone is the most common variety of compact bone among terrestrial vertebrates, and Haversian bone is associated with rapid growth.

As mentioned above, the top layer of acellular bone or aspidin is made of small, teeth-like structure called denticles. These are in fact the precursors of true teeth. Each rounded denticle is capped with an enamellike material closely resembling the enamel of teeth. A second kind of toothlike structure, the placoid scales that cover the bodies of sharks and their relatives, bridges the morphological gap between denticles and teeth. Placoid scales, ultimately derived from denticles, have a well-developed outer layer of enamel and a pulp cavity furnished with a good blood supply. True teeth also have an outer layer of enamel and an inner pulp cavity, as well as a middle layer of dentine. Thus, the teeth of vertebrates had their evolutionary beginning as the denticled surface of aspidin in ostracoderms.

Other types of dermal scales besides placoid have evolved during the course of vertebrate history. They are cosmoid, ganoid, cycloid, and ctenoid, all of which are elaborations of the basic structure of dermal bone.

Cartilage is a protein material composed of specialized cartilage-forming cells called chondrocytes embedded in a gelatinous matrix containing protein fibers. Cartilage does not contain bone-forming osteocytes. Although cartilage is usually softer and more pliable than bone, it can undergo a process of calcification in which spherical aggregations of apatite crystals become embedded in the gelatinous matrix. Although the resulting calcified cartilage is harder and stronger than regular cartilage, it lacks the advantages of bone's structure and composition.

GEOLOGIC TIME

Mention of the word Paleozoic and reference to a time span of such tremendous duration as six hundred million years bring us to the subject of geologic time. This is a consideration of prime importance to the paleontologist, because one of the great advantages that the study of fossils has over other branches of natural history is the possibility of projecting ourselves back through the fourth dimension of time into the past ages of the earth.

It is not easy at first to think in the immense units of geologic time. We are used to thinking in terms of years or centuries or millennia; geologic time is measured in millions of years. It seems almost incredible that there were rains and winds and volcanoes and the cycles of life and death on the earth as far back as one hundred million years or five hundred million years or a billion years ago, yet from the study of the decay of radioactive elements we know that such great time spans are necessary to measure the sequence of events that make up earth history. Careful studies of the disintegration of uranium into lead indicate that there are rocks on the earth almost five billion years in age, and there is good reason to think that the earth had existed for a long

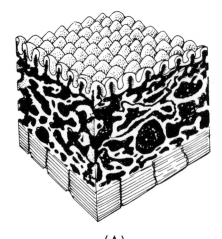

(A)

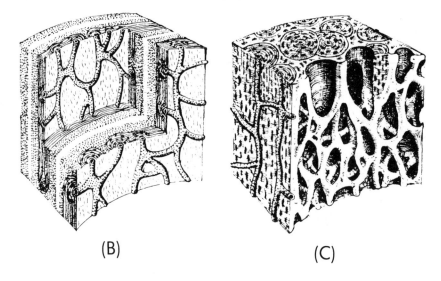

(B)   (C)

Figure 1-4.   Different types of bone tissue: **(A)** aspidin, **(B)** laminar compact bone, and **(C)** Haversian compact bone.

time before these rocks were formed. In terms of human experience this is a very old planet.

It is not possible at this place to go into the evidence for dating the earth or for drawing up the geologic time scale, but a few salient facts can be presented. As already mentioned, the first rocks that can be dated are almost five billion years old. Fossils first appear in abundance in rocks of Cambrian age, and they can be dated as almost six hundred million years old. Although this is the beginning of an adequate

fossil record, it is by no means any indication of the beginning of life on the earth, for there was probably a long time span before the Cambrian period when plants and animals were evolving as very primitive organisms. By the beginning of Cambrian times life was sufficiently advanced so that animals were highly organized, with hard parts capable of fossilization. Almost all the major groups of invertebrate animals and the earliest chordate are present in Cambrian rocks—an indication of the

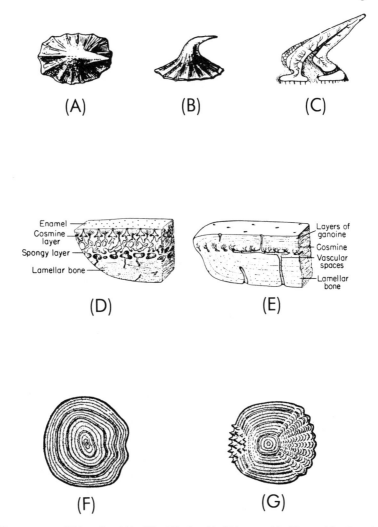

Figure 1-5.   Different types of fish scales: **(A)**, **(B)**, **(C)** placoid, **(D)** cosmoid, **(E)** ganoid, **(F)** cycloid, and **(G)** ctenoid. (C), (D), (E) are cross sections.

incredibly long evolutionary sequence during which life was differentiating and specializing to the comparatively high degree that is characteristic of the earliest Cambrian faunas. Perhaps a few other dates will help to indicate the long evolutionary history of life on the earth. The first vertebrates appeared in the Ordovician period of earth history, nearly five hundred million years ago. The dinosaurs began their long evolutionary history more than two hundred million years ago, and they continued for more than one hundred-fifty million years. They became extinct about sixty-five million years ago, and soon after mammals became the dominant animals on the earth. Human beings, as the genus *Homo*, appeared perhaps two million years ago. In the last few hundred thousand years they have gone far.

Two geologic periods, Cambrian and Ordovician, have been mentioned in the preceding paragraph. Early history has been divided into a number of time divisions called *periods*, and these in turn have been grouped into longer time divisions known as *eras*. Periods have been subdivided into *epochs*.

Several eras mark the course of earth history prior to the beginning of the fossil record, but since this portion of geologic time is rather difficult to interpret, the general practice is to refer to it as "Precambrian times." The Precambrian portion of earth history is long, extending through more than 4500 million years of time. It need not concern us here.

With the beginning of the fossil record earth history can be measured and followed in considerable detail. As a result of cumulative studies carried on during the last century or so, three great eras of earth history are recognized by the sequence of the fossil record. They are, in the order of their age, the Paleozoic, the Mesozoic, and the Cenozoic eras respectively. The Paleozoic era, the time of ancient life, has been divided into seven periods, which in order from oldest to youngest are Cambrian, Ordovician, Silurian, Devonian, Mississippian, Pennsylvanian, and Permian. It is general practice among paleontologists outside North America to combine the Mississippian and the Pennsylvanian into a single period, the Carboniferous.

What are the bases for these names of geologic periods? For an answer to this question we have to go back to the early pioneers of geological science, who laid the foundations for the study of rock strata and their included fossils. The first four of the Paleozoic periods were studied and named by English scholars. The name Cambrian comes from Cambria the ancient name for Wales, where rocks of this age are extensively exposed. Ordovician and Silurian are names based on ancient tribes once living in southern England and Wales, the Ordovices and the Silures, in whose onetime tribal territories there are extensive sequences of rock belonging to these two ages. The name Devonian comes from the English county of Devonshire. As might be expected, the next two names in the geologic column are based upon North American regions where Paleozoic rocks are exposed, along the Mississippi River and in the Allegheny Mountains of Pennsylvania. The older term, Carbonif-

erous, comes from the typically carbonaceous character of rocks of this age; this was the period of great coal deposits over many continental areas. The name Permian is based upon the province of Perm, in northern Russia, where rocks of this age are especially well developed.

The Mesozoic, the time of middle life in earth history, is subdivided into three periods—the Triassic, the Jurassic, and the Cretaceous. The Triassic is so named because it was subdivided into three units in central Europe, where it was first studied. The name Jurassic comes from the Jura Mountains of the Alpine region. And Cretaceous is based upon the Latin word Creta, meaning chalk, in reference to the white cliffs of Dover.

Finally the Cenozoic, the time of recent life in earth history, consists of two periods under which are grouped six subdivisions, generally regarded as of lesser consequence, and therefore classified as *epochs*. The periods are the Tertiary (originally considered as the third great time division in earth history) and the Quaternary (the fourth original time division). The epochs, from oldest to youngest, are the Paleocene, the Eocene, the Oligocene, the Miocene, the Pliocene, and the Pleistocene, names formulated in part by Sir Charles Lyell, the great English geologist, as the result of his studies on mollusks in the Cenozoic rocks of Europe. Of these epochs all but the last, the Pleistocene, belong to the Tertiary period. The Paleocene is so named because it is the ancient (palaeos) division of recent (caenos) geologic history. In Lyell's classification, Eocene was the oldest division—and the name, broadly translated, means "dawn of recent times." Then there come, in order, Oligocene, a name meaning "few recent" (in reference to the few mollusks of that age similar to modern forms), Miocene, "less recent," Pliocene, "more recent," and Pleistocene, "most recent times."

Such was the manner in which the names for the time divisions of earth history became established. Let us now arrange them in order, following the usual geological practice of placing the oldest period at the bottom of the list (since the oldest strata are at the bottom of the rock column), with each successively younger period above its predecessor in the sequence. The material can be set forth in tabular form, as shown in Figure 1-6.

THE SEQUENCE OF VERTEBRATES THROUGH TIME

As mentioned above, the first vertebrates are known from rocks of Ordovician age. However, the fossil evidence for these earliest known vertebrates is very fragmentary, and it is not until we reach the sediments deposited during late Silurian times that fossils are complete enough to give us some idea as to the diversity and relationships of the early vertebrates. Even here the evidence is scanty, so it is actually in Devonian rocks that the fossil record of the vertebrates becomes truly representative. Therefore, for our purposes the story does not begin until about the middle of the Paleozoic era. From then on, however, the history of the vertebrates is well known, as it is revealed in successively younger layers of the earth's crust. As will be shown in the succeeding pages of this book, we can see from the fossil record in the rocks how all the major groups of fishes had appeared by middle and late Devonian times, some of them to continue to the present day, one group to become extinct before the close of the Paleozoic era. We can see how during the transition from Devonian to Mississippian times the first land-living vertebrates, the amphibians, made their appearance as descendants of certain advanced fishes. The amphibians had their heyday, especially during the final stages of the Paleozoic era, after which they continued to develop, but on a small scale as compared to the broad variety of their development at the time of their dominance.

In the course of their evolution the amphibians gave rise to the reptiles, an event that probably took place during Mississippian or Pennsylvanian times, and the reptiles were destined to rule the earth for many millions of years. During the Mesozoic era some reptiles, known as dinosaurs, became the dominant animals on the land, evolving along diverse lines that carried them to all the larger land areas of the earth and into most of the different environments that then existed. The dinosaurs were indeed the rulers of the earth for more than a hundred and fifty million years, but eventually they became extinct. Today the reptiles that survive, although numerous and widely dispersed throughout the world, are but a remnant of the hordes that once ruled the earth.

Before the dinosaurs had become extinct, in fact during the earlier part of their long evolutionary history, two other groups of vertebrates arose from reptilian ancestors. One of these groups, the mammals, appeared during Triassic time, and the other, the birds, arose during the Jurassic period.

By Cretaceous times the birds had become highly specialized; by the beginning of the Cenozoic era they were populating the continents and the islands of the world in essentially their modern form. As long as the reptiles were dominant the evolution of the mammals was comparatively slow. But with the transition from the Cretaceous period into the beginning of the Tertiary period new opportunities were opened to the early mammals. The reign of the reptiles was at an end, and the "Age of Mammals" began. It has continued to this day, when one group of mammals, humans, have developed within a comparatively short span of time to unprecedented heights. This is now the age of intellect (no matter how despairing some people may feel about modern world trends), and because of their intellect the mammals known as human beings are able to look back through time to study the evolutionary history of their forerunners and forebears.

SOMETHING ABOUT CLASSIFICATION

People have been classifying plants and animals, in one way or another, since long before the dawn of history. Primitive people and

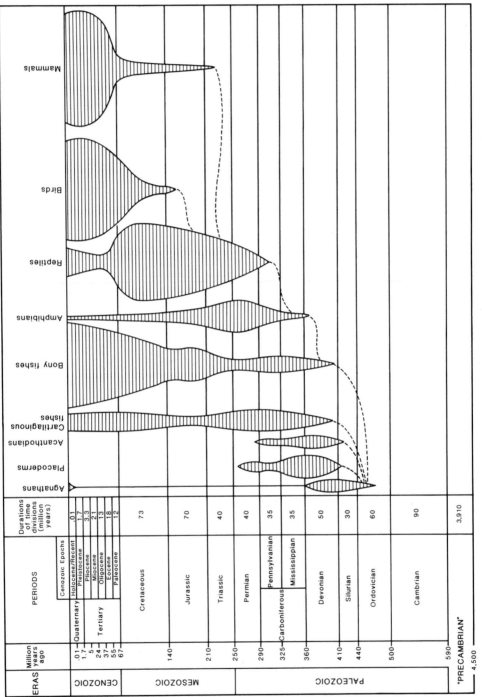

Figure 1-6. Chart of geologic time and the ranges and relative diversity of vertebrate classes through time.

people of ancient cultures commonly have classified organisms according to their suitability as sources of food. For example the Book of Leviticus, in the Biblical Old Testament, instructs the Israelites as to what animals they may or may not eat, the instructions being based upon a classification of sorts: "Whatsoever parteth the hoof, and is cloven footed, and cheweth the cud, among the beasts, that may ye eat," and so on, through a considerable series of animals living in the Middle East.

An early logical classification of organisms was attempted by the Greek scholar Aristotle (384–322 B.C.). Subsequent works dealing with the relationships of plants and animals appeared through the centuries, notably the *Historiae Naturalis* by the Roman Pliny the Elder (23–79 A.D.) and the *Historiae Animalium* by Conrad Gesner of Zürich (1515–1565 A.D.).

Our modern understanding of plant and animal relationships begins with the Swedish naturalist, Carl von Linné, generally known as Linnaeus (1707–1778 A.D.). It was Linnaeus who devised the *binomial* system of nomenclature, whereby every plant and animal is designated by two names, generic and specific. For example, the horse is known scientifically as *Equus caballus*, the first name being that of the genus, the second that of the species. There are other species of the genus *Equus*, such as *Equus burchelli*, a zebra.

Genera and species are grouped into larger categories, in ascending sequence, the family, order, class and phylum, these (with the genus and species) being the obligatory categories essential to the Linnaean system of classification. Other categories such as subclass and suborder are optional.

Besides the traditional Linnaean system, there are two other methods of classification used currently, and both are associated with a branch of systematics called cladistics. One method essentially follows Linnaean classification but adds many more categories, including extra prefixes and suffixes to the hi-

erarchy. Not only does this require one to learn a greater number of category names and their hierarchical rank, but, unlike the Linnaean system, there is no standardization of the extra categories. Different workers use different names and ranks for the extra categories. The second classification method of cladistics does not use categories at all, except for genus and species. Instead, the names of the groups of organisms (not categories) are simply listed in order horizontally. The largest, most inclusive group is given first on the left end of the list, the least inclusive group (usually genus or species) on the right, and all intermediate groups are in between. Two simplified classifications of modern human beings are given below, the first using the traditional Linnaean system, the other using the cladistic method just described:

Class Mammalia
  Order Primates
    Suborder Catarrhini
      Superfamily Hominoidea
       Family Hominidae
        Genus *Homo*
         Species *H. sapiens*

Mammalia, Primate, Catarrhini, Hominoidea, Hominidea, *Homo sapiens*.

The second method is based on cladograms, graphic representations of the phylogenetic relationships of all the groups of organisms. Thus, to fully comprehend a horizontally listed classification as shown above, one must always refer to the cladogram on which it is based. For example, one must look to the cladogram to determine what lesser groups, besides those already listed, are included in the group Primates. In a traditional Linnaean system, each group of organisms is placed within a classification category of established and stable rank in the hierarchy. Therefore, it is easier to learn the classification and to compare the ranks of different groups of organisms using this system.

For this book, a traditional Linnaean system of classification is used for several reasons. It has been in use the longest, and it is probably the system most familiar to the target audience of readers, those seeking a broad introduction to the history of vertebrates. Because it requires the least amount of memorization and advanced knowledge, it is easier to learn. The traditional Linnaean system incorporates conclusions based on uniquely derived characters shared by groups, as does cladistics, as well as overall degree of morphological differentiation, which is omitted from cladistics. For example, in cladistics birds are considered a subgroup of a larger reptilian group, as their phylogenetic relationships indicate. In traditional Linnaean classification, however, even accepting the close relationship between them, birds are elevated to a category rank equal to that of reptiles (i.e., class) because of the great differentiation between the two groups. Finally, the traditional classification provides a conservative and relatively stable foundation with which one can evaluate new, sometimes radical reinterpretations of vertebrate phylogenetic relationships and classification.

A list of the categories of classification used in this book are, in descending order, as follows (obligatory cagetories are in capital letters):

PHYLUM
  Subphylum
    Superclass
     CLASS
      Subclass
       Superorder
        ORDER
         Suborder
          Infraorder
           Division
            Superfamily
             FAMILY
              GENUS
               SPECIES

The names of the higher categories of classification—phylum down through division—are variously written. The names of lesser categories above the rank of genus, however derived, generally have uniform endings. Superfamily names usually have the suffix -oidea, family names -idae, and subfamily names -inae.

The names of the genus and species (as well as subgenus and subspecies) are universally italicized. The initial letter of the generic name is capitalized, but that of the species and subspecies are not.

The classes of vertebrates used in this book (and possible superclasses) are represented in the diagram on page 15. The first possible division of vertebrates into superclasses is based on the type of paired appendages present in the various classes. Vertebrates of the superclass Pisces have fins or similar appendages for movement in water, whereas tetrapods have legs for locomotion on land. The second set of superclasses reflects the presence or absence of jaws. The third scheme indicates that members of a class reproduce by means of a shelled egg or its derivatives (amniotes), or by laying unshelled eggs in water or a moist area (anamniotes). The amnion is one of the special membranes inside a shelled egg.

The species is the basic unit of classification. It represents a distinct, usually continuous population of animals, the members of which breed freely with each other but do not often interbreed with members of other species populations. Two such populations that are closely related to each other, but which do not interbreed, can be regarded as two separate species of a single genus. Thus the sharp-nosed mackerel shark of the Atlantic, *Isurus oxyrhynchus*, and the Pacific mako shark, *Isurus glaucus*, are distinct species of a single genus. They show many similarities. Nonetheless they exist as separate populations, each with its own range, and the members of the two populations are genetically isolated from each other. Because of this there are constant, recognizable differences between them, the differences by which we distinguish and classify these two types of sharks as species.

To the student of modern animals the species is of prime importance, and much of the

SUBPHYLUM   POSSIBLE SUPERCLASSES   CLASSES

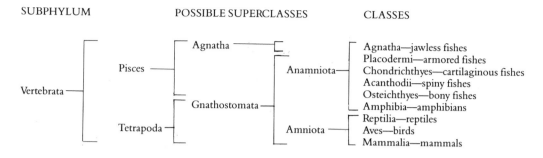

significant work being done on recent vertebrates is concerned with species. But to the student of fossils, the species is of less importance. By the very nature of the materials, fossil species are generally very subjective; their definition cannot be established with so much certainty as the delineation of living species. Therefore the paleontologist, especially the student of fossil vertebrates, is generally interested more in genera than in species. Genera in fossil materials are usually well defined, and serve as the common basis for evolutionary studies. In this book, the emphasis will be upon genera as the lowest category of practical value.

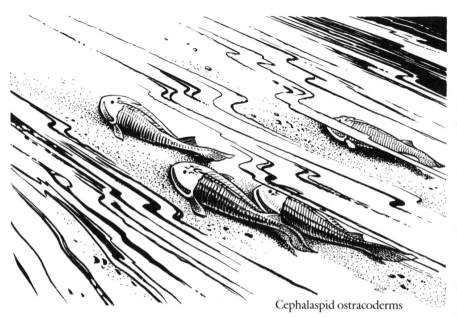

Cephalaspid ostracoderms

Jawless Vertebrates

2

INTRODUCING THE LAMPREY

The lamprey is an elongated, naked, eel-like animal with a long fin on its back and around the tail, but without any paired fins, with a nostril located on top of the head between the eyes, with seven gill openings behind the eyes on each side, and with a peculiar round, jawless mouth, containing sharp teeth. In the earlier stages of its life the lamprey lives on the bottom of streams and lakes, feeding upon small animals that it sucks up through its disc-like mouth, but as it reaches its full growth it becomes a parasite. It attaches itself to a fish by means of its vacuum-cup mouth, and then it proceeds to bore its way into the body of the victim with its sharp, rasping teeth, to suck the blood of its hapless host. The lamprey may abandon its host before the fish dies, or it may stay with the fish to the end.

This rather unpleasant animal is the highly modified survivor of some of the first vertebrates that lived on the earth. Even though the lamprey is specialized in many respects, such as its lack of a bony skeleton, perhaps the result of "degeneration" through the development of parasitic habits, it is nevertheless primitive enough in its general characters to give us some idea of what the first vertebrates were like. The lamprey and its relative, the hagfish, are jawless vertebrates, belonging to the class Agnatha, as did the first vertebrates to appear in the fossil record. Moreover, the lamprey, in particular, shows some close resemblance to some of the early vertebrates that are known to us as fossils. When we look at a lamprey, we get a partial glimpse of the ancient vertebrates that lived almost half a billion years ago.

THE EARLIEST VERTEBRATES

The first bona fide vertebrates known in the fossil record are agnathans, or jawless "fish", recently found in middle Ordovician sediments of Bolivia and Australia. The Bolivian fossils include complete, three-dimensional specimens named *Sacabambaspis*. The Australian material is less impressive, consisting of fragmentary natural molds of the external and internal surfaces of the dermal armor of *Arandaspis* and *Porophoraspis*. The only other vertebrates of Ordovician age are bony scraps assigned to the two genera *Astraspis* and *Eriptychius*, from the Harding Sandstone of western North America. Many bone-like fragments resembling agnathan dermal armor or scales have been reported from upper Cambrian and lower Ordovician deposits in North America, Greenland, and Spitsbergen. These pieces, given the name *Anatolepis*, were originally thought to represent agnathans, but paleontologists who specialize in jawless vertebrates have recently concluded that the fragments cannot be positively identified as vertebrate, and instead, probably represent early arthropod-like organisms.

It is fortunate that one of the earliest known vertebrates, *Sacabambaspis*, is represented by many complete and well preserved specimens. The animal was up to 18 inches long and six inches wide, with a torpedo-shaped body that

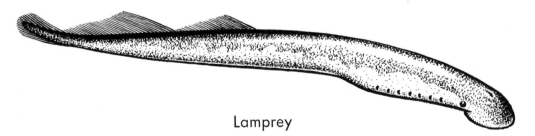

Lamprey

Figure 2-1. A modern jawless vertebrate, the lamprey *Petromyzon*. Notice the nostril on top of the head, and the laterally placed eyes, behind which are the gill openings. About one-fifth natural size. Prepared by Lois M. Darling.

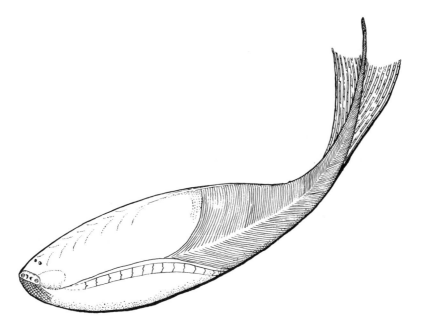

Figure 2-2.   One of the earliest vertebrates, *Sacabambaspis* from the middle Ordovician of Bolivia. About two-thirds natural size.

tapered to a sharply pointed tail with a vertical tail fin. The front third of the body was covered by a bony shield, while the back part had a series of long and narrow V-shaped scales that interconnected to cover the entire surface. The mouth was located on the underside of the front of the head, and two rows of multiple gill openings ran along the flanks of the head, one row on each side. The eyes also seem to have been located on the side of the head, just in front of the first gill opening of each row. Two small openings, which may be paired nostrils, pierced the top of the head above and slightly behind the eyes. *Sacabambaspis* was found in marine deposits containing invertebrates. It is thought that it lived in nearshore salt waters and fed by filtering mud through its mouth and out the gill openings.

Further evidence of early vertebrate life comes from England, in rocks of middle Silurian age. Two genera, *Jamoytius* and *Thelodus*, are represented by complete skeletons which, because of the manner of their preservation, are not easy to interpret. *Jamoytius* is of special interest because it appears to be a very primitive jawless vertebrate, perhaps oc-

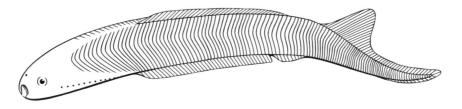

Figure 2-3.   A restoration of the Silurian jawless vertebrate, *Jamoytius*, an anaspid ostracoderm. Notice the lateral eyes, with gill openings behind, the reversed heterocercal tail, and the continuous fin fold. About two-thirds natural size.

cupying a position close to the ancestry of the lamprey and its relatives. The few, rather enigmatic fossil remains of *Jamoytius* show that this animal was small, elongated and tubular. It had a terminal, suctorial mouth, and, on each side of the head region behind the eye, there was a row of circular gill openings. There was a tail fin, with a long lower lobe and a shorter, deep upper lobe, and possibly there were lateral fin folds and a long dorsal fin, for maintaining balance. *Jamoytius* and *Thelodus* are found together in marine sediments.

THE OSTRACODERMS

Although *Jamoytius* is an interesting and, as now known, an all too incomplete record of a middle Silurian chordate, it is not the only indication of chordate life in that particular stage of earth history, because in rocks of late Silurian age the record of the jawless vertebrates as preserved by the fossils of armored types becomes for the first time reasonably well documented. Here the jawless vertebrates appear as fairly adequate and identifiable fossils, and from late Silurian times on the evidence of the backboned animals in the rocks becomes increasingly complete and complex. Although the record of fossil vertebrates in upper Silurian rocks is good, it is in the sediments of Devonian age that the early vertebrates become really abundant. In Devonian times there was truly a great "explosion" in evolutionary development, a flowering of the vertebrates that established many evolutionary lines. Indeed the Devonian was a crucial period in the history of vertebrate life.

The earliest known vertebrates were, as we have noted, jawless animals belonging to the class Agnatha. They are designated collectively as ostracoderms, but in fact they can be classified in several orders that lived from the middle Ordovician to the end of the Devonian period. The orders of agnathous vertebrates, including the modern forms, are:

Cephalaspida (or Osteostraci): *Cephalaspis* is a characteristic genus; *Hemicyclaspis*

Galeaspida: Resemble cephalaspids generally, but with several distinctions.

Anaspida: *Birkenia* is a characteristic genus; *Pterygolepis; Jamoytius.*

Petromyzontida: The modern lampreys.

Heterostraci (or Pteraspida): *Pteraspis* is a characteristic genus.

Thelodontia (or Coelolepida): Poorly known forms of which *Thelodus* is typical.

Myxinoida: The modern hagfish.

The ostracoderms, being agnathous vertebrates, lacked jaws, they either lacked paired fins or had a single pair of fins behind the head, they lacked bony axial skeletons or vertebral columns, and they were typified by a rather well-developed armor of bony plates or scales. Except for these general resemblances, the several orders of ostracoderms were very different from each other; and it seems probable that they represent independent lines of evolutionary development, the result of a long period of phylogenetic divergence before their first appearance in the fossil record.

Among the best-known ostracoderms are the cephalaspids, typified by such upper Silurian and Devonian genera as *Cephalaspis* and *Hemicyclaspis*. These were small animals, generally not more than a foot in length. They were heavily armored, the head being protected by a strong, solid shield, and the body by vertically elongated, bony plates. The head shield was rather flat, but at its back margin it increased in depth where it joined the body. The body was elongated and of general fish-like form, terminating in a tail fin and bearing upon its dorsal surface a small median fin. At the corners of the head shield was a pair of lateral fins.

The head shield appears to have been formed by a solid piece of bone, covering the top and the sides of the head and folded under around the margins of the shield. The front of the shield was rounded in outline as seen from above, and at the back it flared into two lateral projections or horns that pointed posteriorly. In some of the cephalaspids these lateral horns were very long, in others they hardly existed at all. On its dorsal surface the head

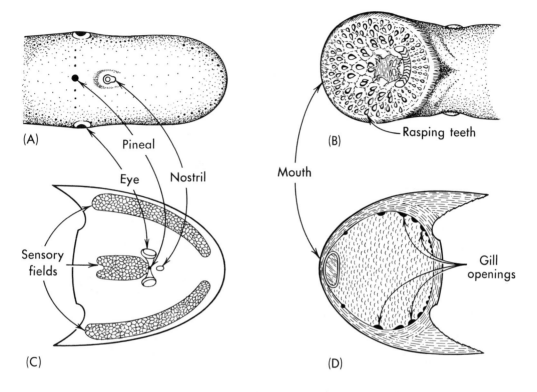

(A)

Pineal

Eye   Nostril

Sensory
fields

(C)

Mouth

Rasping teeth

(B)

Gill
openings

(D)

Figure 2-4. A comparison of the upper and lower surfaces of the head in jawless vertebrates. (**A** and **B**), the modern lamprey. (**C** and **D**), the Devonian cephalaspid ostrocoderm, *Cephalaspis*.

shield was pierced by the openings for the eyes, situated close together and staring straight up at the heavens, while between the eyes and slightly in front of them was a single nasal opening. This is the relationship of eyes to nostril that is seen in the modern lamprey. Furthermore, immediately behind the nostril and between the eyes in the cephalaspids was a well-developed pineal opening, a median communication from the external surface to the brain that may have had a function as some sort of light receptor. There were three areas on the head shield that were depressed and covered by small, polygonal plates. One of these areas was an elongated surface between and behind the eyes, the other two were elongated surfaces on either side of the shield and more or less parallel to its borders. The significance of these areas of plates is unknown, but it has been suggested that they were electric or sensory fields of some sort.

The under surface of the head in the cephalaspids was protected, not by a single shield but by a pattern of fine plates that in life must have been rather flexible. They enclosed the lower surface of the head completely, except for a ventrally placed mouth opening at the front end of the shield, and a series of small ventral gill openings on each side that followed the line of junction between the solid margin of the head shield and the ventral armor plates. In *Cephalaspis* there were on each side ten of these gill openings, corresponding to ten gill pouches that occupied each side of the head shield. Except for *Amphioxus* this is a high number of gills among vertebrates. In the more advanced animals that we commonly designate as fishes, the gills range from seven in the sharks to five in the bony fishes.

The internal structure of the head shield of the cephalaspids has been carefully studied. It indicates that the brain in these early ver-

Figure 2-5. The cephalaspid ostracoderm, *Hemicyclaspis*, a very early jawless vertebrate of late Silurian age. About one-third natural size.

tebrates was of a primitive type. As in the modern lampreys, there were only two semi-circular canals in the ear region of the cephalaspids, in contrast to the three canals in higher vertebrates. Perhaps the most remarkable feature in the cephalaspid brain was the presence of large nerve trunks that radiated out on either side, to supply the supposed sensory fields on the lateral surfaces of the head shield.

As mentioned above, the back of the head shield joined the body armor; perhaps it might be said that the head flowed into the body with no joint between them. Immediately behind the head the body was rather triangular in cross section with a flat base, corresponding to the flat under surface of the head, and with the sides rising to a narrow dorsal apex. The body tapered posteriorly to the base of the tail, where it bent upward to form a strong dorsal support for the tail fin, which was formed of soft rays. Such a tail, known as a heterocercal tail, is characteristic of many primitive aquatic vertebrates. By analogy with modern fishes it is evident that the body and tail in the cephalaspids provided the force that drove these animals through the water. The body plates were arranged in vertical rows, thus giving the ostracoderm a great deal of side to side flexibility. Rhythmic waves produced by muscular action passed alternately down either side of the body, and were transmitted to the tail, swinging it back and forth. The combination of these movements pushed against the water and drove the ostracoderm ahead. In addition, the heterocercal tail drove the animal down as well as forward in the water.

To prevent the body's rolling from side to side there was a single dorsal fin along the middle of the back, just in front of the tail. The front of this fin was stiffened by a spine. In the cephalaspids there were also paired fins that joined the body on either side immediately behind the "horns" of the head shield. These paired fins were scale-covered flaps, and nothing is known of their internal structure. Although it is doubtful that these fins are to be homologized with the true pectoral or shoulder fins of the higher fishes, they probably served much the same purpose as do the pectoral fins in the fish, in that they very likely helped to control balance while the ostracoderm was swimming. It is quite possible, however, that they might have been used to help control the direction of movement as well. For instance, they may have functioned in part as elevator planes, to counteract on occasion the downward push of the tail. The shape of the head shield would also have provided lift. In this respect, the cephalaspids enjoyed an advantage over most of the other ostracoderms, which lacked paired fins.

No internal skeleton has been found behind the head shield in the cephalaspids, or in the other ostracoderms for that matter. Such skeleton as existed in these animals must have been cartilaginous, and therefore has not been preserved in the fossil record.

These are the salient characters of a cephalaspid. What were these ostracoderms like in life? It seems evident from the flattened head shield, from the dorsal eyes, from the flat under surface of the body, and from the ventrally placed mouth that the cephalaspids were bottom-dwelling vertebrates. They probably lived in the shallow waters of streams and lakes, possibly at times in estuaries, where they groveled in the bottom mud, sucking up small food particles through their vacuum-cleaner type of mouth. The food being passed through the throat region by the force of the water that entered the mouth was evidently directed into the esophagus, and from there into the digestive tract, while water and probably a great deal of detritus escaped through the gill openings.

There were many kinds of cephalaspids characterized especially by the wide range of differentiation in the shape of the head shields. *Cephalaspis* itself was a more or less central type. As contrasted with this genus, there were cephalaspids with elongated head shields, others with very short, broad shields, some with tremendous lateral horns on the head shields, others with no horns at all, some with long rostral spikes directed anteriorly from the front of the head shield, others with dorsal spikes sticking up from the back margin of the shield. But except for these differences the cephalaspids were essentially similar in their basic structure.

Having had a glimpse of the Osteostraci, as exemplified particularly by *Cephalaspis,* we turn now to a brief survey of some of the other ostracoderms.

The Galeaspida is a newly recognized order of agnathans based on specimens recently collected from upper Devonian sediments of China. Galeaspids generally resembled cephalaspids, in that they had a large bony head shield and an ossified braincase. However, they lacked the sensory fields of cephalaspids, and the pineal opening was not present on the dorsal shield. Also in contrast to cephalaspids, galeaspids appear not to have possessed paired fins, and the axis of the tail turned down ventrally. Ostracoderms of the order Anaspida, like the cephalaspids, were covered with bony plates or scales. Among the anaspids the eyes were placed laterally, and between the eyes was the single opening for the nostril, and behind that the small pineal opening. Furthermore, there was a slanting row of about eight gill openings on each side in the anaspids, running down from the back of the head in the pharyngeal region; these can be compared with the gill openings of the cephalaspids.

In other respects, however, the anaspids were quite different in appearance from the cephalaspids. *Birkenia* and *Pterygolepis*, which can be described as typical upper Silurian anaspids, were tiny animals of rather fishlike form. *Birkenia* was not flattened for bottom feeding, as was *Cephalaspis*, but rather was comparatively narrow and deep bodied, as if it were adapted for active swimming. Instead of the solid head shield, so characteristic of *Cephalaspis*, the head region in *Birkenia* was covered with a complex pattern of small scales, many of them, especially those covering the throat, shaped somewhat like grains of rice. The mouth was terminal in position, not ventral, and was a transverse slit instead of a round, sucking opening. It had something of the appearance of a mouth, as we are generally accustomed to think of a mouth, yet even so it lacked true jaws. Behind the head the body was covered with a series of vertically elongated scales or plates, arranged in several longitudinal rows.

There were no paired fins in *Birkenia*. On each side, however, there was a small spine that projected from about the region where a pectoral or should fin normally might be placed, and behind, on the ventral surface of the body, there were some additional spines. In addition to these appendages, there were in *Birkenia* several median dorsal spines, running in a row along the back. There was a fairly well-developed anal fin. The tail is of particular interest in that it was fishlike in form, but with the lower lobe of the tail larger than the upper lobe.

It was pointed out above that a fish tail having the upper lobe longer than the lower is known as a heterocercal tail, and such a tail pattern is primitive in many of the aquatic vertebrates. The tail in *Birkenia*, with the lower lobe more prominent than the upper, is accordingly known as a reversed heterocercal tail. The heterocercal tail of *Cephalaspis* would have tended to drive that animal down, as we have seen, and this would have been important for a bottom-feeding ostracoderm. Conversely, the reversed heterocercal tail of *Birkenia* would have driven the animal up, and this would have been an advantage for such a heavily armored vertebrate if it were an active swimmer. It would seem very likely that *Birkenia* swam near the surface of the water, feeding upon plankton there, and for such a mode of life the reversed heterocercal tail stood it in good stead. But one wonders how *Birkenia* managed to swim with any degree of efficiency, even with its well-developed reversed heterocercal tail acting as a propeller. The dorsal spines could not have been very efficient as stabilizers to prevent rolling (some sort of fin membrane is really needed for this), and the two "pectoral" spines were certainly not of much use as stabilizers to control pitching or up and down motion, and yawing or side to side motion. Perhaps *Birkenia* was not a very efficient swimmer, but nevertheless was good enough for the times in which it lived. We might say that the mechanics of vertebrate locomotion in the water were still in an "experimental" stage during late Silurian times.

In some of the anaspids there was a marked reduction of the body armor. For instance, the little Devonian anaspid, *Endeiolepis*, was almost devoid of external bony scales. It may be that anaspids such as this were trend-ing in the direction that led eventually to the lampreys.

Among the earliest known ostracoderms are members of the Heterostraci. This order differs from the ostracoderms that have been described in that the body armor, though well developed, lacked true bone cells. There is quite a range of adaptive radiation in the Heterostraci, some of them being rather small, fusiform, free-swimming types, others being large, flattened, bottom-living animals. Of the former, *Pteraspis* of Silurian and Devonian age is typical. It was a small ostracoderm, heavily armored. There was a heavy head shield, but, instead of being flattened as was the head shield in the cephalaspids, it was rather rounded in cross section. At the front it merged into a separate element, a long, sharp beak or rostrum that extended far beyond the ventrally located mouth. This mouth was a transverse slit, placed near the front on the lower surface of the head shield, and stretched across it was a transverse series

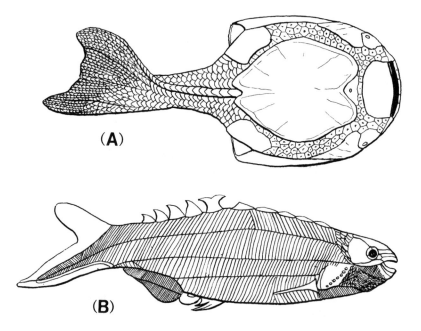

(A)

(B)

Figure 2-6. (**A**) the heterostracan *Drepanaspis* and (**B**) the anaspid *Birkenia*. (A) is about one-third natural size; (B) natural size

of slender plates that may have acted something like jaws. But they can in no way be compared with true jaws. The eyes were situated laterally, on each side of the head shield, and this lateral separation of the eyes is quite typical of the Heterostraci. There was frequently a well-developed pineal opening on top of the head, but there was no visible nasal opening, as in the cephalaspids. At the back of the head shield and on each side was a single exit for the gills, and on the dorsal midline a long spine projected up and back from the rear border of the shield. Cornual plates also projected laterally behind the gill openings and certainly had a function in providing lift and control in the absence of paired fins.

The body in *Pteraspis* was covered with a pattern of small scales; and there was a reversed heterocercal tail, an indication that this ostracoderm, like *Birkenia*, may have been an active swimmer, feeding near the surface. But there were no well-developed median fins, and no paired fins at all. Once again, we are inclined to wonder how efficient a swimmer this little ostracoderm may have been.

The bottom-feeding Heterostraci are typified by such forms as *Drepanaspis*, a lower Devonian genus. These were among the largest of the ostracoderms, ranging up to a foot or more in length. *Drepanaspis* was very flat indeed and very broad. The eyes were set far apart on the sides of the broad head shield, and there was a wide mouth, located on the front of the shield and not in a ventral position. The head shield was covered by a series of plates of varying size, with a very large, oval plate taking up most of the middle region of the dorsal surface. The tail region was comparatively small as contrasted with the broad, flat head shield, a development that is characteristic of many bottomliving vertebrates.

The one other order of vertebrates that has been placed among the ostracoderms is the upper Silurian and lower Devonian Thelodontida, typified by such genera as *Thelodus* and *Lanarkia*. Very little can be said about these early vertebrates, because the fossils are very diffi-cult to study, being little more than impressions in the rock. It is evident, however, that the thelodonts were flattened, with laterally placed eyes. The tail was forked, and apparently of the heterocercal type. The thelodonts differ from all other ostracoderms in that the body covering consisted of minute denticles, rather than flat plates. These denticles, though small, are to be considered as much-reduced plates and not dermal denticles such as those found on the sharks. Very little is known of the internal skeleton in the thelodonts. It is quite possible that these enigmatic little vertebrates were specialized ostracoderms in which there was a fragmentation of the head and body armor to form a sort of shagreen covering of minute denticle-like plates. Such ostracoderms would have been mobile as compared with their heavily armored contemporaries, squires in chain mail rather than knights in plate armor.

THE EVOLUTIONARY POSITION OF THE OSTRACODERMS

The geologic history of the ostracoderms is comparatively restricted, ranging from the middle Ordovician period through the Devonian period. It may be that these vertebrates arose in Cambrian times, it may be at an earlier date, from unarmored ancestors, of which *Jamoytius* is a possible relict. By Devonian times the ostracoderms had become adapted to varied modes of life and various ecological niches, as indicated by the wide divergence in the several orders now known from the fossil materials. And for a time they were successful.

In the end, however, they were unable to compete with the more advanced jawed vertebrates that were evolving rapidly and along many different lines during the Devonian period. So it was that at the close of the Devonian period the ostracoderms disappeared, and except for the lamprey and hagfish, the agnathous vertebrates succumbed to competition from their more efficient contemporaries. A vertebrate without jaws was efficient after a

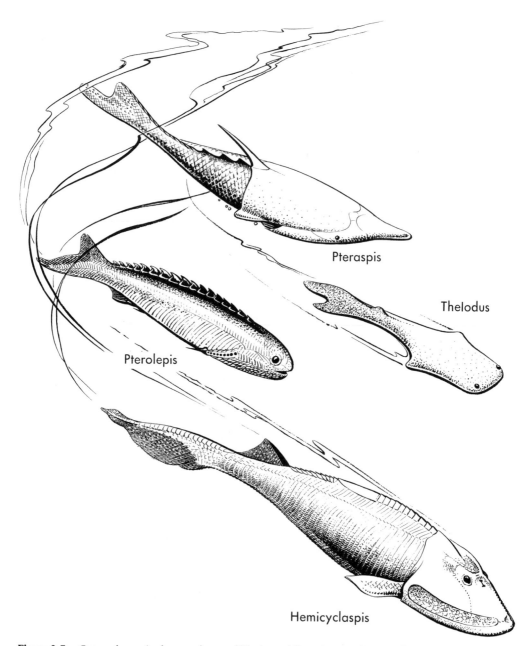

Pteraspis

Thelodus

Pterolepis

Hemicyclaspis

Figure 2-7.   Ostracoderms, jawless vertebrates of Silurian and Devonian age, drawn to the same scale. *Hemicyclaspis* is a cephalaspid, *Pterolepis* an anaspid, *Pteraspis* a heterostracan, and *Thelodus* a thelodont. Prepared by Lois M. Darling.

fashion, but unless it became adapted to certain very specialized habits it was not well enough equipped for survival in a world where a pair of upper and lower jaws had evolved as a food-gathering mechanism.

Were any of the known jawless vertebrates of Ordovician to Devonian times ancestral to the jawed vertebrates or gnathostomes? We would expect that among the great variety of ostracoderms there would be some group containing annectent forms, linking the Agnatha with the higher fishes. Yet the fossil record fails to show any such link. This is not surprising when it is remembered that the Ordovician to Devonian ostracoderms were specialized and widely divergent animals. They were, in effect, the products of millions of years of evolutionary development, and in them we see animals approaching the end of their several evolutionary histories. Consequently it is to an earlier geologic age, possibly to a time as early as the Cambrian period, that we must look for the ancestors of the jawed vertebrates. So far the fossil record has yielded no information on this very important problem.

Acanthodians

Acanthodians and
Placoderms

3

THE ORIGIN OF JAWS

The history of life, like human history, has been marked by certain great developments rising above the general level of events that collectively make up the record. These outstanding evolutionary developments are in the nature of revolutions, affecting profoundly the phylogenetic trends that follow them, just as great historical revolutions, like the American Revolt against Britain or the French Revolution have affected the subsequent histories of the peoples concerned with them. A better comparison might be with the peaceful revolutions in human arts and industries, such as those brought about by the development of the printing press or the application of steam power to machinery.

One of the great events or revolutions in the history of the vertebrates was the appearance of jaws. The importance of this evolutionary development can hardly be overestimated, for it opened to the vertebrates new lines of adaptation and new possibilities for evolutionary advancement that expanded immeasurably the potentialities of these animals. The jawless vertebrates were definitely restricted as to their adaptations for different modes of life, and it is possible that the ostracoderms of middle Ordovician to Devonian times had explored and virtually exhausted the evolutionary possibilities and the ecological niches open to animals of this type. Animals without jaws can evolve as bottom feeders, as did some of the ostracoderms, or they can develop movable plates around the mouth opening that serve after a fashion as weak "jaws," as did some other ostracoderms. Or they can become parasites, like the modern lampreys and hagfishes. Yet even under the most favorable conditions they are denied the possibilities of development that are open to animals with jaws; they simply do not have the structural mechanisms to exploit the opportunities that are available to jawed animals. So it is that the appearance of jaws marked a major turning point in vertebrate evolution.

The appearance of the jaws in vertebrates was brought about by a transformation of anatomical elements that originally had performed a function quite different from the function of food gathering. Here we see the working of a process that has taken place innumerable times in the course of animal evolution; indeed, much of the progress from earlier to later forms of animals has been brought about through the transformation of structures from one function to another. The origin and evolution of the jaws are an excellent example of this evolutionary principle.

The jaws were derived originally from gill arches. It will be remembered that the ostracoderms had a large number of gills, as many as ten in *Cephalaspis*, and it seems probable that the presence of comparatively numerous gills with cartilaginous or bony supports was typical of the primitive vertebrates. At an early stage in the history of the vertebrates at least one and probably two of the original anterior gill arches were eliminated, while another arch, probably the third one in the series, was changed from a gill support into a pair of jaws. This transformation actually was not as radical a shift as on first sight it may appear to have been. Each gill support or arch in the primitive vertebrate was formed by a series of several bones, arranged somewhat in the fashion of a V turned on its side, with the point directed posteriorly. Imagine several such lazy V's in series, thus: > > >. Imagine also the first of these (morphologically the third of the original gill arches) supplied with teeth, and hinged at the point of the V, and you have the primitive vertebrate jaw, in alignment with the gill arches behind them. It is immediately apparent that the transformation of a gill arch into jaws was a natural evolutionary development, perhaps the simplest possible solution to the basic problem of developing a pair of vertebrate jaws.

There are various facts that support such an origin for the jaws in the vertebrates. For instance, a study of the embryological development of certain modern fishes indicates this

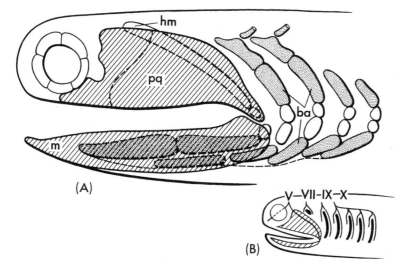

Figure 3-1. The head and gill region in primitive jawed vertebrates. **(A)** Diagram to show the relationships of the primary upper and lower jaws to the branchial arches in a primitive gnathostome, the acanthodian fish, *Acanthodes*. Notice how the elements of the primary jaws are in series with the branchial arches behind them. **(B)** Relationships of certain cranial nerves to the primary jaws and the gill slits in a shark. Here we see the first gill opening reduced to a small spiracle. For abbreviations, see pages 447–449.

origin very strongly. Moreover the arrangement of the nerves in the head region of sharks shows that the jaws are in series with the gill apparatus. Thus the branching of the fifth cranial nerve in the fish, known as the trigeminal nerve, whereby one branch runs forward to the upper jaw and another down to the lower jaw, is identical with the branching of certain other cranial nerves, whereby one branch runs forward in front of each gill opening and one down behind the gill opening. Finally, a simple inspection reveals that in various primitive jawed vertebrates the jaws are in series with and similar to the gill arches.

THE ACANTHODIANS

The earliest known gnathostomes—vertebrates with true jaws—were small forms known as acanthodians, or "spiny sharks," this latter name being based entirely on the general superficial resemblance of these primitive gnathostomes to sharks rather than on any real relationships. Unfortunately, the oldest acanthodians are not known from complete skeletons, but instead are known from numer-

ous teeth, scales, and spines, found in rocks of late Silurian age. Although these ancient remains are incomplete, they give us a clue as to the time when jawed vertebrates entered the evolutionary scene. The acanthodians show very nicely the primitive upper and lower jaws in early vertebrates, as can be seen in the figure of *Acanthodes*, a middle and late Paleozoic genus. In these ancient fishes there was an enlarged upper jaw known as the palatoquadrate, opposed by a well-developed lower jaw, or mandible provided with sharp teeth. Immediately behind these scissorlike primary jaws was a gill arch—the hyoid arch—of which the upper bone, designated as the hyomandibular bone, was enlarged (with its upper end articulating against the palatoquadrate) to form a prop or a connection between the braincase and the jaws. This condition foreshadows the structure of the skull and jaws in the higher fishes, in which the jaws crowd back on the hyomandibular, so that the gill slit between the palatoquadrate and the hyomandibular becomes reduced to a small opening, or disappears completely as it does in most of the bony fishes. The acanthodians are particularly important because of their primitive jaw structure, which nevertheless is basically the same as that of the bony fishes. Here, we see in the fossil record evidence of the first steps in gnathostome evolution.

The acanthodians form a readily recognizable group of extinct fishes, but their position relative to other major groups of early gnathostomes has never been satisfactorily determined. By some students they are recognized as an independent category having the rank of a class, by others they are grouped with the sharks, and by still others they are considered to be bony fishes. Here they are considered as a separate class of vertebrates, but such recognition of their independent status is perhaps an acknowledgment of our lack of understanding of their actual relationships.

The early acanthodians are typified by the genus *Climatius*, of late Silurian and early Devonian age, and it is a good example of these ancient gnathostomes. They were small vertebrates, only a few inches in length, with fishlike bodies tapering from the front portion to the tip of the tail. In the posterior region the body was turned up, and there was a fin beneath this uptilted portion forming a heterocercal tail—the type of tail that seems to be so commonly developed in primitive swimming vertebrates. In addition to the tail fin, *Climatius* was well supplied with median and paired fins. Thus, there were two large, triangular dorsal fins on the back, each consisting of a web of skin supported along its front or

leading edge by a strong spine. Below the more posterior of the dorsal fins, and balancing it, was an equally large and similarly shaped anal fin, also supported along its front edge by a spine. Then there was a pair of anterior or pectoral fins immediately behind the skull and another pair of posterior or pelvic fins in front of the anal fin, while between these two there were five pairs of smaller fins, running along either side of the ventral portion of the body.

Climatius was protected by a dermal armor of small rhombic or diamond-shaped scales that covered the entire body and continued over the head, where they took the form of regularly arranged plates of small size. The head plates were never expanded into large units like the skull bones or head shields in most other vertebrates.

Climatius had large eyes, each surrounded by a ring of bony plates, and the eyes were placed very far forward, so that there was a very restricted nasal region in front of them. It would therefore seem evident that the sense of sight was dominant in these ancient fishes, whereas the olfactory sense played but a minor role in their life. The upper jaw or palatoquadrate was generally ossified in three separate pieces, and it commonly was devoid of teeth. The mandible, as we have seen, was well supplied with teeth. On the sides of the head were gill coverings or opercular flaps, one to each of the five gill arches, and over these there was a larger opercular covering of stiff, bony rods. In effect, therefore, *Climatius*

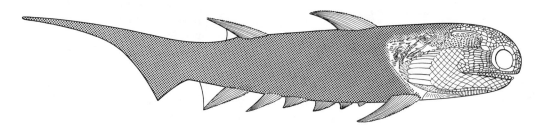

Figure 3-2. The acanthodian, *Climatius*, about natural size. In this restoration the scales are cut away behind the operculum, to show some of the elements of the branchial arches. Note the several pairs of spines along the ventral surface of the body.

had an operculum, somewhat similar to the operculum in the higher fishes but quite different in structure and origin.

This description of *Climatius* gives a picture of the early and generalized acanthodians. Evolution in these fishes, through the remainder of Paleozoic times to the close of the Permian period, was marked by variations on the primitive pattern as exemplified in *Climatius*. The later acanthodians evolved by changes of proportions so that some of them became long and slender, others rather deep bodied. In some there was a loss of fins, both median and paired; in others, like the Devonian form *Parexus*, there was a tremendous exaggeration of the dorsal fin spines, especially the anterior one, to form great spikes on top of the body, stretching far back over the tail region.

The significance of these spines in the acanthodians is not clear; perhaps they were to some extent protective characters that helped their bearers to survive the attacks of large vertebrate and invertebrate predators. However that may be, it would seem that the spines (and the opercular cover) were rather specialized features in animals otherwise comparatively primitive.

It is not surprising that these vertebrates were primitive in many respects, for the geologic record indicates clearly that the acanthodians were among the first of the known gnathostomes, appearing in late Silurian times. By early Devonian years the acanthodians had reached the peak of their evolutionary development and from then on until the close of the Paleozoic era theirs was a declining history. These were freshwater animals, living in rivers, lakes, and swamps of middle and late Paleozoic times, and although they were doomed to eventual extinction they were for the time of their existence characteristic elements of the ancient vertebrate faunas of the continental regions.

APPEARANCE OF THE PLACODERMS

The ancient gnathostomes known as placoderms evolved along varied and divergent lines during early Devonian to earliest Carbonifer-

ous times, and for a time some of them were the dominant vertebrates of the waters in which they lived. But the dominance of the placoderms was of limited duration, because by the end of Devonian period most of them became extinct, and by the end of the early Carboniferous period all of the placoderms had vanished from the earth.

The placoderms were a rather heterogeneous lot, and it is quite possible that, taken together, they are not a truly natural or monophyletic assemblage of vertebrates. It is convenient, however, to combine in one group the several orders of predominantly Devonian jawed vertebrates that were evolving along lines distinct from those of the acanthodians, early sharks and bony fishes. We might regard the placoderms as ancient "experiments" in the evolution of the jawed vertebrates. Perhaps the sharks and the bony fishes of Devonian times were experiments too, but at least they were successful, whereas the placoderms, like the acanthodians, were evolutionary experiments that ultimately failed.

Since the placoderms were so remarkably diverse, exhibiting disparate and often seemingly grotesque adaptations, the classification of these ancient gnathostomes is frustratingly difficult. One arrangement of placoderm orders is as follows:

Stensioellida   Early primitive
Pseudopetalichthydia   placoderms from
Acanthothoraci   lower Devonian rocks.
Rhenanida: "Shark or ray-like" placoderms.
Petalichythida: A restricted group of specialized placoderms.
Arthrodira: For a time, the dominant vertebrates of the Devonian period.
Ptyctodontida: Small, "chimaera-like", mollusk-eating placoderms.
Phyllolepida: Flat, "degenerate" placoderms.
Antiarchi: Small, bottom-living, heavily armored placoderms.

In spite of the diversity that characterizes their evolution, the placoderms shared certain basic features. For instance, in addition to the lower jaws, typical of the ancient gnathostomes, the placoderms had strong upper jaws that were firmly fused to the skull. The gills of placoderms were situated far forward beneath the skull, there was a joint between the skull and the shoulder girdle, and dermal bones covered the head and shoulder girdle. In these early jawed fishes there were always paired fins.

In many respects the various placoderms were specialized; therefore, we cannot look at any genus or even at any larger category as actually ancestral to the higher fishes, although some authorities think that those shark-like fishes known as the Holocephali may be related to the placoderms. Yet, it is evident that the placoderms evolved from very primitive gnathostomes and, when taken together, they indicate many of the characters that we would expect to find in the ancestral jawed vertebrates. Therefore by projecting back from the known placoderms it is possible to visualize certain stages of evolution, some of them evanescent, that vertebrates attained in their first steps beyond the jawless condition.

EARLY PLACODERMS

One might expect the more generalized of these vertebrates to have appeared at the beginning of placoderm evolutionary history. Such is not the case; many of the earliest placoderms, of early Devonian age, would seem to be among the most "aberrant" members of the group. Consequently the first stages of the placoderm record reveal a confusing array of divergent genera, for the most part contained within two orders, the Rhenanida and the Petalichthyida.

The Rhenanids

The group of placoderms known as the rhenanids show certain resemblances to the sharks, and for this reason it has been suggested by some students that these ancient Devonian vertebrates were early relatives of the sharks. The genus *Gemuendina*, from Devonian sediments of central Europe, is the best known of the rhenanids and may be taken as a typical representative of the order. This ancient fish was flattened, with a broad head, and a tapering body covered with small tubercles, much like the denticles that cover the body in modern sharks. However, these tubercles represent the surface ornamentation of plates that were small over much of the body, but became enlarged in the forward region. There was a terminal mouth, equipped with pointed teeth, and the eyes and the nostrils were located on the dorsal surface of the head. Perhaps the most striking feature of *Gemuendina* was the enlarged pair of pectoral fins, giving to this animal an appearance remarkably similar to that of some of the modern rays and monkfishes. Yet the jaw structure of *Gemuendina* is of the primitive placoderm type, so that there is every reason to think that the resemblances of this vertebrate to rays were the result of evolution along similar lines in two separate animal groups. One might say that *Gemuendina* anticipated in Devonian times the *habitus*, the mode of life and adaptations to environment, that was followed in later times by the bottom-living skates and rays. This is an excellent example of *convergence* in evolution—the similar development of unrelated animals for a particular mode of life.

The Petalichthyds

The order now under consideration is based to a large degree upon the genus *Macropetalichthys*, although some other genera, notably *Lunaspis*, belong to the group.

Macropetalichthys had a strong, bony head shield, generally comparable to the head shield in the arthrodires. The ossified braincase within the head shield has been described within recent years, and from this description it has been possible to reconstruct the brain of *Macropetalichthys*, which was similar to the brain of the arthrodires. Behind the head shield there were thoracic plates and a pair of pectoral spines. In *Lunaspis*, there were well-

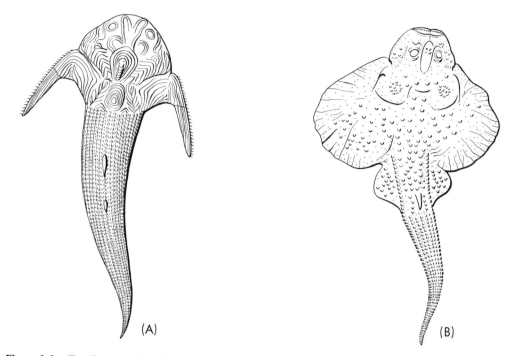

Figure 3-3. Two Devonian placoderms from Europe. **(A)** *Lunaspis*, about one-half natural size. **(B)** *Gemuendina*, one-half natural size.

ossified head and thoracic shields, the two separated by a joint but not by a highly mobile hinge. The thoracic shield bore very large spines. In this animal there was external armor behind the thoracic shield, consisting of large scales. The macropetalichthyds obviously were related to the arthrodires, but it is evident that the two groups diverged at an early stage in their histories to follow separate evolutionary paths.

THE ARTHRODIRES

Certainly the best known, the most varied, and the most spectacular of the placoderms were the arthrodires, often known as the armored fishes or the joint-necked fishes of the Devonian period. These placoderms achieved a position of dominance during late Devonian times, and for a brief spell they were in some respects the most successful of all vertebrates. But their success was short lived, and with the close of the Devonian period the arthrodires became extinct.

A rather typical arthrodire was *Coccosteus*, found in the Old Red Sandstone beds of northern England and Scotland. This was a comparatively small vertebrate, ranging from about one to two feet in length. The body in *Coccosteus* was of fishlike form. But here ends the resemblance to fishes as we know them, for in front the head and shoulder region of this early gnathostome was heavily armored, whereas in the back the body would seem to have been completely naked.

The skull in *Coccosteus* consisted of a series of large bony plates, firmly joined to one another along sutures. These plates can in no way be compared directly with the skull bones of the bony fishes; consequently they have been designated by a series of names that is peculiar to the arthrodires. In many of the arthrodires the skull was quite deep, and strongly arched across its top from side to side. The eyes were large and situated near the front of the skull, and each eye socket contained a ring of four plates, the sclerotic plates, that protected the eyeball. The nostrils were small and placed at the very front of the skull.

There was a very strong lower jaw, hinged to the back of the skull. This lower jaw consisted of a single bone on each side, designated as the inferognathal bone, and anteriorly the upper edge was scalloped into a series of points, superficially like teeth. Opposing the lower jaw were two plates of bone on each side, attached to the front margin of the skull; and these bones, known as the anterior and posterior supragnathals, were also toothlike in shape. Thus this arthrodire had exposed bony plates that functioned as teeth. In some of the more advanced arthrodires, such as *Dinichthys*, these plates were nicely shaped to form scissorlike cutting edges. They must have been very efficient shearing mechanisms.

At the back of the skull on each side was a strong plate known as the paranuchal plate, and on its free surface it carried a strong socket. Behind the skull was a ring of heavy shoulder or thoracic plates, encircling the front of the body, and on one plate on either side of this ring there was a strong "ball" or condyle that fitted into the socket of the paranuchal plate of the skull. In this manner the skull was hinged on each side to the shoulder armor, and since the two hinges were aligned from side to side the skull could be moved up and down on a horizontal axis. In *Coccosteus* as in many of the arthrodires there was a considerable gap between the back of the skull on its dorsal surface and the front of the body armor, which would seem to indicate that there was freedom for strong upward movements of the head. It is generally assumed that these arthrodires opened the mouth, not only by dropping the jaw, as is normal among jawed vertebrates, but also by raising the skull with relation to the body, which would have allowed for a very wide gape. A widely gaping mouth is most advantageous to aggressive carnivores, such as the arthrodires would seem to have been. It is probable, however, that the ability to raise the skull originally evolved as a mechanism to

force water through the gills, crowded as they were beneath the neurocranium.

There was no gill cover or operculum in *Coccosteus*, so it would seem probable that the gills were located entirely in the "cheek" region of the skull on either side and that they opened at the back of the skull.

The heavily armored head and thoracic region of *Coccosteus* is in striking contrast to the naked back portion of the fish. No dermal covering is known from the posterior body portion or the tail in any of the arthrodires. There was a well-developed supporting column, with a cartilaginous notochord, bordered above and below by spines, but without any vertebral discs or centra. The tip of the column was turned up somewhat, so it appears that the tail was of the heterocercal type. In addition to the tail fin there was a median dorsal fin, as indicated by fin rays projecting up from the spines of the vertebral column. Paired fins were not well developed in the arthrodires, but in *Coccosteus* there is evidence that a pair of pectoral fins was situated close behind the skull, and posteriorly there was a pair of pelvic fins, much smaller than the pectoral fins.

The earliest known arthrodires were small and somewhat flattened vertebrates. They had the well-armored head shield hinged to a thoracic shield that is typical of the arthrodires, but in some respects they were very different from the more advanced forms, such as *Coccosteus*. These early arthrodires, as exemplified by the Devonian genus *Arctolepis*, frequently bore very long, strong pectoral spines solidly attached to the body armor. The jaws were rather weak. All in all the evidence would seem to indicate that these earliest of the arthrodires were bottom-living animals, perhaps competing to some extent with the early ostracoderms. The greatly enlarged pectoral spines of the early arthrodires are puzzling; it has been suggested that they served as anchors, to hold the animal steady against rapidly flowing currents in the streams that they may have inhabited.

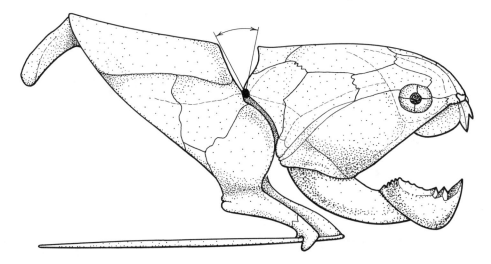

Figure 3-4. The gigantic upper Devonian arthrodire, *Dinichthys*. The head and thoracic shield, shown here, may be eight or ten feet long. Notice in this placoderm the body plates in the upper and lower jaws that functioned as cutting blades. The hinge between the head shield and the thoracic shield allowed the head to be raised as the lower jaw was dropped, thus making possible a large bite.

From such a beginning the trend in arthrodiran evolution was generally toward an increase in size and an increase in mobility. By middle and late Devonian times the trend had progressed to such a degree that the arthrodires had become fast-swimming, aggressive predators, of which *Coccosteus* was typical.

However, the peak of arthrodiran evolution was reached by the giant genera *Dinichthys* and *Titanichthys*, found in the upper Devonian Cleveland shales of Ohio. These were enormous fishes with huge skulls and strong jaws, equipped with large cutting plates. *Dinichthys* attained lengths of thirty feet. The giant of its day, it probably preyed upon any other upper Devonian fishes that it could catch. *Dinichthys* was the ruler of its environment.

PTYCTODONTS AND PHYLLOLEPIDS

Before the placoderms became extinct during the transition from Devonian into Mississippian times, several aberrant lines had evolved. The ptyctodonts, typified by such genera as *Rhamphodopsis*, were very small placoderms with reduced head and thoracic armor and with a large dorsal spine in addi-

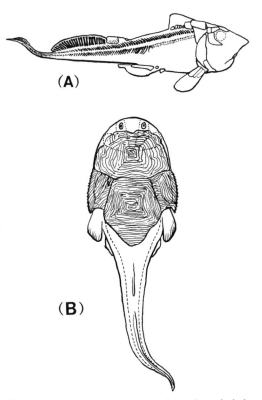

(A)

(B)

Figure 3-5. Representative ptyctodont **(A)** and phylloledid **(B)** placoderms. (A) is one-half natural size; (B) one-fifth natural size.

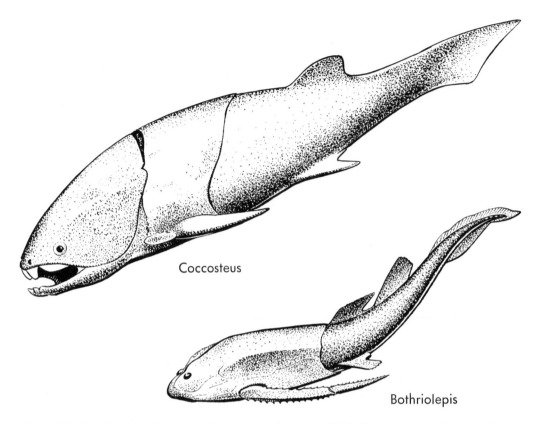

Figure 3-6. Two Devonian placoderms: *Coccosteus*, an arthrodire, and *Bothriolepis*, an antiarch. Drawn to the same scale, each about one-half natural size. Prepared by Lois M. Darling.

tion to the pectoral spines. The dental plates were heavy, and it would appear that these little placoderms were adapted to a diet of mollusks. The phyllolepids such as *Phyllolepis* were medium sized, but very flattened, arthrodires, with ornamented plates. They were probably bottom-living forms.

THE ANTIARCHS

In the upper Devonian rocks of the Gaspé peninsula, Quebec, skeletons of a small placoderm known as *Bothriolepis*, a representative of the antiarchs, are found, sometimes in considerable numbers. *Bothriolepis* was heavily armored anteriorly, having a head shield, with a body shield behind it. The head shield was short, and was composed of large plates; in contrast, the boxlike body shield was rather long.

In *Bothriolepis*, as is usual among the antiarchs, the eyes were situated very close together on the top surface of the head shield, and between them was the pineal opening. On the bottom of the head shield was a small mouth, equipped with very weak lower jaws. With dorsally located eyes and with a ventral mouth *Bothriolepis* superficially resembles some of the ostracoderms, the result of convergent evolution toward similar modes of life in the two classes of vertebrates.

Impressions of portions of the soft anatomy that have been preserved in *Bothriolepis* show, among other things, that this early vertebrate had well-developed and probably very functional lungs, opening from the pharynx. Here

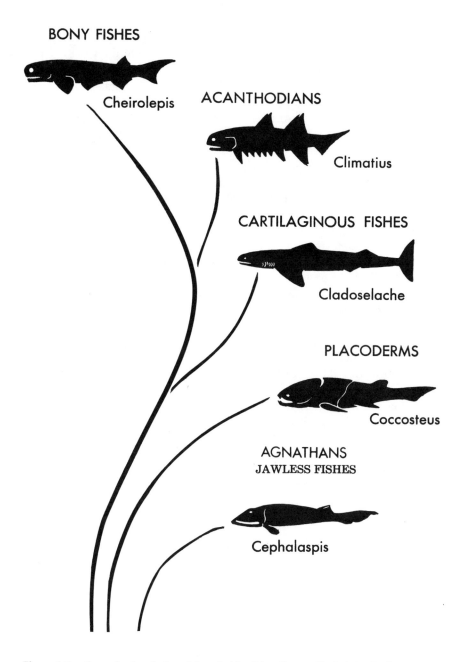

BONY FISHES

Cheirolepis

ACANTHODIANS

Climatius

CARTILAGINOUS FISHES

Cladoselache

PLACODERMS

Coccosteus

AGNATHANS
JAWLESS FISHES

Cephalaspis

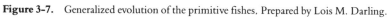

Figure 3-7.   Generalized evolution of the primitive fishes. Prepared by Lois M. Darling.

is one of the many lines of evidence to show that lungs were primitive in the early vertebrates and were common to almost all the principal lines of piscine evolution in the Devonian period.

This antiarch had long, pointed, armored pectoral appendages, attached to the front portion of the body shield by a joint that permitted free motion. Moreover, these "arms" in *Bothriolepis* were hinged in the middle so that they might have been bent to some degree. Evidently they were useful to the animal in getting about; possibly they served like movable hooks to pull *Bothriolepis* around on stream or lake bottoms.

Behind the body shield the body was naked in *Bothriolepis*, although in some other antiarchs, such as *Pterichthyodes* from the Old Red Sandstone of England, the posterior body region and the tail were covered with scales. In *Bothriolepis* the body was elongated and slender behind the thoracic shield, tapering posteriorly to form a sort of heterocercal tail. In addition to this tail there was a median dorsal fin. It is evident that the general plan here is similar to that of the arthrodires, and it would seem logical to suppose that the antiarchs represent a branch separating from an ancestry held in common with the arthrodires, and evolving along a trend toward bottom living and bottom feeding.

INTERRELATIONSHIPS OF PRIMITIVE FISHES

Prior to late Silurian time, no jawed vertebrates are known in the fossil record. But in the relatively short period of time from late Silurian to early Devonian, four major groups of jawed vertebrates first appear: acanthodians, placoderms, cartilaginous fishes, and bony fishes. Thus, no clear indication of the relationships among gnathostomous fishes can be determined from their stratigraphic order of occurrence in the rocks. Instead we must look to comparisons of their anatomy to give us clues concerning their evolutionary connections.

Acanthodians possess several features that indicate their close relationship to bony fishes and, to a lesser degree, cartilaginous fishes. The connection of placoderms to any of the other three groups, however, is not clear. They have been allied with cartilaginous fishes on the one hand, or with bony fishes and acanthodians on the other. However, recent studies of the structure of gnathostome jaws and teeth suggest that placoderms evolved from very primitive jawed vertebrates separately from the lineage that led to the three other groups of jawed fishes. Thus, placoderms may represent a distinct line of gnathostomous vertebrates.

Sardines

The Success of Fishes

DESIGN FOR SWIMMING

Many of the primitive vertebrates that we have surveyed up to this point have been designated as early experiments in the evolution of animals with backbones. We might compare them in a general way with the first weird and wonderful automobiles that managed to get along and survive because there were at the time no automobiles of better design to compete with them. The early vetebrates, like the early cars, followed various lines of development, of which many were doomed to quick failure. For a time the ostracoderms were reasonably successful vertebrates, as were the early gnathostomes, the acanthodians and the varied placoderms, but as other patterns of vertebrate organization became established, and evolved, these earlier vertebrate patterns were suppressed. The ostracoderms became extinct, and today only two types of agnathans survive as relicts of the jawless vertebrates that once had been so abundant throughout the world. The acanthodians and placoderms also became extinct, so that today there are no survivors of these once briefly successful classes of vertebrates.

Many factors possibly were contributory to the disappearance of the ostracoderms, acanthodians and placoderms, but probably the rise and development of the first bony fishes and sharks, the vertebrates that we generally think of when we speak of fishes, was the principal cause for the decline of the pioneer vertebrates. Among the "higher" fishes designs for life in the water evolved that were definitely superior to any of the plans developed among the ostracoderms or early gnathostomes with the result that these fishes prevailed.

What are the factors of design and of life processes that have made the sharks and the bony fishes so eminently suited for life in the water? A simple answer to this question is that these fishes, from the beginning of their evolutionary history, have been primarily very active swimmers. Of course there are many of them that have departed from the active life,

but the central types have been and are good swimming vertebrates. It was probably because of their superior design for swimming that the early sharks and bony fishes took a lead over their ostracoderm, acanthodian and placoderm contemporaries.

The typical shark and bony fish is streamlined, as were many ostracoderms, acanthodians and placoderms, with a body well adapted to rapid movement through the water. The head of the fish acts as an entering wedge, to cleave through the dense medium in which the animal lives, whereas at a point not far behind the head the body reaches its maximum size. From here back the body decreases in its dimensions to the stem or peduncle of the tail, and this decrease in body height, width, and girth from front to back allows the animal to slip through the water with a minimum amount of disturbance, or turbulence. So far we can see little evidence that such a fish is better adapted for active swimming than were ostracoderms and early gnathostomes; it is only when we consider the fins that the superiority of this fish becomes apparent. The ostracoderms, acanthodians, and placoderms had fins variously developed, but in none of the early vertebrates was there a complete array of mobile fins so well perfected as in the higher fishes.

In a typical modern fish there is a large and efficient tail fin or caudal fin, the function of which is to help drive the animal through the water by a back and forth sculling action. Rhythmic and alternate muscular waves pass down the sides of the body and are transmitted to the tail, and these pulsations push back against a column of water, thereby thrusting the fish forward. In addition to the caudal fin there is a median dorsal fin, or frequently two dorsal fins, on the back of the fish. These fins are stabilizers that prevent the animal from rolling as it swims through the water, and they also function as a dorsal keel to prevent side slip. Likewise, there is generally a median ventral fin, the anal fin, also functioning as a stabilizer and keel.

Finally there are the paired fins, consisting of an anterior or pectoral pair and a pelvic pair that may be either anterior or posterior in position. These fins are mobile in the advanced fishes and are important for controlling movements. They may act as planes or elevators to assist the fish in going up or down through the water, they may act as rudders to assist the fish in turning sharply right or left, and they may act as brakes enabling the fish to stop quickly. Indeed, the paired fins can be used in many fishes for "backing up."

The origin of paired appendages such as the fins of modern fishes is an unsolved problem. An early theory held that paired appendages evolved from gills and their supporting gill arches. Studies in the development of modern fishes have shown, however, that the fins of embryos begin as a pair of elongate outpocketings or folds along the sides of the body. These folds develop further into a continuous, fin-like structure that later divides into separate fins through the death of cells between them. These observations and the presence of paired ventrolateral ridges or metapleural folds in the primitive chordate *Amphioxus* (*Branchiostoma*) support the fin fold theory, which is generally accepted today. It holds that paired appendages evolved from a pair of continuous, ventrolateral fin folds in the adults of primitive vertebrates. A single, continuous fin fold is thought to have run along the dorsal surface as well. Local breakdown of tissues in specific places along the fin folds resulted in separate fins. It is interesting to note that the fins of the ostracoderm *Jamoytius* (Fig. 2-3) very closely resemble the continuous fin folds hypothesized for primitive vertebrates.

So with a combination of median and paired fins the fish is well adapted for active life in the water. In addition, modern fishes have evolved along many lines for efficient feeding and for protection, and are generally abundantly endowed for reproducing themselves. Consequently these are at the present time the most numerous of the vertebrates, and, if success is to be measured in numbers,

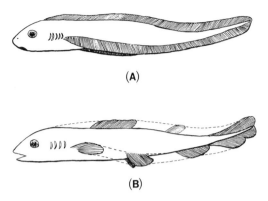

Figure 4-1. Illustrations showing hypothetical stages in the development of paired appendages in vertebrates. **(A)** An early vertebrate (fish) with a single, continuous fin along the dorsal surface and paired continuous fins along the sides of the body. **(B)** A later evolutionary stage in which the continuous fins have been broken into discrete fins, with the lateral ones being paired.

fishes are undoubtedly very successful animals. The species of modern bony fishes outnumber by far all other species of recent vertebrates combined, and in numbers of individuals some species of marine fishes reach population figures of almost astronomical magnitude.

THE HYOID ARCH

The rise of the higher fishes was marked among other things by an important anatomical development that contributed to the advancement of these vertebrates. Early in the evolution of fishes there was a specialization of the first arch behind the jaws, the hyoid arch, so that the upper bone of the arch was transformed into a sort of prop or connecting element, to join the jaws with the braincase. This transformed bone is known as the hyomandibular, and it has played an important role in the evolution of the fishes and of the land-living animals that evolved from the fishes. In the most primitive jawed fishes the upper jaws, designated as the palatoquadrate, were directly articulated with the braincase. This type of articulation is known as the autostylic jaw suspension. In primitive sharks the hyomandibular formed an additional brace

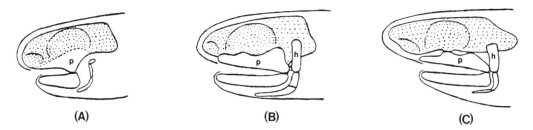

Figure 4-2.   Three basic types of jaw suspension in fishes: **(A)** autostylic, **(B)** amphystylic, and **(C)** hyostylic. h = hyomandibula, p = palatoquadrate.

for the upper jaws, this being the amphistylic type of jaw suspension. In later sharks and many of the bony fishes the hyomandibular formed the entire support from the braincase to the upper jaws, this being the hyostylic type of jaw suspension. Since the hyomandibular was connected to the back of the skull at one end and to the back of the jaws at the other end, the gill slit that originally had occupied the space between the skull and the hyoid arch was greatly restricted. In the more primitive of the modern fishes this restricted gill slit remains as the spiracle, a small, dorsally placed opening in front of the first full gill slit, but in the highly advanced fishes the spiracle disappears completely.

THE CLASSES OF FISHES

The higher fishes are divisible into two large classes, the Chondrichthyes or sharks and other cartilaginous fishes; and the Osteichthyes or bony fishes. Both these groups appeared in the late Silurian period, and it is possible that they may have originated at some earlier time, although there is no fossil evidence to prove this.

The sharks evolved rapidly in Devonian times and continued their expansion through the Carboniferous and Permian periods of earth history. At the close of Paleozoic times many lines of shark evolution died out, but these fishes have nevertheless continued to the present day in comparatively restricted varieties. Throughout their entire evolution the sharks have been primarily marine fishes. The bony fishes also evolved rapidly through the

Devonian and the later Paleozoic periods, but, whereas the history of sharks was circumscribed after the end of Paleozoic times, the evolution of the bony fishes continued in an ever-expanding fashion. Many lines of bony fishes evolved through the Mesozoic era, and toward the close of the Mesozoic, in the Cretaceous period, one group of bony fishes, the teleosts, began a remarkable expansion in their evolutionary history that has continued from that time to the present. In evolving through time, the bony fishes have become adapted for many environments, both fresh water and marine.

EVOLUTION OF THE SHARKS

Sharks are generally considered to be "primitive" fishes, but it is doubtful whether they are more truly primitive than the bony fishes. They appear in the fossil record in the late Silurian period, at about the same time as the first bony fishes. Perhaps the allocation of sharks to a primitive position in the sequence of vertebrate life has grown out of the view that the cartilaginous skeleton, so typical of the sharks, is more primitive than the bony skeleton found in other fishes. Yet it is quite reasonable to think that the opposite is true—that the cartilaginous skeleton of the sharks is a secondary development and that the bone seen in the skeletons of ostracoderms, placoderms, and the first bony fishes is truly primitive.

However that may be, the fact is that the sharks are and have been cartilaginous through the extent of their history. The teeth and various spines are the usual "hard parts" in the shark skeleton, and most of the fossil sharks are known from such remains, although occasionally there has been sufficient calcification in the braincase or the vertebrae for these skeletal elements to be preserved as fossils. In spite of such qualifying statements it is still correct to speak of the sharks as having a cartilaginous skeleton.

Certain other characters, generally, are typical of the sharks as a large group. Internal fertilization of the eggs is usual in these fishes; consequently the males bear clasping devices on their pelvic fins. Furthermore, the sharks are characterized by an absence of lung or air bladders; and in this respect they differ markedly from the other Pisces, for, as we have seen, lungs were early developed in the history of aquatic vertebrates. Most sharks, but not all of them, have separate gill openings, with a small spiracle or reduced gill opening in front of the first full gill slit. Finally, as mentioned above, the sharks have been predominantly marine fishes throughout their history.

One of the first sharks known to us from fossil evidence is the genus *Cladoselache*, found in the upper Devonian Cleveland shales, along the south shore of Lake Erie. This shark was fortunately fossilized in black shales, derived from fine-grained mud, and as a result it is unusually well preserved. The body outline is frequently indicated, and even such soft parts of the anatomy as the muscle fibers and the kidneys are fossilized. From this remarkable material it has been possible to derive a rather accurate picture of *Cladoselache*.

In certain aspects *Cladoselache* was similar to some of the living sharks that are familiar to us. It was rather small—three feet or so in length—and it had a typical "sharklike" or torpedo-shaped body. There was a large heterocercal tail, but in which the two lobes were of equal size. In addition there were two dorsal fins, pectoral and pelvic fins, and a pair of small, horizontal fins, one on either side of the base of the tail. The paired fins had very broad bases whereby they were attached to the body, and because of this the fins could not

have been especially mobile. The pectoral fins were very large, however, and must have been of great importance to the fish as balancing and steering controls.

Cladoselache had very large eyes set far forward in the skull. The upper jaws of this ancient shark were attached to the braincase by two articulations: one immediately behind the eye, the postorbital articulation, and one at the back of the skull, in which region the hyomandibular bone formed a connecting rod between the braincase and the back of the upper jaw. Such a jaw articulation, the amphistylic method of suspension, is relatively primitive among the jawed fishes. In this shark the upper jaw, thus supported at its front and back from the braincase, consisted of a single element, the palatoquadrate, and opposed to this was the single element of the lower jaw or mandible. Each tooth of *Cladoselache* consisted of a high central cusp, with low lateral cusps on each side. This is an early form of tooth structure among the sharks, and is found among many of the ancient fossil forms. Behind the jaws were six branchial arches or gill bars.

The structural pattern seen in *Cladoselache* is in most ways primitive for the sharks, so that this fish may be considered to approximate the central stem from which later sharks evolved. From such a beginning the sharklike fishes evolved in various directions. These are in general: the xenacanth sharks, the "typical" sharks, the skates and rays, the rather enigmatic bradyodonts, and the chimaeras or rat-fishes. As a result of these evolutionary developments the cladoselachians were destined to be replaced in time by the more advanced sharks that sprang from them, yet even so they were sufficiently well adapted to their environment that they could survive until almost the end of the Paleozoic era, their last representatives disappearing during Permian times. Here we see a phenomenon that is common in the evolutionary history of life—the persistence of primitive forms; the grandfather living on in association with his descendants.

There should be mentioned at this place the peculiar whorls of fossilized teeth (obviously

Cladoselache

Figure 4-3. The primitive sharks of late Devonian times were generally similar to many modern sharks. Prepared by Lois M. Darling.

chondrichthyan) which, in former years, were classified among various groups of sharks according to the individual views of the students who worked on them, but which are now generally assigned to the order Eugeneodontida. These whorls, commonly designated as edestid teeth, from the upper Carboniferous genus *Edestus*, are in the form of flat spirals, with the smallest teeth on the inner curves of the spiral, the largest on the outer curves. It is supposed that these are whorls from the symphyseal portion of the jaws, that is, where the two halves of the jaws join, and that they represent a peculiar type of tooth replacement. They are beautiful and striking fossils, and as such have attracted much attention.

The xenacanth sharks, named from the typical genus, *Xenacanthus*, evolved from early Devonian to Triassic times. They were in a sense an aberrant offshoot from the evolutionary tree of sharks, for in contrast to almost all the other sharks these were freshwater fishes, living in shallow rivers and lakes of late Paleozoic and early Mesozoic times. The xenacanths were like their cladoselachian ancestors in that they had the primitive amphistylic type of jaw suspension, but in various other respects they evolved in rather special-

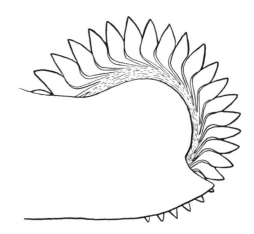

Figure 4-4. A tooth whorl of the lower Permian edestid shark, *Helicoprion*. About one-half natural size.

ized directions. Thus these sharks had an elongated body, with a long dorsal fin running along most of the length of the back. Posteriorly the tail extended straight back in line with the body, to terminate in a point. Such a caudal fin is known as a diphycercal tail, and this tail, in the xenacanths, was obviously secondarily derived from a primary heterocercal tail. The paired fins, too, were unlike the paired fins in other sharks, being composed of a central axis from which fin rays radiated on each

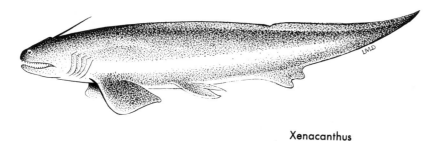

Xenacanthus

Figure 4-5. A Permian freshwater xenacanth shark. About one-seventh natural size. Prepared by Lois M. Darling.

side. One of the most striking features of these strange sharks was the presence of a long spine, projecting backward from the back of the skull. Each of the many teeth was composed of two long divergent cusps or blades, with a small cusp in between them.

While the xenacanth sharks were invading the fresh waters of the continental regions, the more successful and persistent lines of shark evolution were developing in the oceans around the globe. The earliest and most primitive representatives of the modern sharks were the hybodonts, which first appeared in late Devonian times and continued their evolutionary history through the long years of the upper Paleozoic, through the Mesozoic, and even into the beginnings of Cenozoic times. The hybodonts were an intermediate stage in the evolution of the sharks—primitive in that they retained the amphistylic type of jaw suspension (which, as we have seen, was typical of the cladoselachian sharks) but advanced in that the paired fins had narrow bases and, therefore, were rather flexible appendages that were much more useful for control of the fish when swimming than were the broad-based, relatively rigid, paired fins of *Cladoselache*. Moreover, claspers on the pelvic fins first appear in the hybodont sharks, in which respect these fishes show their affinities with all later sharks and show their advance beyond the primitive cladoselachians, in which claspers had not yet evolved.

The hybodont sharks had teeth that were clearly derived from cladoselachian teeth, but were commonly specialized so that the teeth in the back of the jaws were not sharply pointed like the anterior teeth. Instead, they were broad and low-crowned, possibly as an adaptation for crushing the shells of mollusks.

Hybodus, widely distributed throughout the world in sediments of Upper Permian and Mesozoic age, is a typical genus.

Two groups of sharks in the order Galeomorpha, each represented by a small number of genera, seem to be connecting links between the hybodonts and the hosts of varied sharks that live in today's oceans. These are the heterodontoid and hexanchoid sharks, the former exemplified by the recent Pork Jackson shark, *Heterodontus*, of Australia, and the latter by the frilled shark, *Chlamydoselache*, found in northern Atlantic and Pacific waters. *Heterodontus* evidently is a slightly modified descendent of the hybodonts. It closely resembles the hybodonts in its general appearance and in its hybodont-like teeth, with which it crushes and devours mollusks. But it has advanced toward the modern condition in the form of its jaw suspension. *Chlamydoselache* and its related genus, *Hexanchus*, are elongated, predaceous sharks. The jaw suspension in these sharks is of the amphistylic type.

It appears that the hybodont sharks reached the pinnacle of their evolutionary development during late Paleozoic and early Mesozoic times. They were replaced, in effect, during the Mesozoic by the heterodontoid and hexanchoid sharks (perhaps never very numerous) and, more particularly, by the varied sharks and rays that have evolved in large numbers from that time until the present.

With the advent of the Mesozoic era and particularly during the Jurassic period, the modern sharks began a series of evolutionary steps that established the foundation for their success in later geologic periods. In these advanced sharks the primitive amphistylic method of jaw suspension was replaced by the hyostylic suspension, whereby the jaws were attached to the skull only at the back by means of the hyomandibular bone. This type of suspension allows for increased mobility in the development and the functioning of the jaws, and it is characteristic of the advanced fishes.

From Mesozoic times to the present the evolution of the higher sharks has followed two general lines. On the one hand there are the "typical" sharks, streamlined, elongated, fast, aggressive, and highly predaceous. The head is pointed and provided with a widely gaping, ventrally placed mouth, equipped with sharp teeth. On each side there are five separate gill slits, with a round spiracle in front of the first gill. The torpedolike body tapers back into the strong, heterocercal tail. On the back are one or two dorsal fins. The strong paired fins have narrow bases, making them rather mobile and effective in controlling the motion of the animal through the water. On the pelvic fins of the males are claspers. In this group of sharks are the common sand sharks, the mackerel sharks, the tiger sharks, and the dreaded white sharks or man-eaters. Also there are certain divergent types, such as the huge basking sharks and whale sharks and the flattened angel fishes.

The other line of evolution among the higher sharks is that of the skates and rays. These are highly specialized sharks, adapted for life on the bottom. In these fishes the pectoral fins have become greatly enlarged and are used somewhat like wings, so that the fish "flies" through the water. The tail is generally reduced to a small pointed appendage, often a mere whiplash. The gills are ventrally placed in the skates and rays, and water is taken in through a greatly enlarged, dorsally situated spiracle. The teeth are highly modified to form large crushing mills, well fitted for cracking the shells of the mollusks upon which these fishes feed. In this group of sharks we find the banjo-fishes, the sawfishes, the skates, the various rays, and the torpedoes, these last provided with electric organs in the head.

One other group of sharklike fishes, which has never been particularly numerous, can be set apart from the sharks so far considered as a separate subclass—the Holocephali. This subclass includes the chimaeras or rat-fishes, which persist at the present time in deep oceanic waters. The chimaeras are active fishes, with an elongated, pointed rostrum on the head, and with the upper jaws firmly fused to the braincase in an autostylic type of jaw suspension. The pectoral fins are large and fanlike, and the tail is elongated into a long whiplash. Various chimaeras are known in the fossil record, which extends as far back as early Carboniferous times.

Here, we briefly mention the bradyodonts, late Paleozoic sharks that are known mainly from their tooth plates, which occur in great numbers. These "pavement teeth" (as they are often called) obviously indicate adaptations for crushing mollusks. It is probable that the bradyodonts do not constitute a single order of sharks; instead they probably represent a number of parallel groups of mollusk-feeding types that evolved from primitive ancestral sharks. Their relationships are at best uncertain.

The history of the sharks as revealed by the fossil record is not broadly documented. In many respects the fossil evidence is tantalizing and disappointing, often consisting only of isolated teeth or spines, and less frequently of poorly fossilized jaws and braincases. Yet from the fossil evidence as we know it, and from our knowledge of modern sharks, it is quite apparent that sharks have been very successful vertebrates. They have never been numerous in genera and species as compared with many vertebrates, but the forms that have evolved have been extraordinarily well adapted to their environments. There were sharks in late Devonian times and there are sharks today, and through the intervening geologic periods, from the beginning of the Mississippian period to the present day, sharks have lived in the ocens of the world, successfully holding their own against, and even dominating the "higher" types of life that have shared this habitat with them. In spite of competition from the bony fishes, from aquatic reptiles such as the ichthyosaurs, and from aquatic mammals such as the whales, the sharks have carried on in a most successful way. The sharks have succeeded because, with the exception of some of the bottom living skates, they have been very aggressive fishes, quite capable of taking care of themselves in spite of earth changes, changes in food supply, and competitors. It looks as if sharks will continue to inhabit the seas for a long time.

BONY FISHES—MASTERS OF THE WATER

Of all the animals that have lived in the water, none have been so successful as the bony fishes, the class Osteichthyes. Even the most completely aquatic and highly developed of the invertebrates, such as the various mollusks, particularly the complexly evolved ammonites of Mesozoic times, cannot be rated as equal to the bony fishes in their adaptations to life in the water. For the bony fishes have invaded all the waters of the earth, from small streams, rivers, lakes, and ponds of continental regions to all levels of the ocean. They have evolved species showing a remarkably wide range in size, from tiny animals a fraction of an inch in length to huge fishes like the tuna. Their diversity of body form and of adaptations is remarkable indeed. The bony fishes are in many respects the most varied, and as mentioned they are certainly the most numerous of the vertebrates. At the present time they have reached the culmination of their evolutionary history.

What are the characters that distinguish the bony fishes? In the first place, these vertebrates are, as the name implies, commonly highly advanced in the ossification of the skeleton. This is true not only of the internal skeleton, the skull, vertebrae, ribs, and fins, but also of the external skeleton, the outer covering of plates and scales. In the primitive bony fishes the scales were heavy and generally of rhombic shape, and were of two basic types, the cosmoid scales typical of the early lungfishes and crossopterygians and the ganoid scales of the early actinopterygians. The cosmoid scale consisted of a basal portion formed by parallel layers of bone. Above this portion was a middle layer of spongy bone, richly supplied with blood vessels. Finally there was an upper layer of hard cosmine formed around numerous pulp cavities, above which there was a thin layer of enamel. In the ganoid scale the same layers were present, but the enamel was very thick and formed a heavy shiny surface on the scale, known as ganoine. As the bony fishes evolved there was a general reduction in thickness of the various layers comprising the scales so that finally they were composed only of thin bone. Nevertheless the body in most bony fishes has been and remains completely covered by an armor of scales.

In the skull the braincase is completely ossified, whereas the outer bones of the skull are numerous and well formed. Generally speaking, these bones form a complex pattern of several series of related elements, covering the top and the sides of the head, extending back over the gills, and comprising the two halves of the lower jaw. This dominance of bone is carried over into the gill region, so that the several gill arches are composed of articulated chains of bones, whereas the entire gill region is covered by a single bony flap or operculum. Consequently there is a single gill exit at the back or free edge of the operculum as contrasted with the separate external gill openings in the sharks. Moreover, the spiracle is greatly reduced or even suppressed in the bony fishes. The hyomandibular bone is an important element in the skull, forming in most bony fishes the hyostylic support for the jaws from the braincase.

Behind the skull the vertebrae are highly ossified, with spool-shaped central bodies or centra that form an articulated shaft for support of the body. From the centra extend spines directed up (the neural spines) and in the tail region spines directed down (the haemal spines), and from the sides of the vertebrae the ribs extend out and down, to enclose the thoracic region of the body.

There is a compound shoulder girdle, often attached to the skull, and to this the pectoral fins are articulated. All the fins, the pectoral and pelvic fins and the median dorsal, anal, and caudal fins, have bony rays for internal supports.

In the primitive bony fishes there were functional lungs, but in most of these fishes the lungs have been transformed into an air bladder, a hydrostatic organ that helps to control the buoyancy of the animal. The eyes are generally large and important in the life of the fish, whereas the olfactory sense is of secondary consequence. Other characters can be listed as typical of the bony fishes, but these are the features of particular importance to the student interested in their evolution as revealed primarily by fossil materials.

Bony fishes are of ancient ancestry, first appearing as isolated scales in deposits of late Silurian age. The first-known, relatively complete body fossils of bony fishes, from the middle Devonian, were of small or modest size, and were characterized by their heavy diamond-shaped or rhombic scales. The skull showed the basic osteichthyan cranial pattern that was to be the starting point from which the complex skulls of later bony fishes evolved. Thus there was a series of rostral bones that covered the nasal region; behind these bones were paired bones forming the skull roof. Around the eyes were several bones, the circumorbitals, and behind them and on either side of the roofing bones were the bones of the temporal region. Around the edges of the skull were the marginal tooth-bearing bones, and these were joined ventrally and medially by the various bones of the palate.

Opposed to the tooth-bearing bones of the skull were the bones that formed the two rami of the lower jaws.

In these ancient bony fishes the eyes were very large and the mouth was often long, extending the full length of the skull. Behind the skull the fusiform body tapered back to the tail, which was of heterocercal type, with a long upper lobe and a much smaller lower lobe. There was a single dorsal fin, far back on the body, and balancing it on the ventral surface an anal fin. In addition there were the paired fins, the pectoral fins in front and the pelvic fins farther back. In these early fishes the vertebrae were incompletely ossified, and the notochord was still strongly developed.

These first bony fishes belong to the chondrostean suborder known as the Palaeoniscoidei (the palaeoniscoids), which are particularly well exemplified by the Devonian genus *Cheirolepis*. From an ancestry illustrated by *Cheirolepis* the main line of bony fishes, the actinopterygians or ray-finned fishes, evolved and in evolving they passed through three general stages of development. These three stages of actinopterygian evolution can be indicated as three main groups:

	Infraclass (Division)	
Primitive	Chondrostei	Lower Devonian to Lower Cretaceous, with a few forms (sturgeons, paddlefish, and bichirs) surviving to the present.
Intermediate	Neopterygii (holosteans)	Upper Permian to Upper Cretaceous, with a few forms (garpikes and bowfins) surviving to the present.
Advanced	Neopterygii (Teleostei)	Upper Triassic to the present, the majority of modern bony fishes.

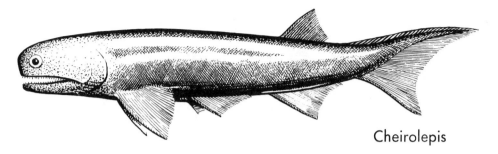

Cheirolepis

Figure 4-6. An ancestral bony fish or palaeoniscoid. About one-half natural size. Prepared by Lois M. Darling.

The general characters of a primitive palaeoniscoid chondrostean have been described above. From such a beginning certain definitive evolutionary trends can be followed in tracing the rise of the actinopterygian fishes through the higher chondrostean fishes, through the holostean grade, and finally through the teleosts. The bony fishes have such a complex history that the details of their evolutionary development include numerous diverse adaptations, too involved for description at this place. It may be useful, however, to set down the general trends that typify the successive stages of evolution in these vertebrates. Please refer to the table below.

Although the chondrosteans have been described as comparatively primitive actinopterygians, it must not be thought that all these fishes were of the generalized type, as exemplified by *Cheirolepis*. In fact there was an early stage of branching out, of adaptive radiation, among the chondrosteans that took place even in the most ancient suborder of the bony fishes, the Palaeoniscoidei. So it was that in evolving through late Paleozoic times the palaeoniscoids followed many lines of adaptation that led to a considerable variety of body forms. The climax of this evolutionary history was attained during the Pennsylvanian and Permian periods, when the palaeoniscoids were perhaps the most abundant of all fishes.

Some of the palaeoniscoids followed the generalized line of adaptation that had been established by *Cheirolepis* and other stem forms. For instance the genus *Palaeoniscus* itself was essentially a very primitive actinopterygian, showing few advances beyond *Cheirolepis*, yet it lived many millions of years later in the Per-

Chondrostei	Holosteans	Neopterygii
		Teleostei
Heavy, rhombic scales	Rhombic scales continued	Thin scales, of rounded shape
Internal skeleton partly cartilaginous	Internal skeleton partly cartilaginous	Internal skeleton completely ossified
Spiracular slit present	Spiracle lost	Spiracle lost
Eye large	Eye large	Eye large
Hyostylic skull	Hyostylic skull	Hyostylic skull
Maxilla fastened to cheek	Maxilla freed from cheek, reduced, and jaws shortened	Maxilla free, transformed into a "pushing" bone; cheek region opened; jaw shortened
Hyomandibular strong	Hyomandibular enlarged	Hyomandibular enlarged
Tail strongly heterocercal	Abbreviated heterocercal tail	Tail homocercal
Pelvic fins usually posterior	Pelvic fins usually posterior	Pelvic fins move forward in many forms
Lungs not transformed	Lungs transformed into air bladder	Air bladder usually completely hydrostatic

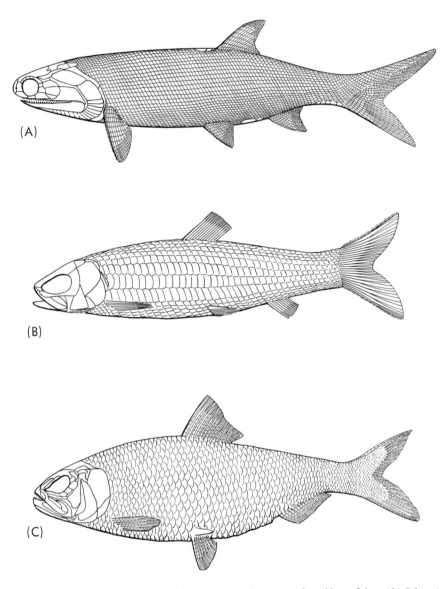

Figure 4-7.   Three stages in the evolution of the actinopterygians or ray-finned bony fishes. **(A)** *Palaeoniscus*, a Permian chondrostean. **(B)** *Pholidophorus*, a Jurassic primitive neopterygian or holostean. **(C)** *Clupea*, a Cenozoic advanced neopterygian or teleost. All one-third natural size. In the sequence from A to C, notice some of the evolutionary changes mentioned in the text.

mian period. On the other hand, some of the Pennsylvanian and Permian palaeoniscoids were highly specialized. *Amphicentrum* was a deep-bodied fish with elongated dorsal and anal fins, and with a tail that was superficially homocercal, although in structure it was heterocercal. *Dorypterus* was another deep-bodied fish,

with fin developments similar to those described above for *Amphicentrium*, but with the anterior part of the dorsal fin tremendously high. It is interesting to note that in this fish the pelvic fins had migrated forward to a position beneath the throat and actually in front of the pectoral fins—a type of fin placement that is especially characteristic of the higher teleost fishes, as we shall see. Other late Paleozoic chondrosteans underwent great reductions in the scaly body covering. In one very aberrant genus, *Tarrasius*, there was not only a loss of scales but also a loss of the pelvic fins and a fusion of the median fins to form a continuous fin along the back, around the tail, and along the ventral surface of the tail.

During their evolutionary development, the palaeoniscoids gave rise to two other orders of chondrostean fishes, which have survived to the present day. One of these surviving orders is the Polypteriformes, now represented by the genera *Polypterus* and *Calamoichthys* of Africa, for many years supposed to be of crossopterygian relationships. The other modern order of chondrostean fishes is the Acipenseriformes, represented in modern faunas by the widely distributed sturgeons and by the paddlefishes of North America and China. In these recent chondrosteans there has been a great reduction of bone, both in the endoskeleton and in the outer armor.

During the Triassic period there were numerous advanced palaeoniscoid fishes inhabiting the waters of the earth. These progressive chondrosteans show an abbreviation of the upper lobe in the heterocercal tail, a reduction in the middle layer of each scale, and a shortening of the jaws. These fishes, of which *Redfieldia* is a typical example, were abundant in Triassic fish faunas, and they can be regarded as intermediate between the chondrosteans and the next higher group of fishes, the Neopterygii.

As the chondrostean fishes declined at the end of the Triassic period they were replaced by the holosteans or primitive neopterygians, which, having arisen from several basic stems, carried farther the specializations already initiated by their immediate ancestors. Holosteans are probably not a natural or monophyletic group, but represent a common stage of anatomical evolution in bony fishes that was attained by several different lines of chondrosteans.

In the holostean fishes the upper lobe of the heterocercal tail was even more abbreviated than it had been in the chondrosteans, there were specializations in the skull and jaws, there were frequent ossifications of the vertebral centra, there was a considerable reduction in the fin rays, and the scales showed structural reduction beyond that of the earlier fishes. The spiracle was lost.

The first holosteans were of comparatively generalized form and are well exemplified by the early Mesozoic genus *Semionotus*, but at an early stage in the evolution of this group specializations took place that imitated in a remarkable way many of the specializations that had occurred earlier among the chondrostean fishes. For instance, deep-bodied genera such as *Dapedius* or *Microdon* evolved during the Jurassic period, showing evolutionary trends similar to those exhibited in the evolution of deep-bodied chondrosteans in late Paleozoic times. Other holosteans were elongated fishes, like the Jurassic form *Aspidorhynchus*.

The culmination of holostean evolution was reached in Jurassic and early Cretaceous times, after which there was a decline. However, two genera of these fishes have persisted into recent times; they are *Lepisosteus*, the garpike of the Mississippi River, and *Amia*, the bowfin of the northeastern United States.

With the progression of the Cretaceous period the holosteans were replaced by the rapidly evolving teleosts or advanced neopterygian fishes, and so began the great development in fish evolution that has continued unabated to the present day. As is apparent in the tabulation presented above, the teleosts advanced many of the evolutionary trends that had already been established among the holosteans. In the skull there was a general shortening of the jaws, with a specialization of the maxilla in the highest types as a toothless bar, and with a concentration of the teeth upon the premaxilla. The heterocercal extention of the vertebral column into the upper lobe of the tail was suppressed, so that the caudal fin became completely symmetrical or homocercal. The internal skeleton became highly ossified. There were varied specializations of the dorsal fin and the paired fins, marked especially by the fre-

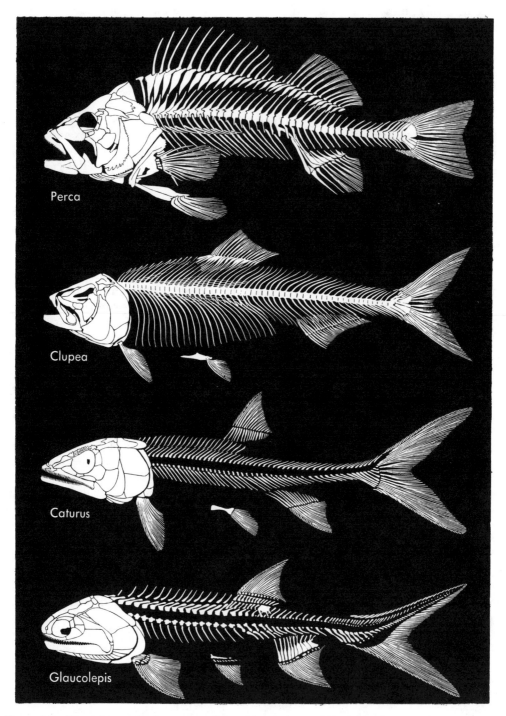

Figure 4-8.   The skeletons of actinopterygian fishes. *Glaucolepis* was a Triassic chondrostean, *Caturus* a Mesozoic primitive neopterygian or holostean, *Clupea* a generalized advanced neopterygian or teleost, and *Perca* an advanced teleost, the last two of Cenozoic age. In the sequence from chondrostean to advanced teleost, notice some of the successive evolutionary changes outlined in the text.

quent migration of the pelvic fins to an anterior position close to the skull. The scales became very thin, and were generally of rounded shape, rather than rhombic.

The first teleosts are represented by the pholidophoriforms, appearing in late Triassic times. Pholidophoriforms are very similar to the better known leptolepiforms, which first appeared in the middle Jurassic. A typical member of this group is *Leptolepis*, from the Jurassic and Cretaceous. This generalized teleost makes a very good intermediate between the holostean fishes and the teleosts, so nice an intermediate form in fact that it has been variously placed by different authorities in both these large categories. By Cretaceous times the teleosts were well established, and from that time on they have followed numerous lines of adaptive radiation that have established them as the masters of all waters, continental and marine. Because of the vast expanse of the oceans, covering the greater part of the earth's surface, most teleosts are marine forms.

A simplified arrangement of the multitudinous array of teleost genera and species may be outlined in large categories in the following manner:

Order Leptolepiformes. These are among the most primitive of the teleost fish, and are, generally, of small size with a "sardinelike" body form. There is a single dorsal fin; the pelvic fins are posterior in position; and the caudal or tail fin is fully homocercal, although there are no expanded hypural bones at the end of the vertebral column to support the tail, as is the case in more advanced teleosts. There are traces of ganoine on the surfaces of the scales. These fishes appeared during middle Jurassic time and continued through the Mesozoic era.

Subdivision Elopomorpha. The elopomorphs, which were particularly abundant during Cretaceous history, were advanced beyond their leptolepiform ancestors, for example, in the development of hypural bones and the loss of ganoine on the scales. Some of the elopomorphs evolved as the tarpons and their relatives, although, oddly enough, others branched off as the highly specialized and morphologically distinct eels.

Subdivision Clupeomorpha. These fish, which appeared at the end of Jurassic times, have continued as highly successful and tremendously abundant denizens of our modern oceans. Fishes such as the her-

ring (*Clupea*), shad, and sardines (only slightly more advanced than the elopomorphs) belong here. Various fishes from the famous Green River deposits of Eocene age, such as the genus *Diplomystus*, are among the well-known fossil clupeomorphs.

Subdivision Osteoglossomorpha. The Cretaceous seas were inhabited not only by numerous elopomorphs but also by the very similar osteoglossomorphs. These fishes were abundant and varied in Cretaceous times, and some of them were of large size. It appears that certain primitive, tropical freshwater teleosts, living today, are descended from early osteoglossomorphs. These include the genus *Osteoglossum*, found today in the rivers of Brazil, North Africa, and the East indies, and the strange, snouted mormyrids of Africa.

Order Salmoniformes. In Cretaceous times, when the teleosts were expanding in many directions, the early salmoniforms appeared. They were among the ancestors of the most advanced, the most varied and numerous, and the most successful of the teleost fishes. The primitive members of this group are exemplified by our modern trout and salmon—fishes of "generalized" body form that are characterized externally by the presence of a soft, fatty dorsal fin, behind and derived from the normal spiny dorsal fin. This basic character (frequently modified by various specializations) is found in all of the advanced teleosts. Various deep-sea fishes come within the boundaries of this group.

Superorder Ostariophysi. Most of the modern freshwater fishes living today can be included within this large group. These are the characins, found in streams and lakes in Africa and South America, the carp, minnows, suckers, and catfish, the most specialized of the ostariophysans. All of these fishes are set apart from other teleosts by the presence in the skull of a small chain of bones, the Weberian ossicles, that connect the ear with the air bladder. These bones, derived from anterior vertebrae, transmit vibrations from the air bladder to the inner ear.

Superorder Paracanthopterygii. The final stages in the evolution of the teleosts seem to have been marked by a dichotomy—into the specialized paracanthopterygians on the one hand, and the parallel acanthopterygians or spiny-finned teleosts on the other. The paracanthopterygians are characterized by a forward migration of the pelvic fins, like the acanthopterygians, and certain specializations in the mouth. This varied group of fishes includes the cod and the haddock, the toadfishes, the deep-sea anglers, as well as others. The variety of form and adaptations in these fishes suggest that they represent a series of parallel adaptations from salmon-like ancestors.

Superorder Acanthopterygii. The spiny-finned teleosts, including a vast majority of modern marine fishes, are the most numerous and advanced of the bony fishes. The common name for these fishes refers to the prevalence of spines (evidently as defense mechanisms), especially, in the anterior portion of the dorsal fin, and in the anal fin as well. The scales are very thin and rounded, usualy with tiny spines on their surfaces. These scales are known as ctenoid scales. The entire endoskeleton is very highly ossified. The pelvic fins migrate to an anterior position beneath the skull—frequently, to a position in front of the pectoral fins. The maxilla is excluded from the mouth and becomes a lever to push the toothed premaxilla forward. These fishes show an enormous range of adaptive radiation and include forms such as the perches, sunfishes, bass, snappers, pipefishes, seahorses, porgies, weakfishes, sailfishes, blennies, parrotfishes, triggerfishes, barracudas, swordfishes, butterflyfishes, molas, flounders, sculpins, and many others.

Body forms and adaptations in the teleosts and particularly in the spiny teleosts are well known and too numerous to consider here. It is enough to say that in form they run the gamut from small to large, from elongated to deep and short, from narrow to round to flat, and in adaptations they range from speedy swimmers to slow swimmers to almost sedentary forms, from dwellers in the open ocean to bottom-living types to lake and river fishes, and from highly carnivorous feeders to scavengers to plant-eating types. It is probably safe to say that no other vertebrates show such wide ranges of adaptation as the teleost fishes.

REPLACEMENT IN THE EVOLUTION OF BONY FISHES

One of the striking features in the evolution of the bony fishes is the predominant role that replacement has played during the long history of these vertebrates. It was pointed out above that the chondrosteans were the first bony fishes to evolve and that they were dominant during late Paleozoic times. Then it was shown that during early and middle Mesozoic times the chondrosteans were replaced by the holosteans, and during the last phases of the Mesozoic era and the Cenozoic era the holostean fishes were replaced by the teleosts. Replacement of one large group by another has been the characteristic pattern in the evolution of the bony fishes.

In the development of this pattern we can see the repetition of many details through time by some remarkable examples of parallel evolution. For instance, the chondrostean fishes evolved some short, deep-bodied forms in Permian times. Then among the holosteans in the Jurassic period remarkably similar fishes were developed. Finally the same pattern of evolution was repeated among the teleosts in Cenozoic times. Examples might be multiplied at great length.

Why should this be? Why should the chondrosteans have waned, to be replaced by holosteans, many of which evolved in ways that were remarkably similar to the adaptations previously established by the chondrosteans? And why did the same process take place as between the holosteans and the teleosts? The answer is probably complex, but it would seem likely that the factor of competition was of primary importance. Competition among the fishes must have been intense through time, as it is today. As new forms evolved by means of genetic processes and natural selection, increasingly efficient mechanisms arose for coping with environments and for meeting competition from other fishes. Thus there were trends toward ever more advanced types, for example, in the structure of the mouth and the development of fins—from chondrosteans through holosteans to teleosts.

But the restrictive qualifications for life in the water are very severe. Streamlining is essential for fast fishes. Deep bodies and essentially homocercal tails are important for fishes that live among coral reefs. Large mouths are important to many carnivorous fishes. Consequently, as the more advanced and more efficient fishes evolved they "swamped" their less efficient predecessors, so that finally the more primitive forms disappeared. Yet all the fishes, whether of primitive or advanced structure, were confronted by similar evolutionary problems that were of necessity solved in similar ways. Consequently the replacement so characteristic of fish evolution was marked by repetition of similar types in successively later geologic ages. Thus, for example, the deep-bodied chondro-

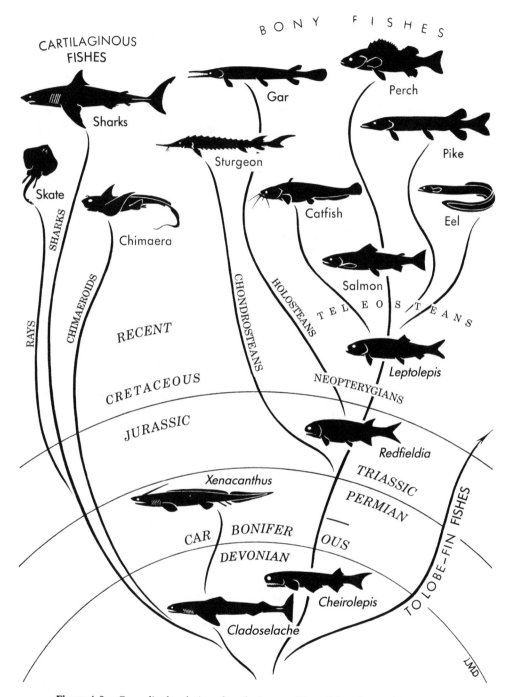

Figure 4–9. Generalized evolution of cartilaginous and bony fishes. Prepared by Lois M. Darling.

steans were well adapted to their environment at one time, but eventually they gave way to the deep-bodied holosteans that were even better adapted to the same type of environment. And the deep-bodied holosteans eventually succumbed to the deep-bodied teleosts.

This is the key to an understanding of fish evolution and, when this key is utilized with discrimination, the story of bony fishes through time loses much of its confusing complexity. It becomes a logical story, and a highly interesting record of evolution.

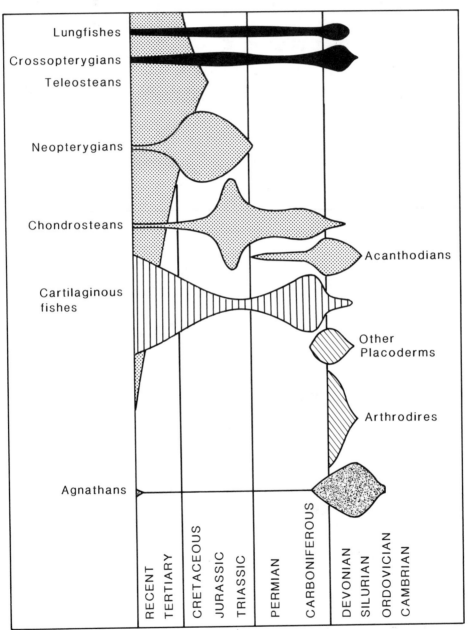

Figure 4-10. Range and relative abundance of fishes through geologic time.

Latimeria

From Water to Land

5

AIR-BREATHING FISHES

With the rise and the almost fabulous increase of the teleosts during the more than two hundred and ten million year lapse of time from the late Triassic period to the present day, the evolutionary history of the fishes has reached its climax. The modern bony fishes that populate the waters of the earth have reached a pinnacle of evolutionary success among the water-living vertebrates, and it seems probable that this culminating phase in the long phylogenetic development of the fishes may continue for an appreciable time into the geologic future. In outlining the story of the fishes we have arrived at the high point, and anything more that is said is likely to be an anticlimax.

Yet when we look at the whole picture of vertebrate evolution we see that the adaptive radiation of the teleosts is in some respects a very complicated side issue. It is a most fascinating side issue, to be sure, for it is important to keep in mind the fact that the modern teleost fishes represent a branch of vertebrate development in many respects as significant in the history of life as any other branch of vertebrate evolution. Nevertheless the teleosts are off the line of evolution that led to the so-called higher vertebrates, so that in order to follow the history of backboned animals beyond the fishes it is necessary to turn to lines of phylogenetic development quite separate from any that have so far been considered. These are the lines of evolution represented by the lobe-finned or sarcopterygian fishes—the lungfishes and their cousins, the crossopterygian fishes.

The lungfishes or dipnoans and the crossopterygians comprise a subclass of the Osteichthyes or bony fishes set apart from the subclass of actinopterygians, which we have reviewed in the preceding chapter. Moreover, the sarcopterygian fishes can be regarded as making up a phylogenetic group equal in rank and importance to all the other bony fishes, even though they have always been restricted in numbers of genera and species as compared with the vastly multiplied actinopterygians.

The first sarcopterygian fishes appeared in early Devonian times, and in their early manifestations these fishes were in many respects rather similar to the first actinopterygians. We might say that *Cheirolepis*, an early actinopterygian described above, had various characters in common with *Dipterus* and *Osteolepis*, representing rather generalized ancestral types of dipnoans and crossopterygians, respectively. For instance, all these fishes were of fusiform shape, and all of them were covered with heavy scales. They all had the primitive heterocercal type of tail, and in all of them the paired fins were located in their primitive positions, with the pectoral fins close behind the head and the pelvic fins far back on the body. In these fishes the skull was covered by a pattern of bony plates.

But there were significant differences that divided these early fishes into two groups, with the ancestral actinopterygians in one group and the ancestral sarcopterygians in the other. Even though the early crossopterygian and dipnoan each had a heterocercal tail, it differed from the tail of *Cheriolepis* in that there was a small epichordal lobe above the body axis, a feature not present in the primitive actinopterygian. Again, the internal structure of the paired fins was basically quite different in the two types. In the primitive actinopterygian the fins were supported by parallel fin rays, as we have seen, but in the early sarcopterygian there was an arrangement of supporting bones that consisted of median or axial elements with lesser bones radiating either on the sides or distally from these central members. This type of fin has been called the archipterygium. The primitive actinopterygians had but a single dorsal fin; the early sarcopterygian fishes had two dorsals. There were important differences in the skull, too. In the early sarcopterygians a pineal opening, a median light receptor, was present on the top of the head, between the two parietal bones, whereas in the early actinopterygians the pineal was generally absent. The two groups can be additionally contrasted in that whereas the

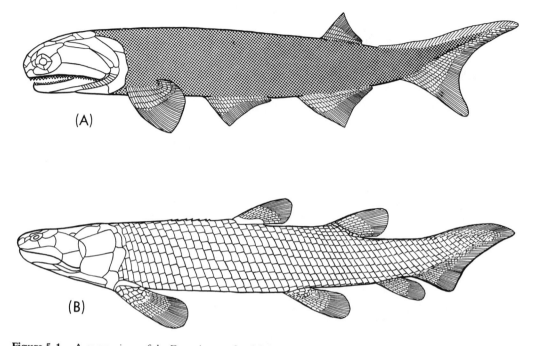

Figure 5-1. A comparison of the Devonian ray-finned fish or actinopterygian. **(A)** *Cheirolepis*, with the Devonian lobe-finned fish or sarcopterygian, **(B)** *Osteolepis, Cheirolepis* about one-third natural size, *Osteolepis* about three-fourths natural size. This figure shows the differences in the proportions of the skull bones and the contrast in the position and the size of the eyes in the two types of fishes. It also shows the single dorsal fin, the long rays of the paired fins, and the heterocercal tail without an upper or epichordal lobe in *Cheirolepis*, as contrasted with the two dorsal fins, the lobed pairs fins, and the heterocercal tail with an upper or epichordal lobe in *Osteolepis*. On the upper surface of the tail of *Cheirolepis* are heavy ridge scales.

early actinopterygians like *Cheirolepis* had large eyes, the eyes in the sarcopterygians were not particularly large. Of particular significance, the sarcopterygians had internal narial openings, which may have been important for air-breathing vertebrates. No such internal nostrils are found in the actinopterygians. Finally, in primitive sarcopterygian fishes the scales were of the cosmoid type, with a thick layer of cosmine above the basal bony layer of the scale, a contrast to the primitive actinopterygian scale, in which the cosmine was limited and the surface of the scale was covered by a heavy layer of enamel or ganoine. These differences indicate that as early as early Devonian times there was a basic divergence between the two lines of bony fishes, even though they had been closely related to each other at the beginning of their evolutionary development.

LUNGFISHES

There are three genera of living lungfishes: *Neoceratodus* in Australia, *Protopterus* in Africa, and *Lepidosiren* in South America. The Australian lungfish, which closely resembles the lungfishes of Triassic age, is almost certainly the most primitive of the three modern types. It lives in certain rivers of Queensland that become reduced during the dry season to stagnant pools. At such times *Neoceratodus* is able to survive by coming to the surface and breathing air, for which purpose it uses its single highly vascular lung. This fish cannot live out of the water.

The South American and African lungfishes, on the other hand, are able to live for months at a time when the rivers that they normally inhabit are completely dried up. At the beginning of the dry season these fishes burrow

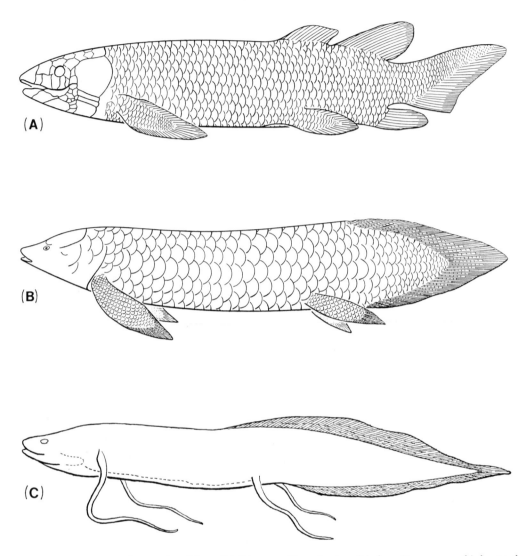

Figure 5-2. Ancient and recent lungfishes. **(A)** *Dipterus*, a Devonian lungfish from Europe, one-third natural size. **(B)** *Neoceratodus*, the modern Australian lungfish. This fish may grow to be five feet or more in length. **(C)** *Protopterus*, the modern South African lungfish.

into the mud and encyst themselves, leaving an opening or openings from their burrow to the outer air; they breathe through these openings. These two lungfishes are characterized by a pair of lungs, in contrast to the single lung of the Australian lungfish.

The ability of the lungfishes to breathe air is certainly suggestive of an intermediate stage between fishes and land-living vertebrates. (In this connection it is interesting to note that the Aus-tralian lungfish is able to "walk" along the bottom of the rivers or pools in which it lives by using its paired fins like legs.) Yet in spite of such specializations in the lungfishes directed toward a method of surviving out of the water, the total evidence points quite clearly to the fact that these vertebrates are not and never have been on the direct line of evolution leading from fishes to the first land-living vertebrates. Briefly the lungfishes show too many specializations, even in the earliest known stages of their evolutionary history, for vertebrates that might occupy an intermediate position along the line from fishes to amphibians.

Early lungfishes are represented by the genus *Dipterus*, as mentioned above. *Dipterus*, a fish of middle Devonian age, possessed many of the generalized sarcopterygian characters that were outlined above for the primitive air-breathing fishes, such as a long, fusiform body terminating in a strong, heterocercal tail, paired fins of the archipterygial type, with a strong central axis down the middle of each fin, and with subsidiary bony rays diverging on either side of this axis, and two dorsal fins. The large, heavy rounded scales were of the cosmoid type.

Contrasted with these primitive characters there were various specialized features that indicate even in as early a form as *Dipterus* the trends that were to take place in the evolution of the dipnoans. For instance, there was considerable reduction of bone in the internal skeleton of this fish, and such a development is found in all of the later lungfishes. The braincase, too, was poorly ossified, although in those Devonian lungfishes in which the braincase has been preserved a certain amount of bone is present. Subsequent to Devonian times the ossification of the braincase was to be completely suppressed. The jaws were partially ossified, yet even here a process of chondrification was beginning that was to become typical of later dipnoans.

The skull was composed of numerous bony plates. In general there was a great multiplication of bones covering the head in *Dipterus*, and because of this it is almost impossible to indicate any homologies between the bones of the skull in this fish and the skull bones in other bony fishes. Likewise, the dentition in *Dipterus* had become highly specialized. The marginal teeth were suppressed in both the upper and lower jaws, and mastication of the food was effected by large, tooth-bearing plates, those above being formed by the pterygoid bones of the palate and those below by

the prearticular bones of the lower jaw. On these plates the teeth were arranged in a fan-shaped fashion, a pattern that was to be carried on through the evolutionary history of the lungfishes. Obviously such teeth were adapted for crushing hard food, and it is probable that the food of the Devonian lungfish, *Dipterus*, was rather similar to that of the modern Australian lungfish, consisting of small invertebrates and vegetable matter.

From *Dipterus* the lungfishes evolved as inhabitants of continental waters during the long lapse of time between the Devonian period and the present day. The lungfishes seemingly were never very numerous, but from the evidence at hand it would appear that they reached their greatest variety of forms in late Devonian and late Paleozoic times. The earlier lungfishes were for the most part variants of the *Dipterus* theme, but as time progressed these fishes evolved along the major lines that have already been indicated, following evolutionary directions marked by progressive chondrification of the skeleton and by modification of the median and paired fins.

The central line of dipnoan evolution led to *Ceratodus*, a genus that became widely distributed during the Triassic and subsequent periods of the Mesozoic era, when this lungfish inhabited most of the continental regions of the world. The modern Australian lungfish *Neoceratodus*, a direct descendant of *Ceratodus*, has changed very little from its Mesozoic progenitor. It is a fair-sized fish with a rather pointed head. The body is covered with large, rounded scales. There is a single, pointed posterior fin running from the dorsal surface around the end of the body to the ventral surface, and obviously formed by a fusion of the original dorsal fins, the caudal fin, and the anal fin. This secondarily simplified, symmetrical tail is known as a gephyrocercal fin. The paired fins are rather elongated and leaf shaped, and in each fin the fleshy portion is covered with scales. Internally the skeleton is greatly reduced; there are no vertebrae but rather an unrestricted notochord, and the bones of the skull are reduced.

The South American and African lungfishes have evolved as side branches from the central stem of dipnoan evolution, as have various fossil forms. Both these modern genera are characterized by gephyrocercal tails, showing evolutionary trends similar to that for *Neoceratodus*. In *Protopterus* of Africa the paired fins are reduced to long, slender whiplike appendages, and in *Lepidosiren* of South America reduction of these fins has progressed to an even more extreme state, so that they are comparatively small.

The distribution of the modern lungfishes on southern continents has been accorded considerable significance by students of earth history, because such a pattern of distribution seems to indicate former close connections between the land areas of the southern hemisphere. This is one of many instances of distributional patterns among fossil vertebrates that seems to be in agreement with the geological and geophysical evidence indicating that there was an ancient, supercontinent, Gondwana[1], composed of the now separate continents of the southern hemisphere, as well as peninsular India. However, it is well to keep in mind the fact that various fossil lungfishes were widely distributed over the earth, not only in Gondwana, but also in the northern supercontinent of Laurasia, composed of modern North America, and Eurasia north of the Himalaya Range, as indicated above for the genus *Ceratodus*, so that the distribution of the modern genera may very well represent the remnants of areas of habitation that were once of broad extent. It is fortunate for us that lungfishes have survived to modern times on several continents because they give us an oblique glimpse of the important vertebrates that formed the link between fishes and the first land-living vertebrates. In the dipnoans we see the collateral uncles of the amphibians.

[1]The concept of an ancient southern hemisphere supercontinent originated with the nineteenth century Austrian geologist Edward Seuss. He named this hypothetical land mass Gondwanaland, after the land of the Gonds, an aboriginal tribe living in central India. This name, although widely used in geological literature, is in a sense tautological, since Gondwana means "land of the Gonds." There is now a trend to discard the older term in favor of the more correct Gondwana.

CROSSOPTERYGIAN FISHES

To trace the direct line of evolution from fishes to land-living vertebrates it is necessary to consider the other group of sarcopterygian fishes, the crossopterygians. These fishes, like the dipnoans, appeared in early Devonian times and, like the first dipnoans, they possessed certain generalized sarcopterygian characters. For instance, the early crossopterygians, as exemplified by the Devonian genus, *Osteolepis*, were fusiform fishes with a strong, heterocercal tail, with lobate, archipterygial paired fins, with two dorsal fins, and with heavy, rhombic scales of the cosmoid type. From this point on, however, striking differences can be seen between the first crossopterygians and their dipnoan cousins.

In the crossopterygians there was no trend toward reduction of the endoskeleton as there was in the dipnoans. *Osteolepis* was characterized by a strong notochord, as were other Devonian crossopterygians, but in this line of evolutionary development the bony elements of the vertebral column were prominent and on their way toward a high degree of functional perfection, as we shall see. The skull and the jaws were completely bony, and the bony pattern was well defined and comparable to that seen in other bony fishes, *as well as that in the early land-living vertebrates*. This is a point of the utmost importance, for it (together with other evidence) indicates the position of the crossopterygians on the direct line of descent between fishes and amphibians. The upper skull bones of *Elpistostege* and *Panderichthys* from the middle and late Devonian are particularly close to those of early amphibians. In *Osteolepis*, on top of the skull between the eyes, there were two large bones that can be homologized as the parietal bones of other vertebrates. On the suture between them there was a pineal opening. In front of the parietals were the frontals and, in addition, a series of small bones covering the rostral portion of the head. Around the eyes were the various circumorbital bones that are seen in the higher bony fishes and the early amphibians, and on the sides of the head were the several temporal bones. Behind the parietal bones and cutting across the skull transversely was a prominent joint, separating the front part of the skull from the postparietal region. It is obvious that there was a certain amount of movement on this joint; the front of the skull could be raised or lowered somewhat in relation to its back portion.

The braincase was highly ossified in *Osteolepis* and other Devonian crossopterygians, and this structure, too, was jointed immediately below the joint in the skull roof. It is possible that the joints in the skull and the braincase of the crossopterygians gave enough flexibility to the head to increase the power of biting when the jaws were snapped shut.

Although there were teeth on the palate in the early crossopterygians, the largest and most important teeth were around the margins of the jaws, both above and below. These teeth were sharp and pointed and well adapted to grasping prey, and it seems obvious that the early crossopterygians were carnivorous fishes. When the teeth of these ancient crossopterygians are cut across and examined under the microscope it can be seen that the enamel is highly infolded to form an exceedingly complex labyrinthine pattern. For this reason the teeth of the crossopterygians are designated as labyrinthodont teeth, and it is an important fact that in the early land-living amphibians the teeth were likewise labyrinthodont in structure.

The crossopterygian fishes had well-developed internal nares or nostrils on the palatal surface of the skull between the vomer and palatine bones, which probably increased the effectiveness of the nose as a sensory organ. Their nasal passages, as in the higher land-living vertebrates, went direct from the external nares or outer nostrils through the internal nares into the mouth or pharynx.

Of particular interest is the internal structure of the paired fins of the early crossopterygian fishes, which are in decided contrast to the paired fins of the dipnoans. In the early crossopterygians there was a single proximal bone in the fin that articulated with the girdle. This is best seen in the pectoral fin, the structure of which is especially well known. Below the single upper fin bone were two bones articulating with it, and beyond these were still other bones radiating toward the distal edges of the fin. Such a scheme of bone arrangement in the fins could very well have formed the starting point for the evolution of the limb bones in land-living animals. There is every reason to believe that the single proximal bone in the paired fins of the crossopterygian fishes is to be equated with the upper bone of the tetrapod or land-living vertebrate limb, namely, the humerus in the front leg and the femur in the hind leg. Likewise, the next two bones of the crossopterygian fin can be homologized with the radius and ulna of the front leg and the tibia and fibula of the hind leg in the land-living vertebrates. Beyond this point homologies are not easy, but it seems probable that the various bones of the wrist and ankle and of the hand and foot evolved from the complex of distal bones seen in the crossopterygian fin. Lungfishes also have bones supporting their lobed fins, but unlike crossopterygians the upper fin bone is followed by only a single bone, not two. Thus, the condition in lungfishes is different from that in crossopterygians and tetrapods.

In many respects, therefore, the early crossopterygian fishes have the characters that we might expect in ancestors of land-living animals, and that is one reason why we can consider these fishes as direct progenitors of the first amphibians. The crossopterygians are to us perhaps the most important of fishes; they were our far-distant but direct forebears.

Thus far we have been concerned with the early crossopterygians of Devonian age as typified particularly by the genus *Osteolepis*. From the basal stock the crossopterygians evolved in two general lines, one being represented by the suborder Rhipidistia, which includes *Osteolepis*, the other by the suborder Coelacanthiformes.

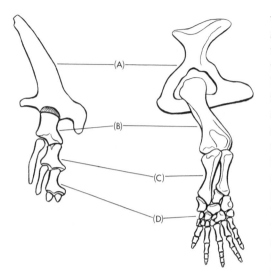

Figure 5-3. A comparison of the pelvic region and hind limb in the Devonian crossopterygian fish, *Eusthenopteron* (left), the Permian labyrinthodont amphibian, *Trematops* (right). A, pelvis; B, femur; C, tibia-fibula; D, pes.

The rhipidistians were primarily freshwater fishes. Beginning in early Devonian times they followed a dichotomous pattern of evolution through the remainder of the Paleozoic era, after which they became extinct. One line of the rhipidistians was that of the holoptychians or porolepiforms, characterized by the genus *Holoptychius* of middle and late Devonian times, an evolutionary side branch which paralleled in an interesting way some of the lungfishes. Thus, in the advanced holoptychians like *Holoptychius*, the body became robust and the scales large and rounded somewhat like those of the Australian lungfish. Also the paired fins were elongated in a manner similar to that seen in the lungfishes, and the median fins, though not fused with the caudal fins, were nevertheless set far back so that they were in close proximity to the tail.

Of more interest to us is the group of rhipidistians known as osteolepiforms, of which the late Devonian genus *Eusthenopteron* is of particular importance. These fishes were obviously on the direct line toward the early amphibians, and they showed various advances over their osteolepid ancestors. *Eusthenopteron* was an elongated, carnivorous fish, with a skull pattern remarkably prophetic of the skull pattern seen in the early amphibians. It was characterized by the advanced nature of the vertebrae. The notochord was strong, but around it at regular close intervals was a series of rings. Above each ring a spine projected up and back, and on the dorsal side of the notochord between the rings were small nubbins of bone. The dorsal spines are to be homologized with the spines of the early amphibian vertebra, whereas the rings may be compared with the intercentra, and the small intermediate bony nubbins with the pleurocentra of the tetrapod vertebra. The tail in *Eusthenopteron* was symmetrical, with the vertebral column extending straight back (not up as in a heterocercal tail) to the tip. The reduction or suppression of a caudal fin of this type would have been a relatively simple matter. In addition to the progressive characters already cited for *Eusthenopteron*, the paired fins were of the structure already described as directly antecedent to the tetrapod limb. It was indeed but a short step from *Eusthenopteron* to a land-living vertebrate.

The other group of crossopterygians, the coelacanths, was far removed from the "main line" of evolution toward the early land vertebrates. These were predominantly marine fishes, often rather deep bodied, with lobate paired fins. There was a reduction of the internal elements of the lobed portion of these fins, while the thin fin supported by the fin rays increased in relative size. The tail, which was symmetrical and of the diphycercal type, possessed an additional small lobe, in line with the body axis and between its main upper and lower lobes. The skull was short and deep with considerable reduction of the skull bones, and of the marginal teeth except for the teeth on the premaxillary bones and on the tips of the dentaries of the lower jaws. The lung or swim bladder was commonly calcified and thus preserved in the fossil materials.

The coelacanths, though beginning in mid-

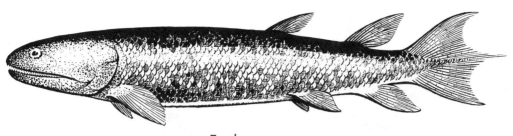

Eusthenopteron

Figure 5-4. A progressive crossopterygian or lobe-fin fish that had evolved in the direction of the early land-living amphibians. About one-fourth natural size. Prepared by Lois M. Darling.

dle Devonian times, were especially characteristic of the Mesozoic era, and they are found in a number of marine deposits belonging to this phase of earth history. The general description for a coelacanth, given above, can be applied most aptly to the Cretaceous genus *Macropoma*, a typical member of the suborder.

Until relatively recently it was thought that the coelacanths became extinct at the close of the Cretaceous period, since no post-Cretaceous forms had ever been found. In the season of 1938–1939 a trawler off the coast of South Africa dredged up a fish that proved to be a living coelacanth. Owing to a series of unfortunate events little more than the skin of this fish was saved. The fish, named *Latimeria*, is rather large, being about five or six feet in length. In form it is extraordinarily similar to *Macropoma* of the Cretaceous period. It has brilliant blue scales, rounded and of large size, and the lobed paired fins are long and strong.

For fourteen years after its discovery, this single specimen was the only known example of *Latimeria*. Then in December, 1952, a second specimen was caught off the coast of Madagascar, an event that set off a train of excited newspaper and magazine articles. This fish was flown to South Africa in a special airplane, and elaborate plans were made for a detailed study of it. But hardly had the excitement died down when several more specimens of *Latimeria* were caught, also near Madagascar. Since then, more coelacanths have been taken on frequent occasions in the vicinity of the

Comoro Islands, between Madagascar and Africa. A considerable series of specimens of this fascinating fish has been intensively studied, and large, monographic descriptions of it have been published. Here we have a valuable link with the past that gives us a glimpse of an important group of vertebrates, hitherto known only from the fossils.

APPEARANCE OF THE AMPHIBIANS

At some time during the Devonian period, possibly during the later phases of Devonian history, some of the crossopterygian fishes came out on the land. Very likely these were osteolepiforms, of the type represented by the genera *Eusthenopteron* and *Panderichthys*. It was a bold step, a venturing of early vertebrates into a completely new environment to which they were only partially adapted. Once having made the step, however, the advanced air-breathing fish soon evolved into a primitive amphibian. With this change vast new possibilities were opened for the evolutionary development of the vertebrates.

What were the factors that led the crossopterygians out of the water and on to the land? Professor A. S. Romer has suggested that it was paradoxically a desire for more water that brought about the first excursions of crossopterygians away from their river and lake environments. According to this idea some of the late Devonian crossopterygians may have been forced by excessive drought to seek new fresh-water pools or streams in which they could

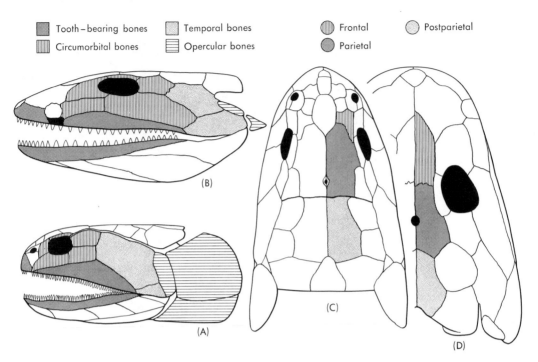

Figure 5-5. (**A** and **C**), lateral and dorsal views of the skull and lower jaw of the Devonian crossopterygian fish, *Osteolepis* and *Eusthenopteron*. (**B** and **D**), similar views of the skull and lower jaw of the Devonian amphibian, *Ichthyostega*. This comparison shows the similarity in the arrangement of the bones in fish and amphibian, but the differences in proportions.

continue to live, and thus they struggled out on the dry ground in an effort to reach the water that was so necessary for their survival. This is certainly a logical explanation of the first stages in the change from an aquatic to a terrestrial mode of life, but there may have been other factors that also contributed to the initial break from life in the water. Perhaps there was a gradual series of changes through time that resulted in increasingly wider excursions away from the water. Perhaps the search for food upon the land may have been as much of a motivating force in this change as the search for fresh bodies of water.

We can only speculate about this. What we do know is that the first amphibians actually appeared at the end of Devonian times, during the transition from the Devonian to the Mississippian period. In upper Devonian sediments in east Greenland fossils of primitive amphibians that are essentially intermediate between the advanced crossopterygians and characteristic early amphibians have been discovered in recent years. These annectant amphibians are known as ichthyostegids, of which the genus *Ichthyostega* is representative.

THE ICHTHYOSTEGIDS

Ichthyostega had a skull about six inches in length. The skull was solidly constructed. Its pattern of roofing bones was closely comparable to the pattern seen in the advanced crossopterygian fishes, yet there were some differences. In the ichthyostegid certain fish bones, such as the operculars and suboperculars that covered the gill region, were lost; but some of the fish bones, like the preoperculars, were retained as reduced elements. Of more importance than any loss of bones were the changes in proportion between the fish

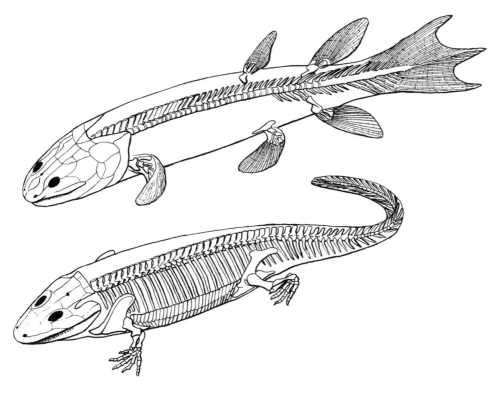

Figure 5-6. A comparison of the skeletons of the crossopterygian fish *Eusthenopteron* and a primitive amphibian.

skull and the amphibian skull. In *Eusthenop-teron*, for example, the rostral portion of the skull in front of the eyes was very short, whereas the postparietal portion of the skull was very long. In *Ichthyostega* this situation was reversed; the skull in front of the eyes had become greatly enlarged, whereas the postparietal portion of the skull had been much shortened, with a consequent reduction in size of the bones in this region. It is interesting to see that *Ichthyostega* was only slightly advanced beyond the crossopterygians in location and development of the internal and the external nares, for the external nasal openings were far down on the side of the skull, and were separated from the internal nares, which were in the front of the palatal region, only by a thin bar of the maxillary bone.

In the postcranial skeleton *Ichthyostega* showed a strange mixture of fish and amphibian characters. The vertebrae had progressed but little beyond the crossopterygian condition, and in the caudal region the *fin rays of the fish tail were retained*. In contrast to the primitive vertebrae and the persistent fish tail, there were strong pectoral and pelvic girdles, with which were articulated completely developed limbs and feet, quite capable of carrying the animal around on the ground.

Osteologically *Ichthyostega* was a fully-fledged tetrapod, in spite of the retention of certain fish characters. Consequently it can be placed among the early amphibians, of which it is a primitive representative. From this point we can follow the evolution of the amphibians as they developed along various lines, one of which led toward still higher types of vertebrates.

Devonian Scene

Early Vertebrate Faunas

6

ENVIRONMENTS OF THE EARLY VERTEBRATES

In a reading of the geological record we see that five classes of backboned animals—the jawless vertebrates, the placoderms, the acanthodians, the Chondrichthyes or sharks, and the Osteichthyes or bony fishes which first appeared between middle Ordovician and early Devonian times—had become well established by late Devonian times. A sixth class, the amphibians, came into being at the very end of the Devonian period. Moreover, for all the fishes, using that term in a very broad sense of the word, the Devonian was a period of evolutionary advancement during which time they were present as dominant aquatic animals throughout the world. Therefore the Devonian period was without doubt one of the most crucial times in the history of life upon the earth, when new lines of evolution were being explored, and when the backboned animals were firmly started on the long and complex evolutionary development that was to terminate in the highly specialized vertebrates of the present day. We might think of a great "evolutionary explosion" as having occurred during the Devonian period, with consequences of the utmost importance to the subsequent history of life on the earth.

The history of the vertebrates during middle Ordovician through Devonian times was, as we have seen, largely a history of animals that lived in continental fresh waters, in rivers and streams and lakes. Therefore it is reasonable to think that the shallow waters of continental drainage systems were the environments in which the primitive pre-Devonian vertebrates enjoyed their evolutionary successes, even though it is quite probable that the ultimate ancestors of the backboned animals, the forms that connected the vertebrates with their invertebrate progenitors, were inhabitants of the seas and the oceans. It is certainly evident that once having begun their several lines of adaptive radiation in fresh waters, the early vertebrates continued in these environments and

did not expand into the seas to any appreciable extent until after the Devonian period.

Many of the middle Ordovician through Devonian vertebrates are found in rocks consisting to a large degree of sandstones and black shales, which would indicate that these animals were living in waters that received great amounts of detritus eroded from nearby lands. In this connection it might be said that the Devonian period was the time when plants were first making a successful conquest of the land, and although Devonian plants were probably numerous, they possibly did not form heavy soil covers as did the plants of later geologic ages. Therefore the land surfaces in Devonian times may have been comparatively bare, so that erosion was very active. Under such conditions, large amounts of sands and muds would have been deposited in the rivers and lakes of those days, and even in the shallow waters along the margins of the continents. These ecological conditions must be kept in mind in making interpretations of Devonian faunas and in drawing comparisons between them.

The early vertebrates of middle Ordovician through Devonian times were generally protected by heavy armor, as we have seen. This armor may have been in part a defense against attacks from some of the large voracious invertebrates that lived in those days, especially in the estuarine and offshore waters of the coasts. But since many of the early agnathans, placoderms, and other fishes were obviously inhabitants of environments where large, predaceous invertebrates were not particularly numerous, and sometimes not even present, it must be supposed that the heavy armor that was so universally present in the ancient backboned animals served largely to protect them from each other. Competition must have been intense, and at an early stage in vertebrate history there were aggressive types that fed upon other backboned animals.

Thus the picture that we get of life during middle Ordovician through Devonian times is one of considerable diversity and active competition between the early vertebrates. There were numerous armored ostracoderms, generally living on the bottom and feeding there, though some of the anaspids were obvious surface-living plankton feeders. There was a great variety of gnathostomes adapted to different modes of life. Some of them, like the acanthodians, were rather generalized animals of very fishlike form and habits. Some, like the antiarchs, were remarkable animals in that they paralleled the ostracoderms to some extent, whereas in certain respects, such as the development of their jointed, paired appendages, they even came to resemble marine arthropods like lobsters. Still other placoderms, like the late Devonian arthrodires, were highly predaceous, and were the first of the vertebrates to evolve into giants. These huge placoderms, unlike the other early vertebrates that have been enumerated, inhabited the oceans, for they were too large to find a living in the restricted waters of rivers and small lakes. Then at this stage of earth history there were the early sharks exploring the possibilities of life in the sea, some of them sharing the environment with the giant arthrodires. Finally there were the primitive bony fishes, including the first actinopterygians. These were for the most part freshwater fishes.

These Ordovician, Silurian, and Devonian vertebrates are widely spread in the sedimentary deposits of several continental regions, and it is probable that if the geologic record were more complete a worldwide distribution of the early backboned animals would be indicated. As the record is preserved, it shows that the early vertebrates lived at many localities in Europe, as far north as Spitsbergen and Greenland, in North America, and in certain parts of Asia, Australia, and Antarctica. It is probable that the continents were more closely connected than they are now, as will be explained subsequently, and that climates and environments were more uniform than the varied climates and environments in which modern vertebrates live. Certainly the freshwater vertebrates of those days indicate essentially similar conditions at widely separated points on several continents, while the marine types, such as they are, point to general uniformity in the oceans of the earth.

It is significant that some of our best-known early vertebrate faunas are found in northerly regions. The presence of various ostracoderms and fishes in Spitsbergen and Greenland and the occurrence of primitive amphibian bones or footprints in Greenland, Eastern Europe, Australia, and South America in sediments representing the final stages of Devonian history show that environments must have been favorable to those cold-blooded vertebrates in what are now boreal zones. Such occurrences are in accordance with modern geophysical and geological evidence indicating that continents have not always been in the positions they now occupy. It would seem that there was a great late Precambrian supercontinent that broke apart in early Paleozoic time, so that land masses were separated by extensive oceans during the Ordovician period. Following this separation there was a collision of North America with northern Europe during Silurian time that closed an ancient proto-Atlantic ocean and formed an immense mountain chain stretching across what are now Scandinavia, Scotland, and New England. This wide uplift, known as the Caledonian Orogeny (mountain building period), created freshwater environments in which the early vertebrates had their beginnings. As we shall see, later continental movements with changed relationships of landmasses had profound influences upon the evolution of the vertebrates.

THE DISTRIBUTION OF EARLY VERTEBRATE-BEARING SEDIMENTS

Although vertebrates first appeared during the middle Ordovician period, the earliest well known vertebrate faunas are found in rocks of late Silurian age. The most abundant fossils occur in northern Europe, especially in England, the Scandinavian region, and Spitsbergen, although some fossils have been found at scattered localities in North America. Vertebrate faunas of this age are unknown in other parts of the world.

In the Ludlow stage of Silurian history we can see in England a passage from older marine sediments to younger estuarine and continental deposits, and it is in these latter beds that the vertebrates make their first well-documented appearance. The succession from the marine to the continental type of deposits was the result of the Caledonian Orogeny, that raised great masses of land above the sea and initiated new environmental conditions throughout this region. The development of continental environments for freshwater vertebrates by this uplift is recorded in the Ludlow rocks of western England by the Ludlow bone bed, a thin stratum of wide extent, filled with the plates and spines of small ostracoderms.

By the subsequent stage of earth history, the Downtonian, the continental uplift was far advanced. Great areas of freshwater deposits were laid down, many of them containing the remains of ostracoderms. These are the so-called Passage Beds of the Downtonian, immediately antecedent to the beginning of Devonian sedimentation, and so intimately linked with Devonian strata in some regions that many authorities have included them in the Devonian sequence. The vertebrates of the Downtonian are much more widely distributed than those of the preceding Ludlovian stage of earth history, and indicate, among other things, the spread of continental conditions favorable to the evolution and geographic expansion of the ostracoderms. Downtonian faunas are found not only in England and Scotland but also in the Ukraine, in various parts of Scandinavia, and in Spitsbergen. At the famous locality of Oesel, an island in the Baltic Sea, vertebrate remains are found in limestones that indicate possible estuarine conditions.

The most famous sequence of Devonian vertebrate-bearing sediments is the classic Old Red Sandstone of England, Scotland, and Wales, first studied and described by Hugh Miller, more than a century ago. The Old Red Sandstone is divisible into three consecutive portions, lower, middle, and upper, representing respectively the three broad stages of Devonian history. Portions of this sequence are exposed at various localities throughout Great Britain, especially along the Welsh border, in the southeastern highlands of Scotland along Moray firth, in northern Scotland, and in the Orkney Islands. The physical expression of the Old Red Sandstone varies, ranging from heavy, red sandstones, as indicated by the name, to coarse conglomerates on the one hand, to gray and black shales and silt-stones and even to limestones on the other. In these rocks are contained numerous fossils, not only of ostracoderms and acanthodian fragments, as in the preceding Ludlovian and Downtonian sediments, but also of complete acanthodians, placoderms, and bony fishes, appearing here for the first time. Indeed, it is from the Old Red Sandstone that we have obtained much of our knowledge of the evolution of early vertebrates and the development of Devonian faunas.

Sediments comparable to various parts of the Old Red Sandstone are found in other parts of the world, and they have yielded many fossils to add to our knowledge of vertebrate evolution during the Devonian period. Rocks of lower Devonian age, comparable in a general way to the lower Old Red Sandstone, are found in the Ukraine, in Spitsbergen, and in central Germany, this last being a marine deposit. In North America lower Devonian vertebrates are found near Campbellton, New Brunswick, at Beartooth Butte, Wyoming, and in the Canadian arctic. Middle Devonian sediments containing vertebrates are found in the European localities already listed; in North America fossil vertebrates of this age are found in several eastern states. Of particular interest are the middle-Devonian vertebrates of Ellesmere Island and eastern Greenland, since they commence a record for the evolution of vertebrate faunas in an area that was to be of great importance during middle and late Devonian times. Middle Devonian vertebrates are found also in New South Wales, Australia. Lower to upper Devonian vertebrates are also known from China.

As might be expected, the record of Devonian vertebrate faunas continues in rocks of late Devonian age in Europe, especially in the British Isles, the Baltic States, and at the locality of Wildungen in Germany. In Ellesmere Island and East Greenland upper Devonian faunas continue the history of vertebrate evolution in that area, and at the top of the sequence are found the first amphibians, the ichthyostegids. Various important upper Devonian faunas are known from North America, especially the fauna of Scaumenac Bay, Quebec, which has yielded fine fossils of *Bothriolepis* and of lungfishes and crossopterygians, and the fauna of the Cleveland shales in Ohio, in which are found giant arthrodires like *Dinichthys* and the early shark, *Cladoselache*. Upper Devonian faunas are also found in the southern hemisphere in Australia, and in the Antarctic.

The sequence of these various upper Silurian and Devonian faunas and their correlative relationships are shown in the accompanying chart.

THE SEQUENCE OF EARLY VERTEBRATE FAUNAS

It would appear that the middle Ordovician through late Silurian vertebrate faunas were dominated by ostracoderms, for at this stage of earth history the jawed vertebrates probably were not very numerous. But ostracoderm supremacy did not last for long, because by the beginning of the Devonian period the jawed vertebrates had become well established. The ostracoderms continued, but their place in nature was being challenged by the early acanthodians and placoderms. At this time,

too, the early arthrodires were prominent in the faunas. Sharks were sparse in early Devonian faunas, and representatives of the bony fishes were scattered and primitive.

In the transition from early to middle and late Devonian history there were great changes in the nature of the vertebrate faunas. The ostracoderms, which had been rather abundant in early Devonian times, were greatly reduced, and continued to the end of the period only as remnants of a once numerous group of vertebrates. The acanthodians also were greatly reduced in abundance, although they continued to the end of the Paleozoic era. On the other hand, the arthrodires expanded through middle and late Devonian times, but with a shift of emphasis from freshwater to estuarine and marine types. It was in the final stages of Devonian history that the arthrodires reached the culmination of their evolution as the giant predators of the Cleveland shales. In addition, some new placoderms, the antiarchs, appeared in middle and late Devonian times.

The most important event in the history of the vertebrates during middle Devonian times was the appearance of whole body fossils of the Osteichthyes, and it is probable that the sudden influx of the new and comparatively advanced bony fishes was instrumental in leading to the decline and final disappearance of the more primitive vertebrates such as the ostracoderms and most of the placoderms. From middle Devonian times on, the bony fishes evolved with great rapidity and variety, first as freshwater types and in subsequent ages as marine fishes. Even in early stages of their evolutionary development in middle Devonian times, the Osteichthyes had differentiated into actinopterygians on the one hand and into lungfishes and crossopterygians on the other. Evidently the initial phases of osteichthyan history, during which the separation of these primary lines took place, were consummated with great evolutionary dispatch, so that almost from the beginning the main lines for the development of the bony fishes had been

CORRELATION OF LOWER AND MIDDLE PALEOZOIC VERTEBRATE-BEARING SEDIMENTS

	England / Scotland	North and Central Europe	Spitsbergen	North America	Other Continents
Devonian — Upper	Old Red Sand-stone — Upper	Baltic States; Wildun-gen; Russia		E. Green-land; Cleve-land shales / Catskill; Scau-menac	New South Wales Victoria; Antarc-tica; Central Asia
Devonian — Middle	Middle		Wijhe Bay Series	Ellesmere Island	
Devonian — Lower	Lower	Podolia, Ukraine; Rhine-land	Wood Bay and Gray Hoek Series	Camp-bell-ton; Bear-tooth Butte	
Silurian	Downton sand-stone / Temeside shales; Ludlow Bone Bed	Oesel Gotland	Red Bay Series		
Orodovician					Australia; Brazil

delineated. Sharks were also present in middle Devonian faunas, but it is not until we reach upper Devonian sediments that we find well-documented records showing an early stage of shark evolution.

The climax of vertebrate evolution during Devonian times was reached with the appearance of the first amphibians at the close of the period. This event ushered in a new phase of vertebrate evolution—the beginning of tetrapod history. The story of the rise and the differentiation of the land-living vertebrates concerns the Carboniferous and later periods of geologic history, and will be the subject of the subsequent chapters of this book.

Ichthyostega

7

Amphibians

THE SPECIAL PROBLEMS
OF LIVING ON LAND

When the descendants of certain crossopterygian fishes ventured out of the water on to the land at the end of the Devonian period and jumped the gap, so to speak, from one vertebrate class to another to become the first amphibians, the backboned animals entered a completely new course of evolutionary development, quite different from anything in their previous history. For the first time the vertebrates were invading a new environment, very different from the one in which they had lived for millions of years. This was, needless to say, a profound step that opened broad avenues for evolution over a tremendous range of adaptations. It was an unprecedented step, involving great transitions in vertebrate structure and a change in the emphasis of evolutionary trends. From this point on, the evolution of the vertebrates other than fishes was directed primarily along lines of adaptation for life on the land (and in the air), although there were secondary reversions to life in the water among various land-living animals.

The most primitive amphibians, the ichthyostegids as we know them in the geologic record, evolved in directions quite different from those followed by their immediate fish ancestors. Even though there are many common characters that connect the higher crossopterygian fishes with the primitive amphibians, there are at the same time great differences between the two groups of vertebrates that result from the specializations of the one group to life in the water and the other to life out of the water. Let us consider these differences, especially in light of the new problems facing the first land-living vertebrates.

One of the important problems with which the early amphibians had to contend was breathing or respiration. Fishes ordinarily obtain oxygen from the water by means of gills, whereas land-living vertebrates secure oxygen from the air by means of lungs. As we have seen, the problem of respiration out of the wa-

ter had been solved for the amphibians by their fish ancestors, the crossopterygians, in which lungs were well developed and probably frequently used. We may say, therefore, that the amphibians had a "head start" on the problem of breathing in the air, so that it was actually not much of a problem for them; they had merely to go on using the lungs that they had inherited from certain crossopterygians. The main point of difference between the fishes and amphibians in this respect was that in most of the fishes that had lungs the gills were still the primary method for respiration and the lungs generally formed an accessory breathing mechanism, whereas in most of the amphibians the situation was reversed. These first land-living vertebrates were primarily air-breathing animals, using their lungs for this function, although in their young or larval stages they continued to breathe by means of gills, as we know not only from the evidence of living amphibian tadpoles but also from fossil evidence.

Another problem to confront the first land-living vertebrates was desiccation or drying-up. This is no problem to fishes, which are continually bathed by the liquid in which they live, but to land-living animals it is a crucial and frequently a severe problem. The first amphibians, therefore, were faced with the necessity of retaining their body fluids when they were no longer immersed in the water. It is probable that the earliest amphibians, such as the ichthyostegids, never ventured very far away from the water and returned frequently to streams and lakes, as indeed do many modern amphibians. Even though such habits would have limited the excursions of ancient amphibians away from water, it is nevertheless probable that these animals at an early stage in their history evolved integuments or body coverings that would protect them against the drying effects of the air. There is now evidence to indicate that some of the first amphibians retained the scales that had covered the body in their fish ancestors. There is also evidence to show that as the am-

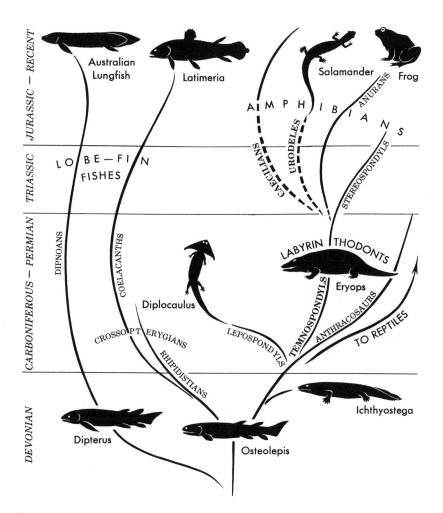

Figure 7-1.   Generalized evolution of the lobe-fin fishes and the amphibians. Prepared by Lois M. Darling.

phibians evolved, especially during the Permian period, they developed tough skins, frequently underlain by ossicles or bony plates. As the outer covering became increasingly efficient for preventing evaporation of body fluids, and served as a tough coat that shielded the animal from its environment, the amphibians became increasingly independent of the water and more able to spend much of their time on the land. This was an important factor in the evolutionary history of these animals, and more particularly of those higher vertebrates, the reptiles, that arose from the amphibians.

To animals living on land gravity is a powerful factor, influencing much of their structure and life, whereas to fishes gravity is of little consequence, because they are supported by the dense water in which they live. Of course the first amphibians had to contend with increased effects of gravity when they were out of the water, and because of this they developed a strong backbone and strong limbs at an early stage in their evolution. The rather simple discs or rings that typically constituted the centra of the vertebrae in the crossopterygians became transformed into interlocking structures that with the aid of muscles and ligaments formed a strong horizontal column for the support of the body. At two points this vertebral column was supported by girdles, the pectoral girdle in front and the pelvic girdle in the back; the girdles in turn were supported by the limbs and the feet.

But land-living animals are not stationary structures like bridges. They move around. Therefore the early terrestrial vertebrates became adapted to a new method of locomotion, and here the limbs and feet were of prime importance. They served not only to hold up the body in counter-action to the force of gravity but also to propel the animal across the land. Here we see a reversal in locomotor functions between the crossopterygian fishes and the early amphibians. In the fishes locomotion was effected primarily by the body and tail, and the paired fins were used for balancing functions, whereas in the early land-living vertebrates the tail was attenuated to become in some degree a balancing organ, and the paired appendages became the chief locomotor organs. The pattern for locomotion initiated by the first amphibians has continued with many variations through the evolution of the land-living vertebrates.

In addition to these new problems facing the first land-living vertebrates, there was the problem of reproduction. Fishes commonly deposit their unprotected eggs in the water, where they hatch. Land-living animals must either go back to the water to reproduce, or they must develop methods for protecting the eggs on the land. The amphibians made several great advances in their adaptations to life on the land, but they never solved the problem of reproducing themselves away from moisture. Consequently these animals throughout their history have been forced to return to the water, or among some specialized forms to moist places, to lay their eggs.

THE BASIC DESIGN FOR LIFE ON THE LAND

How did the early amphibians respond in their evolutionary development to the new requirements imposed upon them as a result of the change from a fully aquatic to a partial land-living mode of life? As has been mentioned above, the ichthyostegids that appeared at the end of the Devonian period took the first steps to meet the basic requirements for life on the land. Among the immediate successors of the ichthyostegids, the amphibians of Mississippian and Pennsylvanian times, adaptations for land life reached a point of stabilization, so that the primary characters of the tetrapods or four-legged vertebrates were established. These earliest tetrapods and many of their descendants are known as labyrinthodont amphibians (from the structure of the teeth), and much of amphibian history is concerned with the evolution of the labyrinthodonts.

In the amphibians of Carboniferous times the adaptations for breathing air were advanced beyond those of their ichthyostegid forebears. For example, in the ichthyostegids the external nostrils were far down on the margins of the skull and separated from the internal narial openings only by a thin bar of bone (see Figure 5-5); in the later amphibians the external nostrils were located on the dorsal surface of the skull, and the internal nares, now in the front part of the palatal region, were quite separated from them (Figures 7-5 and 7-12). Consequently there was a well-defined nasal passage in these amphibians, leading from the external nostrils into the throat, and this was the basic plan for air intake that was to be followed during the evolutionary development of all the higher vertebrates.

In the ichthyostegids, as we have seen, the vertebrae were but little advanced beyond the crossopterygian condition, which means that the spinal column was not efficiently designed for supporting the body of the animal out of the water. However, the successors of the ichthyostegids had specialized vertebrae, of which the bony elements, preformed in cartilage, had progressed to such a degree as to have interlocking joints (see Figure 7-12). These vertebrae constituted in series a strong column, quite capable of holding up the body against the downward pull of gravity. This was an important advance in the evolution of land-

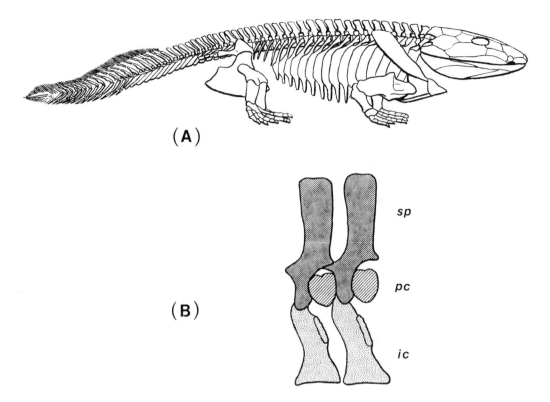

(A)

(B)

Figure 7-2.   The **(A)** skeleton and **(B)** vertebrae of the earliest known amphibian, *Ichthyostega*. sp = spinal process of neural arch, pc = pleurocentrum, ic = intercentrum.

living vertebrates; it provided the mechanism whereby these animals could evolve to considerable size.

In these early vertebrates the girdles became comparatively strong, the pectoral girdle (when seen from in front) taking the form of a broad U in which the fore part of the body was slung by strong muscular attachments, the pelvic girdle being a heavy V, attached by a single vertebra to the backbone. The limb bones, characteristically broad and heavy, were composed in each leg of a strong proximal bone, the humerus of the forelimb and the femur of the hind limb, and two distal bones, the radius and ulna of the forelimb and the tibia and fibula of the hind limb. There were hands and feet with five toes, articulating with the limbs by means of carpal or wrist bones, and tarsal or ankle bones. In conjunction with the perfection of the limbs, the tail became variously reduced.

The skull in the early Carboniferous labyrinthodonts was heavy, and was solidly roofed by skull bones that had been inherited from crossopterygian fishes. Only five openings pierced the skull roof, the two nostrils, the two eyes, and behind the eyes the pineal foramen, an opening that during life contained a median light receptor. At the back of the skull there often was a prominent notch on either side, bounded above by the tabular bone of the skull roof and below by the squamosal bone of the temporal region of the skull. This tympanic notch was for the accommodation of the eardrum, a new character in the amphibians. From the eardrum, a tight membrane that stretched across the tympanic notch, a single bone, the stapes, bridged the gap between the outer portion of the skull and the wall of the braincase. One end of the stapes was attached to the tympanic membrane, and the other end was inserted into a hole, the fenestra ovalis, in the side of the braincase, and by means of this arrangement sound waves that impinged upon the eardrum were transmitted by the stapes to the inner ear. The stapes was the old fish hyomandibular, transformed from a bone originally part of a gill arch, subsequently a prop to hold the jaws and the braincase together, into its new use as part of a hearing device. This is an interesting example of transformation through evolution of a structure from one function to an entirely new function—a process that has occurred time and again during the history of life. It should be added that in most early tetrapods, the stapes was still a large bone that retained its old function as a prop for the jaws, as in crossopterygians. By late Carboniferous times, however, labyrinthodonts with stapes slender enough to function in air-borne sound detection first appeared.

Numerous sharp teeth bordered the edges of the jaws in the early amphibians, and in addition there were other teeth located on the horizontal surfaces of the bones in the front part of the palate. These teeth had the complex labyrinthine structure of the enamel that was so characteristic among many crossopterygian fishes. The palate in the early amphibians was commonly solid or but slightly fenestrated, and there was generally a joint between the pterygoid bones of the palate and the braincase, allowing some movement between the braincase and the skull. On the other hand, there was no joint in the skull itself, as was characteristic of many crossopterygian fishes.

Such was the general plan typical of the early land-living vertebrate, as shown by the fossil evidence.

LABYRINTHODONT AMPHIBIANS— ANTHRACOSAURS (EMBOLOMERES)

Among the early relatives of the ichthyostegids were the amphibians known as anthracosaurs, which appeared during the late Devonian period, approximately the same time that ichthyostegids first appeared. Two major evolutionary trends may be seen within the anthracosaurs, one exemplified by those members of the group known as embolomeres, the other by the seymouriamorphs.

In the embolomeres each vertebra generally was composed of two discs or rings, one behind the other, known respectively as the intercentrum and the pleurocentrum. In very primitive anthracosaurs such as *Proterogyrinus* from the early Carboniferous of North America, the pleurocentrum was larger than the intercentrum. In typical embolomeres, the two bones were of approximately equal size. The neural arch and spine rested on the two vertebral discs, and the three elements together— intercentrum, pleurocentrum, and neural arch—composed the anthracosaurian type of vertebra. Large facets or zygapophyses on the front and the back of each neural arch formed articular surfaces that locked the vertebrae together with strong but movable joints. As a whole, the embolomere vertebral structure was well adapted for supporting a land-living animal.

In many respects, the embolomeres retained the generalized features that have already been

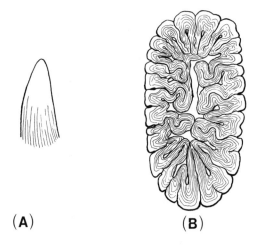

(A)   **(B)**

Figure 7-3. Side view **(A)** and diagrammatic cross-section **(B)** of a labyrinthodont tooth, greatly enlarged. The straight lines on the external surface and the sinuous lines in the cross-section indicate the complex infolding of the tooth's enamel.

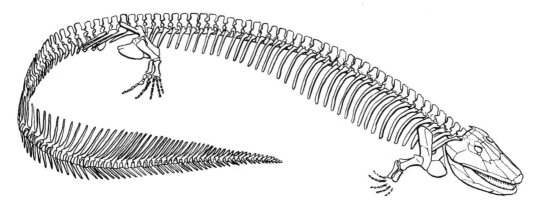

Figure 7-4.   The anthracosaur labyrinthodont *Eogyrinus*. About two meters long.

listed as characteristic of the primitive amphibians. In *Eogyrinus,* a typical genus from the late Carboniferous of Europe, the skull was rather deep, with a primitive type of skull-roof pattern and, on each side of the skull, was a large otic notch for the accommodation of the tympanic membrane. The palate was primitive, almost solid, and pierced only by small openings or vacuities; it articulated with the braincase by means of a movable joint.

The shoulder girdle, so close behind the skull that this animal seemed to have had very little neck, was a rather complex structure, composed on either side of several bones. The front of the girdle consisted of clavicle and cleithrum, and in back of these bones was the scapulo-coracoid with which the upper arm bone, the humerus, articulated. Ventrally the two halves of the shoulder girdle were joined together by the median interclavicle.

The pelvis was a fairly strong, platelike structure, composed on each side of three bones; the ilium, the ischium, and the pubis. The ilium consisted of a narrow upper blade, which was attached to one vertebra, and a broad lower portion. The posterior ischium and anterior pubis were expanded, heavy bones that joined each other and the ilium firmly and, in the region where these three bones met, there was a rounded depression, the acetabulum, forming a socket for the ar-

ticulation of the femur, the upper bone of the hind limb. The two halves of the pelvis were joined along the bottom edges of the pubis and ischium by a heavy symphysis that held the entire structure together in a strong V-shaped support.

Rather weak limbs were attached to the pectoral and pelvic girdles, indicating that *Eogyrinus* (like many of the embolomeres) was to a large degree a water-living amphibian, as might be expected in a very primitive tetrapod that was drifting away from the life lived by its fish ancestors.

LABYRINTHODONT AMPHIBIANS—ANTHRACOSAURS (SEYMOURIAMORPHS)

Seymouria is a small tetrapod found in the upper portion of the lower Permian sediments exposed to the north of the town of Seymour, Texas. This vertebrate is much too high in the geologic column to have been an ancestor of the reptiles, yet in its structure it is almost exactly intermediate between the amphibians and the reptiles. Thus *Seymouria* is a good example of a persistent structural ancestor; it shows how paleontological grandfathers may live on with their descendants.

It seems evident also that this animal and its relatives, known under the collective name of seymouriamorphs, had ancestors in common with other anthracosaurian amphibians. *Seymouria* had a rather deep skull, with a prominent otic notch at the back for the accommodation of the eardrum, as was characteristic of the anthracosaurs. The skull was completely roofed, and all the labyrinthodont skull bones were present, including both a supratemporal and an intertemporal bone, behind the eye. There were sharp, labyrinthodont teeth around the margins of the jaws, and in addition there were some large teeth on the palatine bones, as is typical of the labyrinthodonts. The occipital condyle was single, as it was in the anthracosaurs. Indeed the skull showed many features in common with certain labyrinthodonts.

The postcranial skeleton, on the other hand, displays various progressive characters that link *Seymouria* with the early reptiles. For instance the neural arches of the vertebrae were broadly expanded from side to side and swollen, which is just what we find in some of the earliest reptiles. Moreover, the dominant vertebral element was the pleurocentrum, whereas the intercentrum was reduced to a small, wedge-shaped bone, again a reptilian character. In the shoulder girdle the interclavicle had a long, median stem, typical of the early reptiles and in decided contrast to the labyrinthodonts. The ilium was expanded much beyond the amphibian condition, and a second vertebra was being incorporated into the sacrum. Primitive reptiles had two sacral vertebrae as contrasted with the single vertebra of the amphibians. The humerus was generally similar in form to that of early reptiles and was typified by the presence of a special foramen characteristic of reptiles. Finally, although there were three proximal bones in the ankle, as is typical of the amphibians, *Seymouria* had an arrangment of toe bones similar to that of the early reptiles, with two phalanges in the thumb and great toe, three in the second digit, four in the third, five in the fourth, and three in the little finger, four in the little toe. This is the primitive reptilian phalangeal formula, usually expressed as 2-3-4-5-3(4).

Was *Seymouria* an amphibian or a reptile? The ultimate answer to this question depends on whether *Seymouria*, like modern reptiles, laid an amniote egg on the land, or whether, like modern frogs, it returned to the water to deposit its eggs. Unfortunately, there is no direct paleontological evidence at the present time that gives us a clue about this important and diagnostic attribute. Some years ago, however, Dr. T. E. White, who made a detailed study of *Seymouria*, noted that there appeared to be sexual dimorphism among the known fossils of this interesting tetrapod. In some individuals the first of the haemal arches that project below the tail vertebrae was located at a considerable distance behind the pelvis, whereas in other individuals the first of these arches was rather close behind the pelvis. This may indicate that the animals having a considerable gap between the posterior border of the pelvis and the haemal arch were females that showed an adaptation for the passage of a large amniote egg through the cloaca. Of course, this is only speculation but, in conjunction with various reptilian characters of the postcranial skeleton, it is perhaps significant; it places *Seymouria* on the side of the reptiles.

However, opposed to this, the features of the skull and dentition place *Seymouria* rather securely on the side of the amphibians. And added to this evidence is the clear indication that some genera closely related to *Seymouria*, notably *Discosauriscus* from Europe, were characterized by gill-bearing larvae.

These conflicting bits of evidence in the morphology and the inferred physiology of *Seymouria* and its relatives nicely illustrate the fact that evolution, even within a single species, does not proceed evenly along a broad front. An animal may be advanced in some features and primitive in others, and it is this mixture of advanced and primitive characters which makes *Seymouria* and its relatives such puzzling intermediates between two major classes of vertebrates. Here, we see the effects of a rather well-documented paleontological record. With such a record at hand the divi-

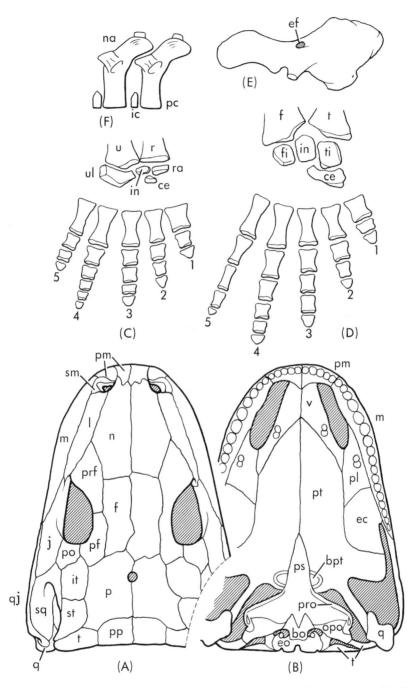

Figure 7-5. *Seymouria,* a tetrapod from lower Permian sediments of Texas. **(A)** Dorsal view of skull. **(B)** Palatal view of skull. **(C)** Fore foot. **(D)** Hind foot. **(E)** Humerus. **(F)** Two dorsal vertebrae in lateral view. All about one-half natural size. For abbreviations, see pages 447–449. Note here the general labyrinthodont-like arrangements of bones in the skull roof and the three separate proximal bones of the ankle: the reptile-like phalangeal formula, the foramen in the humerus, and the swollen arches of the vertebrae.

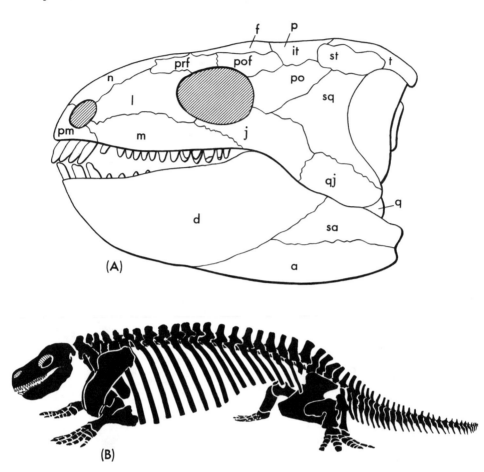

Figure 7-6. *Diadectes,* from the lower Permian of Texas, the skeleton of which shows some primitive reptilian features, although the skull retains some amphibian characters. **(A)** Lateral view of the skull, about two-thirds natural size. Note the presence of both intertemporal and supratemporal bones. For abbreviations, see pages 447–449. **(B)** The heavy skeleton is about six feet in length.

sions between groups break down. The evidence of the seymouriamorphs indicates, too clearly for the peace of mind of those students who wish to categorize animals within neat boundaries of classification, that evolution is a continuum.

Although *Seymouria* shows the mixture of amphibian and reptilian characters described above, it is now generally placed within the anthracosaurian labyrinthodonts, as are its close relatives. It would seem that these amphibians evolved in the direction of reptiles, but were not themselves reptilian ancestors. This role was to be played by some small,

swamp-dwelling tetrapods, the remains of which (described on page 104) have been found in Nova Scotia. Nonetheless it is evidently from an anthracosaurian base that the reptiles had their origins.

One small group of animals, here placed among or close to the seymouriamorphs, are the rather equivocal diadectomorphs, of which the genus *Diadectes,* from the lower Permian beds of North America, is typical and the best known form. Because *Diadectes* has features characteristic of both amphibians and reptiles, it is uncertain as to which of these classes of tetrapods it belongs. The diadectomorphs showed an early trend toward large size. In *Diadectes* the quadrate bone of the skull was pushed forward and, at its upper end, where it joined the bones of the skull, there was a prominent notch. There is some argument whether this notch is the same as the primitive otic notch in the labyrinthodonts, or whether it is a secondary feature, caused by the specialization in the diadectomorph skull from a more primitive, reptilian condition. Regardless of the answer, the effect of the forwardly shifted quadrate was to shorten the jaws, which had peglike teeth in front and transversely broadened teeth in the cheek region. These specializations in the jaws and teeth, together with the capacious nature of the body, suggest that *Diadectes* and its relatives were plant-eating tetrapods. Although *Diadectes* and its relatives show numerous reptilian features, especially in the postcranial skeleton, it has been suggested that because of various primitive, seymouriamorph characters in the skull, they should be classified among the amphibians. But as E. C. Olson, a noted authority on Permian tetrapods, has pointed out, the inclusion of *Diadectes* in the Amphibia is "mostly an assignment of convenience."

LABYRINTHODONT AMPHIBIANS— PALEOZOIC TEMNOSPONDYLS

Early in the history of the amphibians the important order of labyrinthodonts known as temnospondyls arose. These amphibians, originating in the Mississippian period as close relatives of the ichthyostegids, were most characteristically developed in the subsequent Pennsylvanian and Permian periods. The intercentra and pleurocentra of their vertebrae were not equal discs, as in the embolomeres, nor were the pleurocentra much larger than the intercentra as in seymouriamorphs. Instead, the intercentra of temnospondlys were large wedge-shaped elements, and the pleurocentra were comparatively small blocks that fitted in between the intercentra. This condition is essentially the reverse of that seen in seymouriamorphs. In each vertebra the neural arch was supported by both the intercentrum and the pleurocentrum. Together, this association of bones formed the rhachitomous vertebra, probably the basic, central type of vertebral structure among the labyrinthodonts, since its initial stages are indicated in the ichthyostegids and even in advanced crossopterygian fishes. As in the anthracosaurs, the neural arch in the rhachitomous temnospondyls carried well-formed zygapophyses for strong articulations between vertebrae.

Temnospondyls became the dominant amphibians of Permian times, and of these the genus *Eryops* from the lower Permian sediments of Texas can be described as a typical Paleozoic example of the group.

Eryops, six feet or more in length, was a heavily built animal. It had a very large, broad, and rather flat skull in which the roofing bones had become thick and rugose. There was a deep otic notch for the ear. In contrast to the anthracosaurs, the palate in *Eryops* was rather open, with a large fenestra or vacuity on either side. This opening of the palate seemed to go along with the flattening of the skull in *Eryops* and other temnospondyls, and it was probably developed to make room for the large eye and the eye muscles. Another contrast with the anthracosaurs was the solid joint between the palate and the braincase in *Eryops*. There were strong labyrinthodont teeth around the margin of the jaws and some very large teeth in the palate.

The vertebral column was extraordinarily strong, an indication that *Eryops* was well adapted for life on the land. Moreover, the other parts of the skeleton behind the head were robust. The ribs were greatly expanded. The shoulder girdle was heavy, with the scapula and coracoid dominant. The cleithrum and the clavicle were in *Eryops* reduced to rather narrow splints along the front edges of the scapula and coracoid. The pelvis was also heavy. The limbs, though short, were very stout. This characteristic is particularly appar-

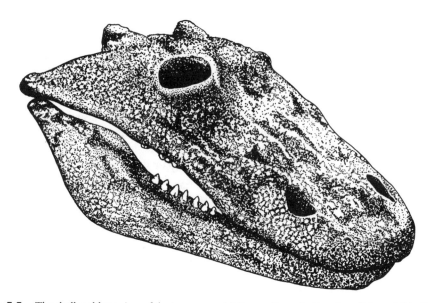

Figure 7-7.   The skull and lower jaw of the temnospondyl *Eryops,* from the Permian of Texas. This illustrates the heavy, solid skull typical of labyrinthodonts. About one-fourth natural size.

ent in the humerus, which was almost as broad at either end as it was long, and in the femur, a very thick bone. The lower limb bones and the hands and feet were likewise strongly constructed. The hand had only four digits, which is typical for temnospondyls. There were bony nodules in the skin that formed a heavy armor, to protect *Eryops* from some of its vicious contemporaries.

It is quite apparent that *Eryops* was a rather well-adapted land animal, even though it probably whiled away many languorous hours in the water. One can think of this amphibian as living a life somewhat similar to the life of a modern alligator, in and out of the water along streams, rivers, and lakes. *Eryops* was probably a fish-eating amphibian, but it must have been rather aggressive, and it may have supplemented its diet by preying upon land-living animals. Certainly there is reason to believe that this big temnospondyl was able to compete actively with many of the reptiles of

Figure 7-8.   The skeleton of *Eryops,* which is about five feet in length.

Eryops

Figure 7-9. *Eryops* represents a high point in the evolution of the amphibians. This big Permian rhachitome was as large as many of the reptiles of the time, and probably was able to compete actively with them. Prepared by Lois M. Darling.

Permian times. In many respects *Eryops* represents a high spot in the evolution of the amphibians.

Although *Eryops* is cited as a typical member of the Paleozoic temnospondyls, it must not be thought that all these amphibians were large, aggressive carnivores, well suited for life out of the water. The rhachitomous temnospondyls evolved along various lines of adaptive radiation during the Pennsylvanian and Permian periods, and although there were numerous large semiaquatic and land-living types like *Eryops,* there were others that became specialized for quite different modes of life. For instance, in one group there were animals of medium size, characterized especially by elongated rather pointed skulls, the jaws of which were provided with numerous sharp teeth. These were the archegosaurs, and it seems obvious that they were primarily water-living, fish-eating animals. Still others were small, and were adapted for ecological niches unavailable to the big predatory forms.

Of these, the genus *Trimerorhachis* from the Permian beds of Texas is especially well known. It was a rhachitomous temnospondyl no more than a foot or two in length, with a much flattened skull, in which there was some reduction of bone and a corresponding increase of cartilage. The body of *Trimerorhachis* was covered with an armor of overlapping scales that appear to be homologous with fish scales. It is reasonable to think, in the light of these features and because of the types of sediments in which *Trimerorhachis* is found, that this was predominantly a water-living amphibian, inhabiting shallow streams and ponds and feeding upon small animals in such a habitat. Another small temnospondyl from the Permian beds of Texas that may be contrasted with *Trimerorhachis* is the genus *Cacops,* characterized by a heavy skull, with an enormous otic notch closed behind by a bar of bone, by a relatively short body and strong legs, by a reduced tail, and by a covering of heavy, bony armor over the back. Evidently *Cacops* was a land-living amphibian, strongly protected by its armor against the depredations of predators. This amphibian must have had an enormous eardrum; perhaps it was a nocturnal animal like many of the modern frogs.

In certain late Pennsylvanian and early Permian deposits in Europe the skeletons of very small labyrinthodonts, commonly known as "branchiosaurs," have been found. It was argued by some paleontologists that these little amphibians constitute a separate group called the Phyllospondyli. More recently, however, it has been demonstrated that moot branchiosaurs were larval forms of some of the larger temnospondyls contemporaneous with them. Branchiosaurs frequently show indications of fossilized gill arches, as might be expected in larval types. Moreover, in a series of branchiosaurs graded according to size, the eye decreases in relative size from the smallest to the largest members of the series, which is what we might expect in a sequence of developing larval amphibians. For these and other reasons, the group of Phyllospondyli has no reality in the history of the amphibians.

The rhachitomous amphibians enjoyed a long and successful sojourn over much of the earth's land areas during Permian times, but with the close of this phase of earth history their period of dominance came almost to an end. A few rhachitomous types known as trematosaurs, which persisted into the first half of the Triassic period, were remarkable in that the vertebrae showed the rhachitomous structure, whereas the skull became advanced beyond the condition so typical of the Triassic labyrinthodonts. Trematosaurs seem to have been marine, fish-eating animals, anticipating in a way the plesiosaurs and other marine reptiles that were so abundant in later Mesozoic times. This seeming incursion of the amphibians into the sea was not, however, very successful, for the trematosaurs died out soon in the Triassic.

LABYRINTHODONT AMPHIBIANS— MESOZOIC TEMNOSPONDYLS

Later temnospondyls of the Mesozoic era, especially characteristic of the Triassic period throughout the world, followed lines of adaptive radiation quite different from those typified by Paleozoic forms, from which they were derived. Whereas the latter were commonly specialized for life on the land (although these amphibians must have spent a great deal of time in the water, as have most amphibians during the long evolutionary history of the class), the Mesozoic temnospondyls were highly specialized aquatic types that probably seldom ventured on to the land. It might be said that these forms reversed the general trend of evolution that had been followed by the labyrinthodont amphibians from the beginning of the Mississippian period to the end of the Permian period, by returning to the environment of their ultimate ancestors. Their pattern of evolution was certainly a successful one for the Triassic period, and these animals became the most common and widely distributed of continental vertebrates. They continued their evolution right up to the beginning of the Cretaceous period, and then became completely extinct. We can think of temnospondyl evolution during the Mesozoic, and particularly during Triassic times, as representing a final vigorous flowering of the labyrinthodonts before their extinction.

Since these animals had largely abandoned the land, they no longer needed the strong vertebral column that was so well developed in the rhachitomous temnospondyls; and, as so often happens among vertebrates that have returned from life on the land to life in the water, there was a secondary simplification of the vertebral column. Instead of the interlocking elements characteristic of the rhachitomous vertebrae the centra in some Mesozoic temnospondyls were reduced to simple blocks composed of the intercentra alone, above which were the neural arches and spines. The pleurocentra were lost. This is the stereospondylous type of vertebral structure.

The Mesozoic temnospondyls, having returned to an aquatic mode of life, were not bothered by problems of gravity as were their rhachitomous ancestors. Consequently there was a common trend among these amphibians toward an increase in size, with the result that some stereospondylous forms surpassed their large Permian ancestors, to become the largest amphibians ever to live. Not all Mesozoic temnospondyls were large, but the general trend was in this direction.

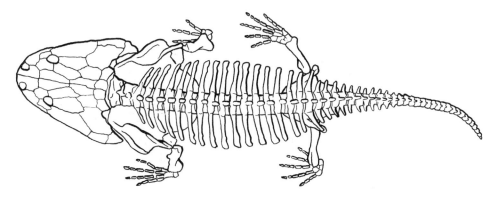

Figure 7-10.   The large-headed temnospondyl *Metoposaurus,* of late Triassic age. About two meters long.

The genus *Metoposaurus* (often designated in the literature as *Buettneria*), found in upper Triassic sediments of the United States, as well as in Europe, northern Africa, and India, exemplifies the results of the evolutionary trends in these last of the labyrinthodonts. In this large stereospondylous labyrinthodont, the skull was inordinately large as compared with the body. Growth factors during evolution were such that as the animals increased in size during time, the skull increased at a greater rate than the body, so that many late Triassic temnospondyls were almost grotesque because of their immense heads. Not only was the skull large in these amphibians, it was about as flat as the proverbial pancake. As might be expected, flatness in the skull was accompanied by flatness in the body.

Along with these changes in proportion there was a general trend among the Mesozoic temnospondyls toward increase of cartilage and reduction of bone in some parts of the skeleton. Thus the palatal region of the skull became very open by reduction of bone, and the braincase became cartilaginous. The elements in the wrist and the ankle were completely reduced to cartilage, so that they are seldom preserved in the fossils.

Strangely enough, the reduction of bone in these amphibians was differential. Although the palate became open and the braincase and the wrist and ankle components were reduced

to cartilage, bone in some other parts of the skeleton was increased, so that the top of the skull became extraordinarily heavy and thick, as did the clavicles and interclavicle—the ventral bones of the pectoral girdle. In surprising contrast, the upper bones of the pectoral girdle were small and weak. So was the pelvis. And as a corollary of these small upper segments in the girdles, the limbs and feet in the later stereospondyls, such as *Metoposaurus,* were astonishingly small.

Thus, evolution in the stereospondylous labyrinthodonts led finally to the production of bizarre animals that peopled the streams and ponds of late Triassic times with queerly proportioned beasts like ones that we might encounter in mediaeval manuscripts. These amphibians showed a strange combination of characters; yet, in spite of this combination, the later temnospondyls were remarkably successful animals.

Even more aberrant than *Metoposaurus* and its relatives were the plagiosaurs, typically represented by a few genera of late Triassic age. These grotesque amphibians were characterized by an extremely flat skull, which was remarkably broad as compared with its length. The body also was flat and broad, and the limbs were small. One genus, at least (*Gerrothorax*), had persistent external gills, like the ones in certain modern salamanders. These labyrinthodonts represent a grotesque sideline

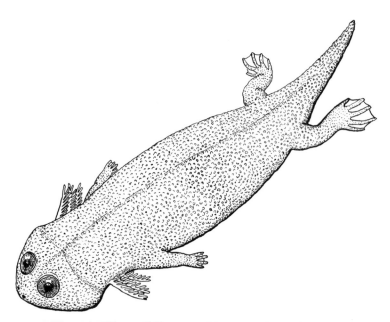

Figure 7-11.   The plagiosaur temnospondyl *Gerrothorax*, of late Triassic age. About one meter long.

of temnospondyl evolution that carries to extremes some of the evolutionary trends seen in other Mesozoic forms.

TRENDS IN LABYRINTHODONT EVOLUTION

It may be useful at this point to summarize the varied courses of labyrinthodont evolution by a chart or a table, in order to show at a glance the trends of development in these amphibians. Of course it is not possible to include all the numerous characters involved during the evolution of the labyrinthodonts, but at least some of the more striking ones are considered. Briefly, the evolutionary trends in the labyrinthodonts may be expressed in the following manner. This table indicates that evolution in the labyrinthodonts was directional in that it generally proceeded along certain courses from the end of the Devonian to the beginning of the Cretaceous periods. In general, it is possible to trace the rise and the development of these amphibians along two evolutionary trends. From fish ancestors to the

	Anthracosaurs		Ichthyostegids	Temnospondyls	
	Seymouria-morphs	Embolomeres		Rhachitomous	Stereospondylous
Habitat:	Terrestrial	←Aquatic	←Aquatic	→Terrestrial	→Secondarily aquatic
Backbone:	Strong	←Weak	←Weak	→Strong	→Secondarily weak
Vertebrae:	Reversed Rhachitomous	←Embolomerous	←Prerhachitomous	→Rhachitomous	→Stereospondylous
Skull:	Deep	←Deep	←Deep	→Flattened	→Extremely flat
Palate:	Closed	←Closed	←Closed	→Open	→Widely open
Braincase:	Ossified	←Ossified	←	→Ossified	→Cartilaginous
Occipital Condyle:	Single	←Single	←Single	→Double	→Double
Limbs:	Strong	←Weak	←Strong	→Very Strong	→Secondarily weak

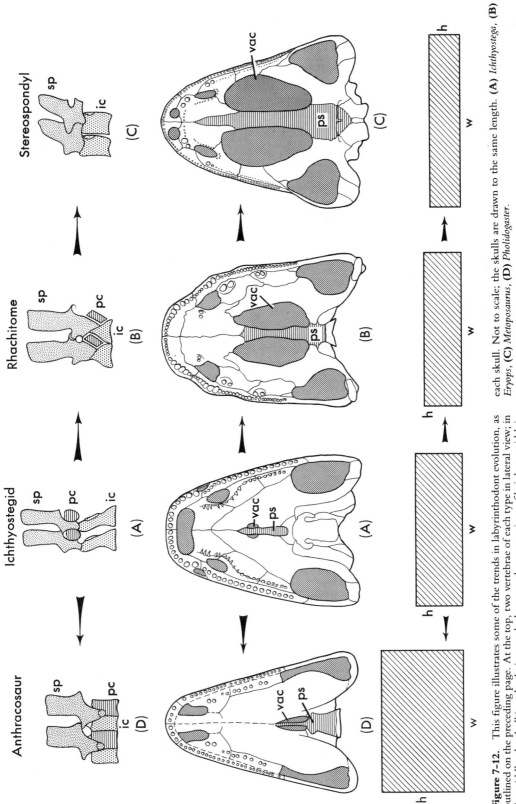

Figure 7-12. This figure illustrates some of the trends in labyrinthodont evolution, as outlined on the preceding page. At the top, two vertebrae of each type in lateral view; in the middle, the skull in palatal view; at the bottom, the proportions of height to width in each skull. Not to scale; the skulls are drawn to the same length. **(A)** *Ichthyostega*, **(B)** *Eryops*, **(C)** *Metoposaurus*, **(D)** *Pholidogaster*.

primitive ichthyostegids and beyond, the labyrinthodonts evolved on the one hand to a high point exemplified by the strongly terrestrial rhachitomous temnospondyls of the Permian, after which there was a secondary return to the water, as seen in the large Triassic stereospondylous amphibians. On the other hand, there was a dichotomous pattern of development, evolving from primitive types, one branch trending toward a permanently aquatic mode of life, as seen in the embolomeres, the second branch trending toward terrestrial forms that showed many reptilian characters, as seen in the seymouriamorphs.

LEPOSPONDYLS

Thus far in this chapter, we have been concerned with the labyrinthodont amphibians—the first of the land-living vertebrates, the ubiquitous dwellers in ancient swamps, rivers, and lakes. We have learned that they spread across the continents to become dominant and very numerous during late Paleozoic and Triassic times eventually to become extinct in early Cretaceous times. As we have indicated, these amphibians were the ones in which the several bony elements forming each vertebra were preformed in cartilage.

But they were by no means the only amphibians that lived in those distant days. They shared the land and the streams and the ponds with other amphibians, which can be called lepospondyls. In the lepospondyls the vertebrae were not preformed in cartilage, but rather were formed directly as spool-like, bony cylinders around the notochord, and generally united with the neural arch. The lepospondyls were a highly varied group that appeared as early as the Mississippian period, reached the height of their evolutionary development in the Pennsylvanian and early Permian periods, and then, before the close of Paleozoic times, became extinct.

The lepospondyls were never large, and through most of their history probably never very numerous. It is obvious that during the golden age of the amphibians, the late Paleozoic, the lepospondyls were not direct competitors with their large and multitudinous cousins, the labyrinthodonts; instead they were adapted to certain ecological niches that were not exploited by the labyrinthodonts. For instance, many of the lepospondyls remained as small, generally primitive amphibians, suited for life in the undergrowth at the edge of the water, or to life in the swamps. Such were many of the genera composing the order Microsauria, of which a typical form was the middle Pennsylvanian genus, *Microbrachis*.

Still other lepospondyls of the late Paleozoic were small, elongated snakelike amphibians in which the legs were suppressed. The genus *Ophiderpeton* of Pennsylvanian age was a characteristic member of this amphibian order, known as the Aistopoda, represented in the fossil record by only a few genera.

The most numerous and varied of the late Paleozoic lepospondyls were the amphibians belonging to the order known as the Nectridea. These lepospondyls followed two lines of evolutionary development in Pennsylvanian times, characterized on the one hand by an elongation of the body to an eel-like or snake-like form (paralleling the aistopods) and on the other by a flattening and broadening of the body and the skull. The first of these two evolutionary trends is exemplified by the genus *Sauropleura*, in which the body was long, but not so long relatively as in the aistopods, the limbs were absent, and the skull was very pointed in front. Evidently animals such as this lived a snakelike type of life in the Carboniferous swamps.

Diplocaulus, from the Permian beds of Texas, is a well-known genus that probably represents the culmination of evolutionary development among those nectridians in which the trend was toward flattening of the skull and body. In this amphibian the bones forming the sides of the skull and the skull roof grew laterally to such an extent that they formed broad "horns" or lateral projections that in the adult caused the skull to be considerably wider

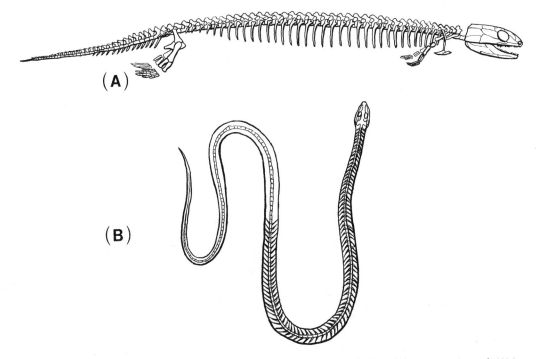

Figure 7-13. The lepospondyl amphibians **(A)** *Microbrachis,* a microsaur, and **(B)** *Ophiderpeton,* an aistopod. (A) is about natural size; (B) about 75 centimeters long.

than it was long. This factor of lateral growth became established not only in the phylogeny or life history of the genus *Diplocaulus* but also in the ontogeny or life history of the individual. Young individuals of *Diplocaulus* had a rather "normal" skull shape, but as growth proceeded the skull grew laterally at a much faster rate than it grew longitudinally. Consequently the skull in the adult *Diplocaulus* was shaped something like a very broad arrowhead. The jaws, which were not involved in this growth, remained small.

The body in *Diplocaulus* was also flat, and the legs were very small and weak. Evidently this animal was a water-living amphibian, probably spending most of its time on the bottoms of streams and ponds. Why the skull evolved to such a grotesque shape is unclear.

MODERN AMPHIBIANS

A large gap, both in time and in morphology, separates the modern amphibians, salamanders, frogs, and the limbless caecilians,

from the varied labyrinthodonts and lepospondyls that were characteristic of the continents of late Paleozoic and early Mesozoic times. Because of this gap, and because there are certain characters that are shared in common by the recent amphibians, these modern denizens of our lakes and streams are usually grouped within the subclass Lissamphibia. But there has been skepticism in recent years as to the validity of grouping the modern amphibians in a single subclass.

The characters that seem to unite the modern amphibians include:

1. Common to frogs, salamanders, and caecilians:
   a. The presence of a zone of weakness between the base and crown of the teeth (a condition termed pedicellate).
   b. Cutaneous respiration, supplementing or replacing the use of lungs.
   c. Simple, cylindrical vertebral centra.
2. In frogs and salamanders, but not caecilians:
   a. Skull with large openings in the orbital and temporal region.
   b. The presence of two ossicles in the middle ear—

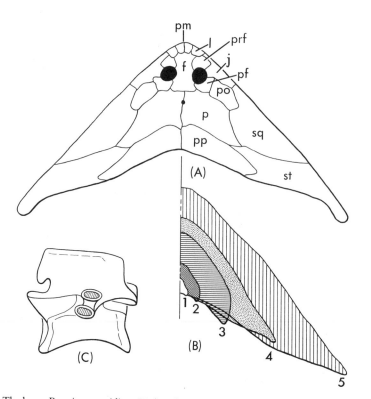

Figure 7-14.   The lower Permian nectridian, *Diplocaulus*. **(A)** The skull in dorsal view, one-third natural size. **(B)** Growth stages in the skull of *Diplocaulus* (here shown by the right halves in dorsal view) from a very young individual, 1, to a large adult, 5. **(C)** Right lateral view of a vertebra, slightly larger than natural size. This is a characteristic lepospondyl vertebra. For abbreviations, see pages 447–449.

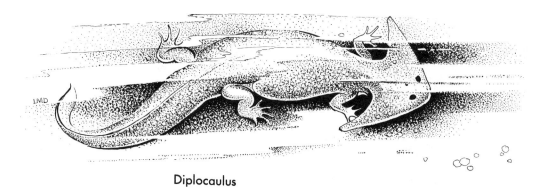

Diplocaulus

Figure 7-15.   The nectridian, *Diplocaulus,* inhabited early Permian streams and ponds in North America. Adult individuals were three feet or more in length. Prepared by Lois M. Darling.

stapes and operculum—rather than the single stapes characteristic of other lower tetrapods.

Despite these similarities, there is no evidence of any Paleozoic amphibians combining the characteristics that would be expected in a single common ancestor. The oldest known frogs, salamanders, and caecilians are very similar to their living descendants. Only the frog ancestor, *Triadobatrachus,* from the Lower Triassic beds of Madagascar, shows evidence of ancestry among Paleozoic amphibians. The specialization of the ear region in frogs, with a large tympanum supported by the squamosal, and a small columnar stapes, is seen in small temnospondyl labyrinthodonts such as the dissorophids, which are likely ancestors of frogs.

The basic structure of the braincase, skull roof, and particularly the area of the middle ear, suggests that caecilians are plausibly derived from elongated, burrowing microsaurs. The cranial anatomy suggests that salamanders also evolved from microsaurs with more normal body proportions, but with emargination of the cheek to accommodate lateral jaw musculature.

In the proto-frog *Triadobatrachus,* the skull bones were greatly reduced in number and area, approaching the condition typical of modern toads and frogs. In addition, the ilium in the pelvic girdle was elongated, the body was moderately short, and the bones of the hands and of the feet were developed to a degree similar to that seen in typical frogs. Ribs were still present, and there was a remnant of a tail.

With the advent of the Jurassic period there was a rather sudden and complete change from *Triadobatrachus* to the modern type of anuran, a development involving some very great changes in the skeleton. The skeleton of *Vieraella,* the oldest known true frog from the lower Jurassic of South America, is essentially like that of living frogs. From Jurassic times on, frogs and toads have been characterized by an open flat skull, with the bones greatly reduced in area, by a short back, by a complete suppression of the tail in the adult, and by

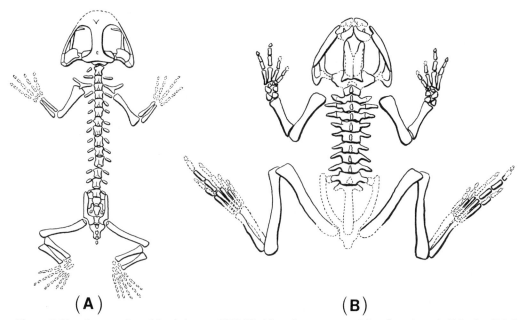

(A)   **(B)**

Figure 7-16.   A restoration of the skeletons of **(A)** *Triadobatrachus,* an ancestral frog from the early Triassic of Madagascar, three-fifths natural size; and **(B)** *Vieraella,* the earliest modern type frog from the early Jurassic of South America, about twice natural size.

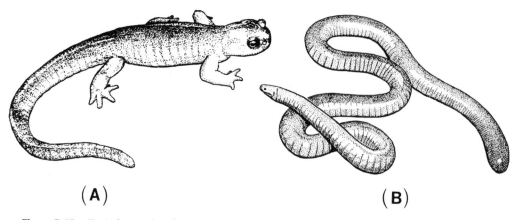

Figure 7-17. Typical examples of modern **(A)** salamanders and **(B)** caecillians. About one-half natural size.

greatly elongated hind legs and short front legs. In line with these developments the vertebrae in the back have been reduced from the primitive number to about eight, and each vertebra is made up of a large neural arch that grows down into the region of the centra. The true centra have been suppressed. Such vertebrae as have remained behind the pelvis are fused into a single, spikelike bone known as the urostyle. The ribs are completely suppressed. In the ankle there is an elongation of certain bones and the development of an extra joint, to give greater power to the leg for leaping. The shoulder girdle is specialized and strengthened to take up the shock of landing, and the forelimbs, though small, are strongly modified as landing gear.

In this way the frogs and the toads have become adapted to a life by the edge of the water, or even out on land away from the water. The skull with its very large mouth is an efficient insect trap. The hind legs give the animal power to make enormous leaps, either on the land or from the land into the water. We are all familiar with the efficiency of these adaptations in the frogs and toads, which all in all have through the years been very successful, even though very noisy, amphibians. They have lived from the Jurassic period to the present, a span of more than two hundred million years, and today they are distributed far and wide over the earth, in a great variety of habitats.

The specializations that so distinguish the anurans are in decided contrast to the morphological features of the salamanders or urodeles. In the salamanders the body has remained generalized, and evolution has been characterized by a secondary loss of bone and its substitution by cartilage in the skeleton. Salamanders are predominantly water-living amphibians, and some modern forms, like the axolotl of Mexico, retain their larval condition and their gills throughout life. The earliest known salamander is *Karaurus* from upper Jurassic sediments of the USSR.

Caecilians, also called apodans or gymnophionans, are wormlike, burrowing amphibians that live in the tropics. The oldest caecilian is *Apodops* from the Paleocene of South America.

First Land Eggs

The Advent of Reptiles

THE AMNIOTE EGG

During Carboniferous times a great forward step occurred in the evolution of the vertebrates; the amniote egg appeared. This was a major innovation in vertebrate history, to be compared with the appearance of the lower jaw, or the migration of the backboned animals from the water on to the land, and like these preceding evolutionary events of great moment, the perfection of the amniote egg opened new areas for the development of animals with backbones. The transformation of gill arches into jaws raised the early vertebrates from comparatively small, bottom-living animals to active and aggressive fishes that ranged the waters of the earth. The transition from water to land made available to the amphibians an entirely new environment. And the appearance of the amniote egg liberated the land-living vertebrates completely from any dependence upon the water during the life history of the individual.

In those animals reproducing by the amniote egg development is direct from embryo to adult. The egg is internally fertilized, after which it is deposited on the ground or in some suitable place, or sometimes retained within the oviduct of the female until the young animal hatches. The egg contains a large yolk that furnishes nourishment for the developing embryo. It also contains two sacs, one being the amnion, which is filled with a liquid and contains the embryo, the other the allantois, which receives the waste products produced by the embryonic animal during its sojourn in the egg. Finally, the entire structure is enclosed by a shell that is tough enough to protect the contents of the egg, yet porous enough to allow the passage of oxygen into the egg and the passage of carbon dioxide out. This egg provides a protected environment for the development of the embryo; in effect it furnishes a little private pool, the amnion, in which the embryo can grow, safely screened from the outside world by the tough shell of the egg.

Animals with an egg such as this can wander freely over the land, without having to return to the water to reproduce, as has been necessary for the amphibians through the millions of years of their history. The first animals to have the amniote egg were the reptiles.

The reptiles were derived from the amphibians, specifically from certain anthracosaur labyrinthodonts, the transition from amphibian to reptile taking place during the Carboniferous period of earth history. Of course the final crossing of the threshold from the amphibians to the reptiles occurred with the perfection of the amniote egg, but of this there is no fossil evidence. The oldest known amniote egg, which is from lower Permian sediments in North America, represents a time long after the reptiles had become well established on the land.

Even though the oldest known reptile egg is far removed in time from the first amniote egg, some speculations may be in order. Since amphibian eggs as we know them lack the amnion and allantois for embryonic support and respiration, they are necessarily small—less than ten millimeters in diameter. The hatchlings are correspondingly tiny. The modern plethodontid salamanders, which are small amphibians, often deposit their eggs in damp places on land rather than in the water, as is the usual amphibian pattern. The young then hatch as miniature adults instead of going through the usual amphibian tadpole stage. Perhaps the first amniote eggs evolved from such amphibian terrestrial eggs and were produced by very small reptiles. Interestingly, the oldest known fossil reptiles were very small, being of such size that they might have laid eggs barely beyond the size limits of amphibian eggs.

REPTILE CHARACTERS

The mixture of amphibian and reptilian characters seen in *Seymouria* is indicative of the gradual transition that took place between the two classes during the evolution of the verte-

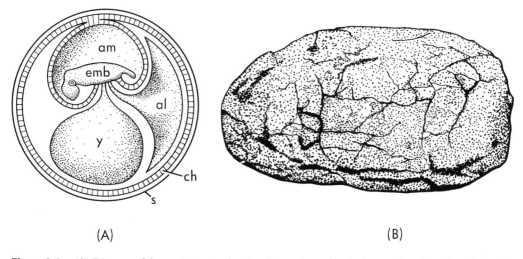

Figure 8-1. **(A)** Diagram of the amniote egg, showing the amnion cavity (am) containing the embryo (emb), the allantois (al), the yolk sac (y), the chorion (ch), and the outer shell (s). **(B)** The oldest known reptile egg, from the lower Permian of Texas. One and a third times natural size.

brates. Because the change was gradual rather than abrupt it is difficult to draw clear-cut distinctions between amphibians and reptiles when all the fossil materials are taken into consideration. There are, however, certain characters typical of reptiles and others generally typical that may be outlined at this place.

Structurally the reptiles are generally characterized by a rather deep skull, in contrast to the frequently flattened skull of the labyrinthodont amphibians. The primitive otic notch has been suppressed in the reptile skull, although there are some reptiles that have an otic notch, probably of secondary origin. Another feature typical of the reptiles is the extreme reduction of the post-parietal elements in the skull, so that the bones behind the parietals are small, or shifted from the skull roof to the occiput, or even suppressed completely. The pineal foramen, so characteristic of the labyrinthodonts, persists in the early reptiles, but in many advanced forms it disappears. The pterygoid bones of the palate are prominent in the reptiles, and in the primitive forms these bones carry well-developed teeth. In the front of the mouth the reptilian palatine bones may have small teeth, but they lack the large tusklike palatine teeth, so common in many

labyrinthodonts. Finally, the occipital condyle is single in most of the reptiles, a character that is foreshadowed in the anthracosaurs.

As in seymouriamorphs, the reptiles are typified by a large pleurocentrum forming the main body of the vertebra, with the intercentrum reduced to a small wedge, or suppressed in the more advanced forms. There are in the primitive reptiles two sacral vertebrae as contrasted with the single amphibian sacral, whereas in many of the advanced reptiles the sacrum includes several vertebrae, sometimes as many as eight. The ilium of course is expanded to form an attachment with the expanded sacrum. In the reptilian shoulder girdle there is an emphasis of the scapula and coracoid, with the cleithrum reduced or suppressed, and with the clavicle and interclavicle often present but much reduced as compared with these same elements in the labyrinthodonts. Among primitive reptiles the ribs form a continuous and generally similar series from the skull to the pelvis, but in the more advanced reptiles there is commonly a differentiation of ribs in the neck, thoracic, and abdominal regions.

The limbs and the feet in the reptiles show certain advances over the appendages in the labyrinthodont amphibians. Even in the primitive reptiles the limb bones are generally more slender than these bones in the labyrinthodonts. In the wrist there are never more than two central bones, as contrasted with the four such elements of the labyrinthodont wrist. In the ankle the proximal bones are reduced to two as contrasted with the three amphibian elements. This reduction has been brought about by a fusion of the inner bone, the tibiale, with the intermedium and also with one central bone, the coalesced element being the equivalent of the astragalus in higher vertebrates. The outer bone, the fibulare, is the equivalent of the calcaneum in the mammals. The reptiles have a basic phalangeal formula of 2-3-4-5-3(4), although this number may be modified in many of the specialized reptiles.

Finally, many reptiles have a horny epidermis that often takes the form of folded, overlapping scales.

CAPTORHINIDANS

The earliest and most primitive reptiles belong to the order Captorhinida, and typical of these is the genus *Hylonomus,* found in Carboniferous swamp deposits at Joggins, Nova Scotia. *Hylonomus* was a small reptile, perhaps thirty centimeters, or about a foot, in length. The skull, about three centimeters, or an inch and a quarter, in length was solidly roofed, somewhat elongated, and rather deep. The eyes were placed laterally. On the skull roof was a well-developed pineal opening (at the junction between the parietal bones)—a character inherited from amphibian ancestors. The postparietal and tabular bones were comparatively small and were pushed back from the top of the skull to the occiput or back surface. There were sharp teeth on the margins of the jaws, and small teeth on the pterygoid bones of the palate. The body was elongated and the limbs were strong but rather sprawling. The postcranial skeleton was characteristically reptilian, with reduced intercentra, a large scapula-coracoid complex, a long inter-

clavicle, a slightly expanded ilium, two sacral vertebrae, and two proximal bones in the ankle (the astragalus and calcaneum) with the primitive reptilian phalangeal formula. The tail was long.

Hylonomus belongs to a major subdivision of the Captorhinida known as the Captorhinomorpha, so named from the characteristic genus, *Captorhinus.* The captorhinomorphs were generally small reptiles. *Captorhinus,* for example, was a Permian genus about a foot in length. In this genus, as in all the captorhinomorphs, the skull was truncated at the back, and the quadrate bone for the support of the lower jaw was vertical in position. The jaws were very long, with numerous sharp, pointed teeth. Evidently, this little captorhinomorph (like all members of this evolutionary line) was a carnivorous animal and probably preyed on small amphibians and reptiles, perhaps even on large insects.

From an ancestry that may be exemplified by *Hylonomus* or *Captorhinus,* the captorhinidans evolved in several directions during late Paleozoic and early Mesozoic times. One of the early offshoots from the captorhinomorph stock was the procolophons—of late Permian and Triassic age, the only captorhinidans to survive beyond the end of Permian time. The procolophons were small reptiles, similar to many modern lizards in size, and perhaps in habits as well. It is quite possible that these reptiles fulfilled the ecological role in the faunas of Triassic times that the lizards were to fulfill in animal assemblages of later geologic ages.

The procolophons were widely distributed during the Triassic period. Ancestral forms are known from the upper Permian sediments of Russia, but it was in the Triassic phase of the Karroo series in South Africa that the group became well established. *Procolophon* was the characteristic genus, a small reptile with a rather triangular-shaped, flattened skull, having large orbits, a large pineal opening, and a limited number of peglike teeth. During the course of Triassic times, the procolophons

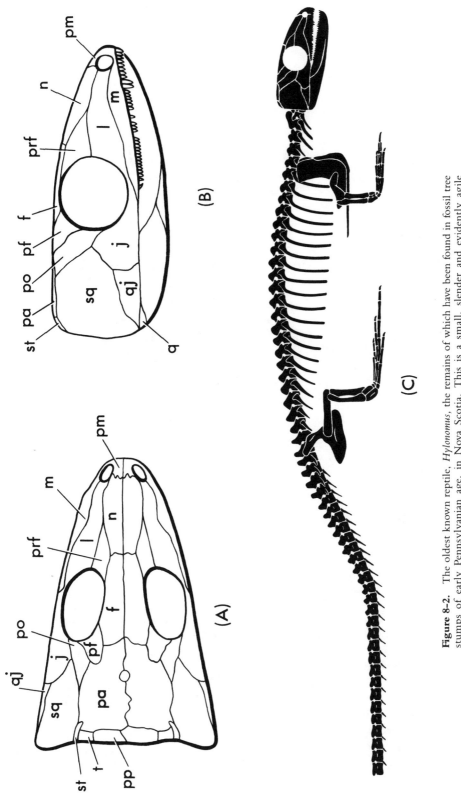

Figure 8-2. The oldest known reptile, *Hylonomus*, the remains of which have been found in fossil tree stumps of early Pennsylvanian age, in Nova Scotia. This is a small, slender and evidently agile captorhinomorph, characterized by the solidly roofed skull of the cotylosaurs. **(A)** Dorsal view and **(B)** lateral view of the skull, about two and a half times natural size. For abbreviations, see pages 447–449. **(C)** The skeleton, about natural size.

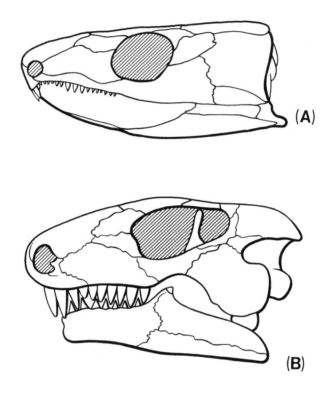

Figure 8-3.   Two captorhinidan skulls. **(A)** *Captorhinus,* from the early Permian of Texas. About one and one-third times natural size. **(B)** *Procolophon,* from the early Triassic of South Africa. About one and one-half natural size.

spread through central Europe, as far north as northern Scotland, and into North and South America.

Of particular significance was the rather recent discovery in Antarctica, at a locality less than six hundred kilometers from the South Pole, of *Procolophon trigoniceps,* a species heretofore characteristic of the Lower Triassic of South Africa. This occurrence of *Procolophon,* added to the evidence of other fossils coincides with geological and geophysical information indicating that East Antarctica was connected with southern Africa in Triassic time. There

was never any notable increase in size among these reptiles, but during the later portions of the Triassic period there was a strong trend among the procolophons for the eye opening to become enormously large, for spikes to develop on the sides and the back of the skull, and for the teeth to become very much reduced in number and specialized in form.

Closely related to the procolophons were the pareiasaurs, found especially in the upper Permian sediments of southern Africa and of Russia. Unlike the procolophons, however, the pareiasaurs were large, heavy reptiles; in-

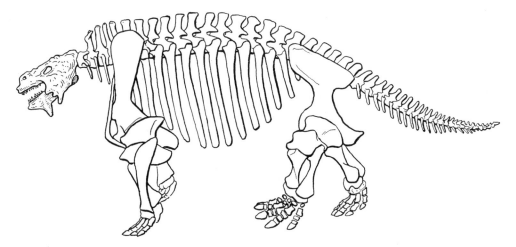

Figure 8-4.   Skeleton of the pareiasaur *Scutosaurus,* from the late Permian of Russia. About two meters long.

deed, they were among the giants of their time. Many of these reptiles were eight to ten feet in length and, since in all of them the body was very bulky and the tail was relatively short, they were necessarily very massive animals. The limbs were remarkably stout in order to support the great weight of the animals, and the feet were short and broad. The body cavity was enormously capacious—an indication that the pareiasaurs probably fed on bulky plant food, a supposition borne out by the form of the teeth, which were small and serrated and evidently well adapted for slicing vegetable matter. There were numerous teeth on the palate. The skull, although of cotylosaurian form, was broad and massive, the skull roof and temporal region usually being ornamented with numerous rugosities and spikes. The presence of bony scutes above the backbone indicates that the pareiasaurs were protected by body armor.

CLASSIFICATION OF THE REPTILES

The establishment of the first captorhinomorph reptiles from a labyrinthodont ancestry was a major step in the evolution of the vertebrates and an event of prime importance to the subsequent history of the reptiles. It seems probable that all other reptiles were ul-

timately of captorhinomorph ancestry. There has been a great deal of paleontological debate about this; some authorities would derive the several large lines of reptilian evolution from separate amphibian stocks, others would derive them variously from within the earliest reptiles. Some authorities view the reptiles as showing a major evolutionary dichotomy, into what E.C. Olson has termed the Parareptilia and the Eureptilia, or into what D.M.S. Watson, in a somewhat different manner, designated as the Sauropsida and Theropsida. There are other variations on this theme, with a consequent lack of agreement as to how the supposed basic division of the reptiles into two evolutionary lines took place.

Perhaps the most practical course to take at the present time is to recognize but avoid this controversial problem by classifying the reptiles in an empirical fashion. Generally speaking, a rather satisfactory grouping of the orders of reptiles can be made according to the development of the temporal region in the skull. This system, like any scheme of classification based upon a single character or a set of closely related characters, is not perfect, but it does give a useful arrangement for the study of reptilian orders.

As we have seen, the captorhinidans resembled their labyrinthodont ancestors in that the skull roof was a solid structure, pierced only by the nostrils, the eyes, and the pineal opening. As the reptiles evolved, the skull roof behind the eyes generally was fenestrated by additional openings, commonly called the temporal fenestrae or openings, the purpose of which was to give room for increased attachment and the bulging of the jaw muscles. Although each type of temporal opening represents a pair of windows in the skull, one on the right side and a matching one on the left, they are usually discussed in the singular, as if referring to only one side of the skull.

In some reptiles there is an upper temporal fenestra high on the skull roof behind the eyes; in others there is a lower opening on the side of the skull. In still other reptiles there are both temporal openings, one on the top of the skull roof and one in the side. The upper opening, or superior temporal fenestra, is bounded along its ventral border by the postorbital and squamosal bones. The lower or lateral temporal fenestra is primarily bounded along its dorsal border by the postorbital and squamosal bones. Finally the two temporal openings, when present, are separated by the postorbital and squamosal bones. Upon these varying relationships of temporal openings (or the lack of openings) the orders of reptiles may be arranged as follows:

Subclass Anapsida
  No temporal openings in the skull behind the eye.
    Order Captorhinida: the earliest and most primiitive reptiles.
    Order Millerosauria: millerosaurs, lizardlike forms.
    Order Mesosauria: mesosaurs, small aquatic (freshwater) reptiles.
    Order Chelonia: turtles and tortoises.
Subclass Synapsida
  A single lower temporal opening, bordered above by the postorbital and squamosal bones.
    Order Pelycosauria: pelycosaurs, primitive mammallike reptiles.
    Order Therapsida: therapsids, advanced mammallike reptiles.
Subclass Diapsida
  Both upper and lower temporal openings separated by the postorbital and squamosal bones.

  Infraorder uncertain
    Order Araeoscelida: the earliest and most primitive diapsids, including *Petrolacosaurus*.
    Order Choristodera: champsosaurs, aquatic diapsids.
  Infraclass Lepidosauromorpha
    Order Eosuchia: early terrestrial and aquatic lepidosauromorphs.
  Superorder Lepidosauria
    Order Sphenodonta: lizardlike forms such as the modern tuatara, *Sphenodon*.
    Order Squamata: lizards and snakes.
  Infraclass Archosauromorpha
    Order Rhynchosauria: rhynchosaurs.
    Order Thalattosauria: primitive marine diapsids.
    Order Trilophosauria: trilophosaurs.
    Order Protorosauria: protorosaurs.
  Superorder Archosauria
    Order Thecodontia: Triassic archosaurs, ancestors of the dominant diapsids of Mesozoic times.
    Order Crocodylia: crocodiles, alligators, and gavials.
    Order Pterosauria: flying reptiles.
    Order Saurischia: saurischian dinosaurs.
    Order Ornithischia: ornithischian dinosaurs.
  Infraclass Euryapsida: modified diapsids which retain the upper temporal opening but lose the ventral bony border of the lower temporal opening, resulting in, at most, a ventral embayment to indicate the former presence of the lower temporal opening
  Superorder Sauropterygia: marine reptiles of Mesozoic times.
    Order Nothosauria: nothosaurs (primitive sauropterygians).
    Order Plesiosauria: plesiosaurs (advanced sauropterygians).
  Superorder Placodontia: placodonts.
  Superorder Ichthyosauria: ichthyosaurs or fishlike marine reptiles.

Until quite recently, the Euryapsida was thought to be a separate subclass of reptiles equal in rank to the Anapsida, Synapsida, and Diapsida. New fossil discoveries and interpretations have shown, however, that the now extinct euryapsids were actually modified diapsids in which the lower temporal opening was reduced and eventually lost.

PRIMARY RADIATION OF THE REPTILES

Although the captorhinidans were the first reptiles, it must not be thought that they stood alone during the early history of the reptiles. Other reptiles that probably descended from

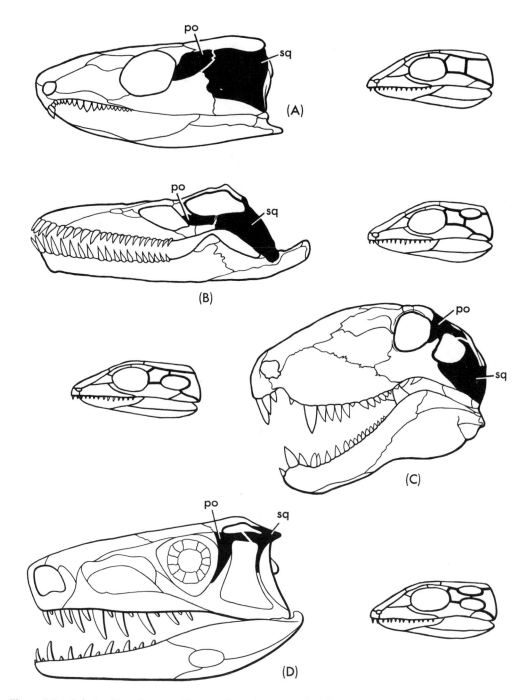

Figure 8-5. Relationships of temporal bones and openings in reptiles, illustrated by the skulls of characteristic genera, and by schematic diagrams. **(A)** Anapsid skull (*Captorhinus*, a captorhinidan) with no temporal opening. **(B)** Euryapsid skull (*Muraenosaurus*, a plesiosaur) with a superior temporal fenestra bounded below by the postorbital and squamosal bones. **(C)** Synapsid skull (*Dimetrodon*, a pelycosaur) with a lateral temporal fenestra bounded above by the postorbital and squamosal bones. **(D)** Diapsid skull (*Euparkeria*, a thecodont) with superior and lateral temporal fenestrae separated by the postorbital and squamosal bones. Not to scale.

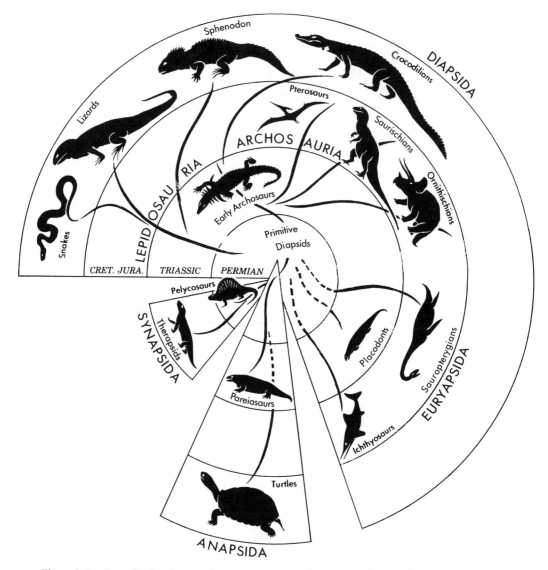

Figure 8-6. Generalized evolution and major classification of the reptiles. Prepared by Lois M. Darling.

the captorhinomorph stem appear at a remarkably early period in the geologic record of these vertebrates, so that it would seem as if there had been an "explosion" of various higher categories soon after the reptiles became differentiated as a class. It has already been shown that this is a common evolutionary phenomenon, repeated time and again through the history of life.

Among the earliest of the known reptiles were the aquatic mesosaurs, appearing during the lower Permian period. The mesosaurs, known from the single genus *Mesosaurus,* were rather highly specialized in spite of their very early debut in reptilian history. They were small and elongated reptiles, with extended jaws at one end of the body, equipped with very long, sharp teeth, and with a long, deep tail at the other end. The slender, toothy jaws were obviously adapted for catching fish or

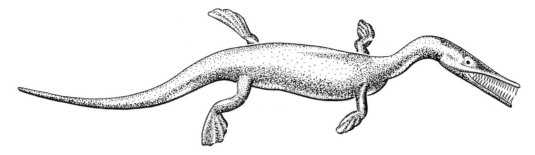

Figure 8-7. The early anapsid *Mesosaurus,* a small aquatic reptile from lower Permian sediments of South Africa and Brazil.

perhaps small crustaceans; the deep tail was plainly a mechanism for swimming. The shoulder and hip girdles were rather small, the limbs were slender, and the feet were enlarged into broad paddles. There can be little doubt but that the mesosaurs spent most of their time in the water, and it is probable that they seldom if ever ventured on to the land. From the nature of the sediments in which these reptiles are found, it would seem likely that they were inhabitants of freshwater lakes and ponds.

The back of the skull in the mesosaurs is difficult to interpret because it is generally crushed as a result of fossilization. Consequently, the precise systematic position of these reptiles has been a matter of debate for many years. There has been considerable sentiment for placing them tentatively among the synapsids, since it appears that there may have been a lateral temporal opening in the skull. It is quite possible, however, that this opening was an independent development, unrelated to the synapsid temporal opening. It is a significant fact that the neural arches in the mesosaurs were swollen in much the same fashion as the neural arches of captorhinomorphs. On the whole it is most probable that the mesosaurs represent a very ancient and independent evolutionary line of reptiles descended from very primitive captorhinomorphs that developed briefly during early Permian times. They are found in only two localities in the world, in southern Africa and in southern Brazil. This fact is one of numer-

ous accumulating lines of evidence to indicate the probability that Africa and South America were closely connected during middle and late Paleozoic time, as constituent parts of the ancient supercontinent of Gondwana.

The synapsids, evidently originating from a primitive captorhinomorph base, became established during Pennsylvanian time, as we shall see. The first synapsids were pelycosaurs.

Even the diapsids, in many respects the culminating group of reptiles, had their beginnings in Pennsylvanian times. The first diapsids were small reptiles; the earliest known is the genus *Petrolacosaurus* from the Pennsylvanian sediments of Kansas. Another early diapsid genus was *Youngina* from the Permian of South Africa.

Petrolacosaurus was a generalized reptile of rather lizardlike form, and probably of lizardlike habits. Of course such a comparison does not imply direct relationship with the lizards. It is a general comparison only, a comparison with something familiar. *Petrolacosaurus* had a slender body and slender limbs, well adapted for scurrying around over the ground.

By Permian times the early diapsids, such as *Youngina,* were small, lightly built animals that were able to run rapidly and thus survive. These were, generally speaking, the founders of a great line—the later diapsid reptiles.

It seems probable that the euryapsids had their beginning soon after the establishment of the early diapsids. The euryapsids had become well differentiated by Triassic times.

So it was that, once having become established through the perfection of the amniote egg, the reptiles quickly diverged in many directions, to initiate at an early stage in their history the main evolutionary lines that they were to follow. The land was open to them; they were free from dependence upon the water for reproduction, and they quickly made use of the available ecological and evolutionary opportunities. They entered upon a long history, extending from Pennsylvanian times to the present, during much of which, until the end of the Cretaceous period, they were the dominant land animals of the earth. It is our purpose now to follow the various ramifications of reptilian history through more than two hundred million years of geologic time.

Dimetrodon

Mammal-Like Reptiles

9

DELINEATION OF THE SYNAPSIDA

Among the early reptiles were the synapsids, which first appear in rocks of late Pennsylvanian age. This remarkable and interesting subclass of reptiles bridged the gap between the primitive reptiles and the mammals, for its early members were very close indeed to the ancestral captorhinomorphs whereas some of its latest genera approached so closely the mammalian stage that it is a moot question whether these animals should be classified among the reptiles or among the mammals. All this development from primitive reptile to mammal took place between the end of the Pennsylvanian period and the close of the Triassic period.

The synapsids, as we have seen, were those reptiles with a lower temporal opening behind the eye. In the more primitive genera the opening was bounded above by the postorbital and squamosal bones, but as the synapsids evolved during late Permian and Triassic times the temporal fenestra was enlarged to such an extent that the parietal bone frequently became a part of its upper border. These were constantly quadrupedal reptiles through the extent of their history, and in this respect they differed markedly from various other reptiles, especially many of those that flourished during the Mesozoic era. Furthermore, the synapsids showed few tendencies toward the loss of bones, which is rather remarkable in a group of animals that, at the end of their history, approached so closely the mammalian grade of structural development. Even the primitive pineal opening was retained through much of synapsid history. On the other hand, there was an early trend among the synapsids toward a differentiation of the teeth into anterior incisors, enlarged canines, and laterally placed cheek teeth; and among the later synapsids this development resulted in a rather high stage of dental specialization. In these reptiles the eardrum seems to have had a low position, near the articulation of the jaws.

The vertebrae were primitively amphicoelous, which means that the ends of the centra were rather flat. Among the earlier synapsids there were small, persistent intercentra. In the synapsid shoulder girdle there were two coracoid elements, the primitive coracoid, often called the procoracoid, and a new bone behind it, designated simply as the coracoid. The limbs were in general similar to the limbs in the ancestral captorhinomorphs, but they frequently showed evolutionary advances through slenderizing and perfection of the individual bones—an indication that the synapsids were active and rather efficient at getting around over the ground.

Two successive orders comprise the Synapsida; the Pelycosauria of Pennsylvanian and Permian age, and the Therapsida of Permian and Triassic age, with some therapsids extending into Jurassic time.

THE PELYCOSAURS

The first synapsids were pelycosaurs of late Pennsylvanian age. These reptiles initiated a line of evolutionary development that is most completely recorded in the upper Pennsylvanian and lower Permian sediments of North America, particularly in Texas, Oklahoma, and New Mexico. Elsewhere the remains of pelycosaurs are fragmentary and scattered.

The early pelycosaurs showed in general the primitive features that have already been listed for the synapsids. Thus the skull had its full complement of bones (lacking the intertemporal bone, which was lost during the transition from labyrinthodonts to early reptiles), there was a pineal opening, there were vertebral intercentra, and the limbs were similar to the limbs of the captorhinomorphs, only somewhat more slender. In many respects the skulls of some of the early pelycosaurs, such as *Archaeothyris,* show sufficiently significant resemblances to primitive captorhinomorph reptiles, as exemplified by *Paleothyris,* as to indicate a very ancient ancestry for the pelycosaurs, dating back almost to the early Pennsylvanian beginnings of reptilian evolution.

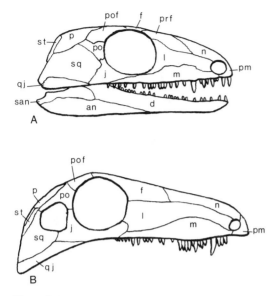

Figure 9-1. **(A)** *Paleothyris,* a primitive anapsid reptile from the Middle Pennsylvanian of North America. Twice natural size. **(B)** *Archaeothyris,* the oldest known pelycosaur, from the Middle Pennsylvanian of North America. About three-fourths natural size. Note the lower temporal fenestra, behind the orbit. For abbreviations see pages 447–449.

OPHIACODONTS

The first pelycosaurs can be included in a suborder known as the Ophiacodontia, a very primitive member of which is the above mentioned *Archaeothyris.* In this small reptile the orbit is large, and set somewhat posteriorly as compared with the position of the orbit in *Paleothyris.* Furthermore, the skull is characterized by a small lateral temporal opening, bounded above by the postorbital and squamosal bones, thereby exhibiting the appearance of a distinctive pelycosaurian feature. Furthermore, the occipital region of the skull slopes back and down to a position about on a level with the articulations between the skull and the lower jaw, again a distinctive feature among the pelycosaurs. Furthermore, there is some differentiation in the size of the teeth in the maxilla, thereby foreshadowing the strong differentiation of the teeth seen in many later pelycosaurs. In many other respects, however, especially in the size and relationships of the various skull bones, *Archaeothyris* clearly shows its derivation from an early captorhinomorph such as *Paleothyris.*

The lower Permian genus, *Varanops,* was a small reptile, three or four feet in length, with a slender body, slender limbs and a long tail. In this early pelycosaur the spines of the vertebrae are somewhat elongated, anticipating the often extreme elongation of the vertebral spines in various specialized pelycosaurs. The skull is rather narrow and deep, the orbit is large, and the lateral temporal opening is enlarged, as is true in the more advanced pelycosaurs. The jaws were very long and the teeth were numerous and sharp. There was no indication of an otic notch, and the ear was located in the vicinity of the jaw articulation.

It was but a short step from *Varanops* to the genus *Ophiacodon,* a large Permian reptile, commonly five to eight feet in length, and in general similar to its predecessor. The skull was very deep, with long jaws, these being provided with many sharp teeth. Evidently *Ophiacodon* was a fish-eating reptile that lived largely along the shores of streams and ponds.

From an ophiacodont stem the pelycosaurs evolved in two directions. One line of pelycosaurian development led to large aggressive, land-living carnivores, the sphenacodonts, the other to large plant-eating forms, the edaphosaurs.

SPHENACODONTS

The sphenacodonts carried forward in Permian times the evolutionary trends that had been initiated among the ophiacodonts, and in the general aspects of the skeleton the two groups show many similarities. In two respects, however, advanced characters developed in the sphenacodonts that went far beyond any of the ophiacodont specializations. In the first place, the sphenacodont dentition was strongly differentiated, a specialization that was correlated with refinements in the skull. There were large, daggerlike teeth in the premaxilla, in the front part of the maxilla,

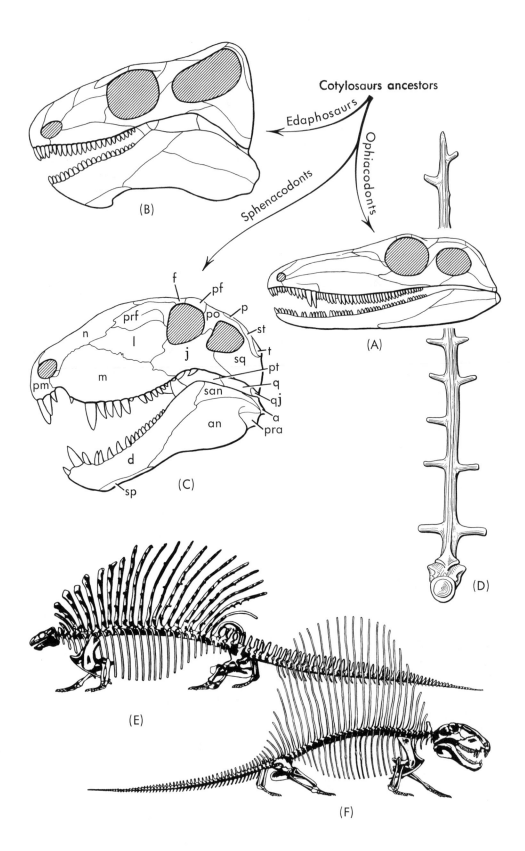

Cotylosaurs ancestors

Edaphosaurs

Ophiacodonts

Sphenacodonts

(B)

(A)

f
pf
prf
po
p
n
l
st
pm
j
t
m
sq
pt
q
san
qj
a
pra
an
d
sp

(C)

(D)

(E)

(F)

and in the anterior portion of the dentary, whereas the teeth along the lateral borders of the dental arcade were considerably smaller than the anterior teeth. The skull was deep and narrow, which was an adaptation for long, strong jaw muscles that allowed the mouth to be widely opened, and closed with a powerful snap. Such specializations were obviously quite advantageous to very aggressive reptiles that preyed upon other large vertebrates.

The other sphenacodont specialization was the elongation of the vertebral spines, which is not so easy to interpret as the development of the skull and jaws. In the genus *Sphenacodon* the spines were tall, but not unduly so, and it is clear that they might have served for the origin and insertion of strong back and neck muscles. But in another genus, *Dimetrodon,* the spines of the vertebrae from the neck back to the sacrum were tremendously elongated, reaching their greatest height in the middle region of the back. These spines almost certainly supported a web of skin so that there was a longitudinal "sail" down the middle of the back in this remarkable reptile.

What is the meaning of the sail in *Dimetrodon*? This question has been debated for many years, and as yet no conclusive answer has been given. Some have argued that the sail was protective, but it is difficult to see how such a structure would give much protection to the animal. Again it has been suggested that the sail was a sort of "psychological warfare" device; that it made *Dimetrodon* look big and impressive, thereby frightening any potential enemies. This explanation is not very convincing. It has been proposed that the sail was an expression of sexual dimorphism—that *Dimetrodon* with the big sail was the male, and *Sphenacodon* without the sail was the female. But the evidence is against this proposal because the two forms are found in different lo-

calities. *Dimetrodon* is found in Texas and *Sphenacodon* in New Mexico, which regions during lower Permian times were ecologically separated. Perhaps the most logical explanation so far made is that the sail was a temperature-regulating device that added a great area of skin surface for warming up or cooling off the animal. This supposition is strengthened by the fact that the area of the sail, as determined by the length of the vertebral spines, varies proportionately to body size in the several species of *Dimetrodon*. Thus the largest of these reptiles have disproportionately tall spines. It seems reasonable to think that a structural modification so extreme as this must have been significant in the life of *Dimetrodon,* that it must have had a considerable adaptive value. Certainly *Dimetrodon* was a highly successful reptile over quite a span of geologic time.

EDAPHOSAURS

The second evolutionary line of Permian age springing from an ophiacodont ancestry, that of the edaphosaurs, was quite different from the sphenacodont line. The edaphosaurs were inoffensive plant-eaters, as is shown by the structure of the skull and teeth. In these animals the skull was remarkably small as compared with the size of the body. It was short and rather shallow, as contrasted with the elongated deep skull of the sphenacodonts. The teeth in the edaphosaurs were not strongly differentiated in size, as were the sphenacodont teeth, but rather were essentially uniform, making an unbroken series around the margins of the jaws. In addition to the marginal teeth, there were in the edaphosaurs extensive clusters of teeth on the palate.

The genus *Edaphosaurus* was characterized by an elongation of the vertebral spines, but the spines were heavier than they were in

Figure 9-2. Pelycosaurs. **(A)** Skull of *Varanosaurus,* one-half natural size. **(B)** Skull of *Edaphosaurus,* one-third natural size. **(C)** Skull of *Dimetrodon,* one-fifth natural size. **(D)** Dorsal vertebra of *Edaphosaurus* in posterior view, one-third natural size. **(E)** Skeleton of Edaphosaurus. **(F)** Skeleton of *Dimetrodon.* These reptiles were about six to ten feet in length. For abbreviations, see pages 447–449.

Dimetrodon, and they were ornamented with numerous short lateral spikes or crossbars arranged irregularly along their length, something like the yardarms on the mast of an old sailing ship. All of which makes *Edaphosaurus* even more of a puzzle than *Dimetrodon.* What could be the meaning of such an adaptation? Surely the growth and nutrition of the sail and the bizarre spines in *Edaphosaurus* must have constituted a serious drain upon the energy of this animal.

Not all the edaphosaurs had sails. *Casea* and the related genus, *Cotylorhynchus,* were massively built edaphosaurs with small heads and barrel-like bodies. *Cotylorhynchus,* the largest of the pelycosaurs, probably played the same role among the early Permian reptiles of North America that the pareiasaurs did in South Africa and Russia.

THE THERAPSIDS

Whereas the pelycosaurs were of late Pennsylvanian to late Permian age and are known largely from North America, the therapsids were of middle and late Permian to Jurassic age and are known from all the continental regions, but especially from the Karroo sediments of South Africa. It appears likely that the therapsids had a pelycosaurian ancestry, but from the beginning of their history these reptiles followed evolutionary trends that led to the attainment of specializations much different from those reached by any other reptiles. These were the reptiles that took the road leading to the mammals, and some of them approached very closely the mammalian stage of organization.

As indicated above, there was a strong trend among the therapsids for the lateral temporal opening to become enlarged, so that in the advanced forms the upper boundary of the opening was formed by the parietal bone, not by the postorbital and squamosal bones. The quadrate and quadratojugal bones were reduced to very small elements, often loosely connected to the skull, as contrasted with the large quadrate in most reptiles. In the more advanced therapsids there was a secondary palate below the original reptilian palate, this new palate being formed by the premaxillary, maxillary, and palatine bones. It served to separate the nasal passage from the mouth, thereby increasing the efficiency of breathing, especially while the animal was feeding. The pterygoid bones were generally solidly fused to the braincase. In the lower jaw the dentary bone tended to enlarge at the expense of the other jaw bones. In most therapsids there was a large notch in the angular bone, bordered below by a prominent flange. This character was inherited from the pelycosaurs, in which it was an adaptation for the insertion of strong pterygoid muscles for closing the jaws. The differentiation of the teeth progressed in the therapsids to high levels of development, with the advanced genera showing sharply contrasted incisors, canines, and cheekteeth, which in some of these reptiles were of complex form, often with accessory cusps or broad crowns. In many therapsids the occipital condyle became double, as in the mammals.

In the postcranial skeleton there was frequently a considerable regional differentiation of the ribs and vertebrae, so that the neck was quite distinct from the body, which in turn might have a distinct lumbar region. The legs were generally "pulled in" beneath the body, with the elbows pointed more or less backward and the knees forward. The body was thus raised from the ground, so that the efficiency of locomotion was increased. This was a departure from the general sprawling pose so characteristic of the pelycosaurs. The scapula was a large bone in the shoulder girdle, and there were two coracoid elements. The ilium was expanded forward, and there was generally an elongated sacrum, making a strong attachment between the backbone and the pelvis. The feet were well formed and adapted to efficient walking and running over the dry ground.

Within this general plan the therapsids developed during Permian and Triassic times,

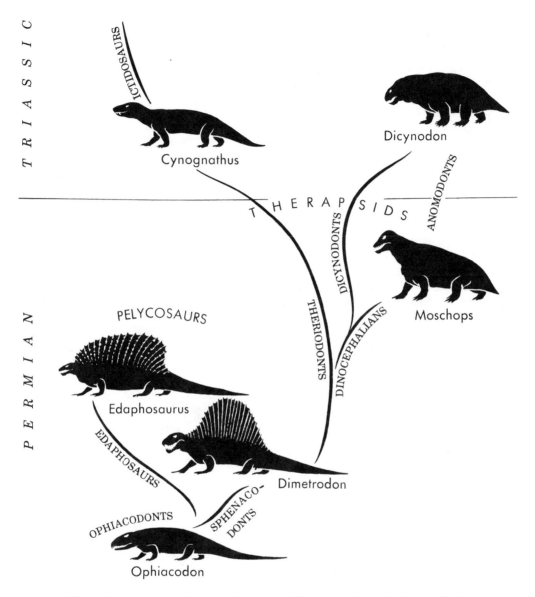

Figure 9-3. Evolution of the synapsid or mammal-like reptiles. Prepared by Lois M. Darling.

and during their evolution these reptiles, originating from Middle Permian ancestors, followed two general lines of adaptive radiation. One of these therapsid lines was that of the anomodonts, containing the large dinocephalians, and the widely spread dicynodonts; the other was the very mammal-like theriodonts.

EOTITANOSUCHIANS

Biarmosuchus, an eotitanosuchian, from the middle Permian of Russia, is one of the most primitive of known therapsids, revealing in most of the features of its skull significant resemblances to the skull in the sphenacodont pelycosaurs. In this early therapsid the orbit was large and set far back in the skull, while behind the eye opening there was a vertically elongated temporal opening, bounded above by the postorbital and squamosal bones—the basic synapsid condition, as has already been mentioned. As in the sphenacodonts, the maxillary bone on the side of the rather deep skull was enlarged and separated the lacrymal bone from the nasal opening. There was a very large canine tooth in the front of the maxilla and there were teeth on the pterygoid bones in the palatal region. The vertebrae are like those of primitive sphenacodont pelycosaurs, but do not have the elongated neural spines so typical of the pelycosaurs. The configuration of the girdles and the limb bones, especially the inturned head of the femur and the rather symmetrical structure of the feet, indicate that this therapsid walked with the feet well beneath the body, in what might be called a semierect posture.

Phthinosuchus, from the Upper Permian of Russia and somewhat more advanced than *Biarmosuchus,* is annectent between the most primitive known therapsids and the more highly evolved members of the order. The skull of *Phthinosuchus* also is remarkably similar to a sphenacodont skull; however, it shows advances toward the therapsids in its large temporal opening, in the depression of the jaw articulation, and in the presence of single large canine teeth in both upper and lower jaws. The skeleton also seems to show advances toward the therapsid condition. From such a nicely intermediate form the therapsids evolved along their several lines of adaptive radiation.

DINOCEPHALIANS

From an eotitanosuchian base, as exemplified by *Biarmosuchus,* the therapsids evolved along three major lines of adaptive radiation. These were the dinocephalians, to be immediately considered, the dicynodonts and the theriodonts, to be treated in subsequent sections of this chapter.

The Permian dinocephalians were in many ways the most archaic of the therapsids and,

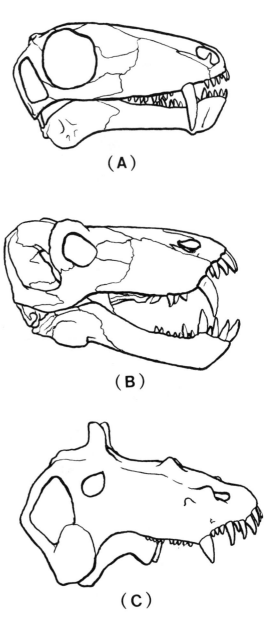

(A)

(B)

(C)

Figure 9-4. Synapsid skulls and jaws from the Permian of Russia. **(A)** *Biarmosuchus,* a primitive therapsid. One third natural size. **(B)** *Titanophoneus,* a dinocephalian. One-fifth natural size. **(C)** *Estemmenosuchus,* a dinocephalian. One-ninth natural size.

although these reptiles showed therapsid adaptations, such as the expansion of the ilium and the general pose of the limbs, they retained various primitive characters of the pelycosaurs. For instance, they had no secondary palate, and their dentary bone was of moderate size. The particular characters distinctive of the dinocephalians were their large size and the pachyostosis, or thickening, of the bones in the skull. Often the top of the head was rounded and elevated into a sort of dome or boss, in the middle of which was lodged a large pineal opening. These reptiles were massive animals that must frequently have weighed a thousand pounds or more in life.

The trend to giantism, so characteristic of the dinocephalians, was established at an early stage in their history; hence, the lower levels of the South African Karroo beds, which in their totality contain such a magnificent record of therapsid evolution, are filled with the bones of giants. Furthermore, there was an early dichotomy among the dinocephalians, with one branch, the titanosuchians, becoming large, ponderous carnivores, and the other, the tapinocephalians, becoming large and equally ponderous herbivores. There is good reason to think that the titanosuchians preyed largely on their tapinocephalian cousins.

One of the oldest yet one of the more specialized of the dinocephalians is *Estemmenosuchus,* a titanosuchian from the lowest Upper Permian zone in Russia. This strange therapsid had horn-like protuberances on the skull—a large upwardly directed pair above the eyes, very large laterally directed jugal protuberances, and small rounded bumps on the front of the skull. Perhaps these weird-looking growths may have served as protective features. There were large canine and incisor teeth, but the cheek teeth of this reptile were remarkably small.

Other titanosuchians are typified by the Permian genera *Titanophoneus* from Russia, and *Jonkeria* from South Africa. The former was a rather medium-sized therapsid, but *Jonkeria* was of gigantic proportions—twelve feet or more in length. In this large lumbering carnivore the snout of the heavy head was elongated and provided with sharp teeth, including large incisors, behind which, on each side above and below were long piercing ca-

nines. The cheek teeth were comparatively small. The body was robust and the limbs were very stout.

In *Moschops,* a typical tapinocephalian, the skull was high and short, and the teeth were undifferentiated and rather peglike. It seems obvious that these teeth, set in comparatively short jaws, were adapted for an herbivorous diet. The shoulders were much higher than the pelvic region in this reptile, so that the back sloped, giraffe-fashion, from neck to tail. The limbs were heavy, and the feet were broad. *Moschops* and its relatives probably wandered over fairly dry uplands during the Permian period, feeding upon the available vegetation.

DICYNODONTS

The venyukoviamorphs, represented by a handful of genera, seem to be intermediate in position between the dinocephalians and the dicynodonts. In these reptiles, as exemplified by the genus *Venyukovia* from the Permian of Russia, there was a reduction in the dentition, and the skull exhibited changes in proportions that were clearly prophetic of the highly evolved skull in the dicynodonts, which we now consider.

The dicynodonts, the most successful of the therapsids if phylogenetic longevity, numbers of individuals, and the extent of distribution over continental areas are criteria of success, appeared in middle Permian times and evolved in a remarkably uniform structural pattern, as seen in *Dicynodon,* through the late Permian and the whole of the Triassic period. During late Permian times they were among the commonest of all reptiles, at least as indicated by the fossil record, and in the Triassic period they spread to all the continents, to enjoy a world-wide distribution.

The dicynodonts ranged in size from reptiles no more than a foot or so in length, such as *Endothiodon* from the Permian of South Africa, to large, massive animals, such as the Triassic genera *Stahleckeria* from Brazil or *Placerias* from Arizona, as big as the largest

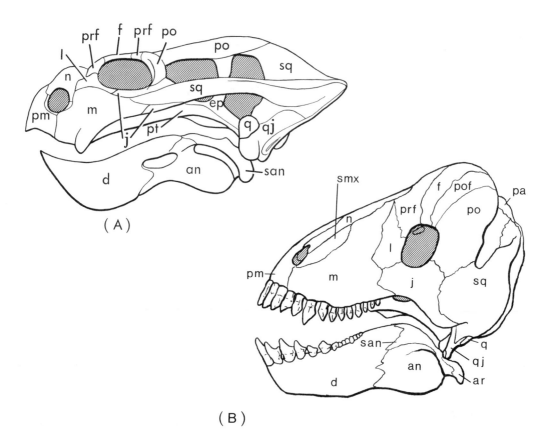

Figure 9-5.   Therapsids or mammal-like reptiles from Permian sediments of the Karroo series of South Africa. **(A)** *Dicynodon,* a dicynodont, one-half natural size or less. **(B)** *Ulemosaurus,* a dinocephalian, one-sixth natural size. For abbreviations, see pages 447–449.

dinocephalians. The body was short and broad, and was supported by strong limbs, so posed that the animal was raised up from the ground in the usual therapsid fashion. The ilium was expanded and strong, and likewise the shoulder girdle was large and strong. The tail was short.

It is in the skull that the dicynodonts show their greatest specializations; certainly it was different from the skull in other therapsids. Although strong, the bones in the temporal region were thinned by an increase of the temporal openings, and by emargination along their lower borders, so that they formed long arches. Indeed the skull in the dicynodonts is noteworthy because of its open construction

and the presence of long, bony bars rather than broad platelike areas behind the eye. The front of the skull and the lower jaw were narrow and beaklike, and except for a pair of large, upper tusks in some of the skulls the teeth were reduced to tiny remnants or were completely absent. There can be little doubt but that the upper and the lower jaws were covered with horny beaks, like the beak in the modern turtles. The presence or absence of tusks in the dicynodonts has been sometimes considered as an expression of sexual dimorphism. This may possibly be true for some species, but evidently is not the case within a large proportion of these very successful therapsid reptiles. All in all this was a rather

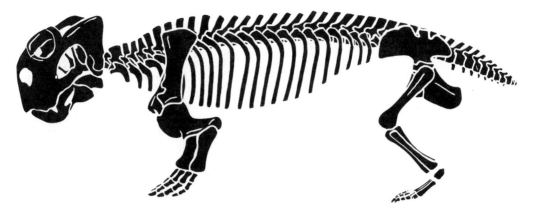

Figure 9-6. *Lystrosaurus,* a lower Triassic dicynodont, found in South Africa, India, Russia, China, and Antarctica. About one-fourth natural size.

strange reptilian design, but nevertheless a very successful one. The dicynodonts were very probably herbivorous. One genus, *Lystrosaurus,* of Triassic age, appears to have been amphibious, living along the edges of rivers and lakes.

One of the important paleontological events of recent years was the discovery of *Lystrosaurus* in Antarctica. This dicynodont, found in association with *Procolophon* (see page 104), is specifically identical with *Lystrosaurus* in South Africa. *Lystrosaurus* is also found in India and Russia, as well as in China. The presence of this reptile on continents now widely separated adds corroborative evidence to the concept of a once great Gondwana supercontinent, stretching across the southern hemisphere during late Paleozoic and early Mesozoic time.

THERIODONTS

While the dinocephalians and the dicynodonts were evolving in their own clumsy ways the theriodonts were developing rapidly in directions that were to lead directly to the mammals. These most mammal-like of the therapsids were small to large carnivorous reptiles, which appeared during middle Permian times and developed through the remainder of the Permian period, continued into the early and middle phases of the Triassic period,

and lingered on, in diminished numbers, to the close of Triassic time. They are found in various parts of the world, but are most numerous and best exemplified in the Karroo beds of South Africa.

The theriodonts, constituting a large and varied suborder of the therapsids, may be divided into a number of infraorders that evolved along more or less parallel lines. Of these the gorgonopsians are perhaps the most primitive and are confined to rocks of Permian age. The therocephalians, somewhat more mammal-like than the gorgonopsians, extend from Permian into Lower Triassic sediments. A third infraorder, the cynodonts, are the most mammal-like of the theriodonts, and some cynodont offshoots, notably the tritylodonts and the ictidosaurs or chiniquodonts, approach so closely the mammalian condition that they may be separated from the mammals only by the use of certain osteological details.

Most of the theriodonts were small to medium-sized therapsids, well developed for comparatively rapid movement in pursuit of their prey. *Cynognathus* can be described as a genus typical of these theriodonts. This lower Triassic therapsid was at its maximum size about as large as a big dog or a wolf. It had a rather large skull that in general form was vaguely doglike, hence the name *Cynognathus.*

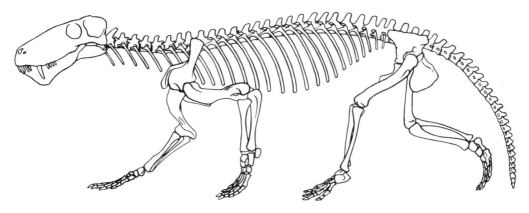

Figure 9-7.   Skeleton of the gorgonopsian theriodont, *Lycaenops,* one-tenth natural size, illustrating the mammal-like characters and pose in an advanced therapsid skeleton.

The skull was elongated and rather narrow. Behind the eye there was an enlarged temporal opening, of which the parietal bone formed the upper portion, and within which very powerful muscles were located for closing the jaws. In the skull the maxillary bone was expanded to form a large plate on the side of the face, whereas in the mandible the dentary bone was so large as to form almost the entire body of the lower jaw, with the bones behind the dentary quite small and crowded.

The teeth were highly specialized and differentiated. In the front of the jaws, above and below, were small, peglike incisors, obviously adapted for nipping. Behind the incisors was a gap, followed in both jaws by a greatly enlarged tooth, the canine. The canines were certainly for piercing and tearing, and they indicate the highly predaceous habits of *Cynognathus*. Behind the enlarged canines and separated from them by another gap were the cheek teeth or post-canines, which in *Cynognathus* were limited in number to about nine on each side and were specialized by the development of accessory cusps on each tooth. These teeth were for chewing and cutting the food, and they indicate that *Cynognathus* must have cut its prey into comparatively small pieces before eating it, rather than swallowing the food whole as do many reptiles. In this reptile there was a well-developed secondary palate, which separated the nasal passage from the mouth. The indication that *Cynognathus* comminuted its food, together with the evidence that the nasal passage was separated from the mouth, show that this animal was very active. Small pieces of meat would be quickly assimilated in the digestive tract, and would quickly replenish an energy output that in this animal must have been proportionately much greater than in the "typical" reptile.

The quadrate bone in the skull and the articular bone in the lower jaw formed the hinge on which the jaws worked, but these bones were very small. The skull articulated to the backbone by a double condyle, formed by the exoccipital bones.

The vertebral column in *Cynognathus* was strongly differentiated into a cervical region with very small ribs, a dorsal region with large ribs, a lumbar region, again with small ribs, a sacrum involving several vertebrae, and a tail. Since the sacrum was elongated, the ilium was enlarged, mainly by a forward growth of the iliac blade, and this gave it a very mammalian appearance. The scapula was also advanced toward the mammalian condition in that its front border was everted or turned out. Here was the beginning of the scapular spine, so characteristic of the mammals.

The limbs were held beneath the body in *Cynognathus,* with the knee pointing forward and the elbow pointing backward to a considerable degree, a pose that increases the

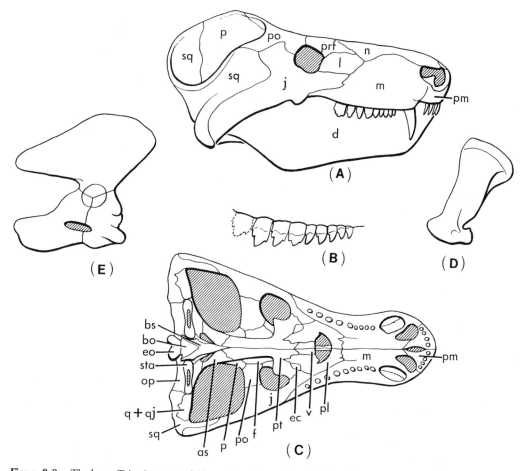

Figure 9-8. The lower Triassic mammal-like reptile, *Cynognathus*. **(A)** Lateral view of the skull. **(B)** Enlarged view of postcanine teeth, showing the cusps. **(C)** Palatal view of the skull. **(D)** Lateral view of the scapula. **(E)** Lateral view of the pelvis. All about one-fourth natural size. For abbreviations, see pages 447–449. This figure shows particularly the enlarged temporal opening of the skull, the differentiated dentition with a large canine and cusped cheek teeth, the large dentary forming the major portion of the lower jaw, the spine along the front of the scapula, and the extended ilium of the pelvis.

efficiency of locomotion in a four-footed animal. The feet were well formed, and adapted for walking by an equal development of the toes. In this animal there was no loss of the primitive phalangeal bones; rather some of the phalanges were greatly shortened. In some other theriodonts, however, there was commonly a reduction of phalangeal elements to the mammalian number of two in the thumb and great toe and three in all the other toes.

The various specializations that have been described for *Cynognathus* show that it was a very active carnivore. For a reptile it was get-

ting close to the mammalian stage of development in many respects, and we wonder whether it was like the mammals in characters that are not indicated in these fossils. Did *Cynognathus* have a coat of hair? Did it have a fairly constant body temperature?

There are various clues among the theriodonts, not necessarily confined to the genus *Cynognathus,* that indicate the close approach of these therapsids to the mammalian condition. For example, the genus *Thrinaxodon* (from the Triassic of South Africa) is a small reptile showing many advanced morphologi-

Cynognathus

Figure 9-9. A cynodont, an advanced mammal-like reptile about the size of a large dog, of early Triassic age. Prepared by Lois M. Darling.

cal features. Skeletons of *Thrinaxodon* are often found in a curled-up position, as if these animals had assumed this posture to conserve body heat. Might they have been endothermic, or "warm-blooded"? It is interesting that, nestled close to the skeleton of an adult *Thrinaxodon,* there was discovered the skull of a tiny individual of the same kind. This looks like the association of a mother and baby—which, if true, would point to the development of parental care among these animals. Of course, we are here indulging in speculation, but certainly it is more than idle speculation.

Thrinaxodon, conspecific with the African form, has been found in Antarctica, associated with *Procolophon* (see page 104) and *Lystrosaurus* (see page 123). The association of these typically African Triassic reptiles in Antarctica furnishes particularly strong evidence indicating that Antarctica and South Africa were closely bound together during early Triassic time.

Cynognathus, and *Thrinaxodon,* which have been described briefly to illustrate the characters of the theriodonts in general, are more particularly representative of the Cynodontia. The cynodonts are perhaps the most numerous and widely spread of the theriodonts; their fossil remains are found in upper Permian and Triassic rocks in Africa, South America, Europe, Asia, Antarctica, and in North America as well. In many respects they represent a peak in theriodont evolution.

The theriodonts known as gorgonopsians,

of which the Permian genus *Lycaenops* is typical, initiated many of the trends that were to culminate in the later theriodonts. Thus, in these earlier theriodonts, the dentary bone was large, but not as large as in *Cynognathus*; the teeth were differentiated but not highly specialized; there was as yet no secondary palate; the occipital condyle was single; and so on.

From the Permian gorgonopsians evolved not only the Triassic cynodonts, like *Cynognathus,* but also, in another direction, the therocephalians of Permian and Triassic age. The earlier therocephalians, like the genus *Lycosuchus,* were in many respects as primitive as the gorgonopsians, but they did show certain advances such as the enlargement of the temporal opening and the reduction of the phalanges to the mammalian formula. From the later therocephalians, arose still another group of theriodonts, the baurias, which carried some characters to a high degree of specialization. For instance, in *Bauria* there was no bar of bone separating the orbit from the temporal opening, which is the condition typical for primitive mammals. There is reason to believe that the most highly developed theriodonts, the ictidosaurians, arose from a bauriamorph stem.

Some very interesting theriodonts, seemingly derived from a cynodont base, were the tritylodonts, of which the genus *Tritylodon* is characteristic. This genus, first discovered in the upper Triassic beds of South Africa many years ago, was for decades considered to be a very

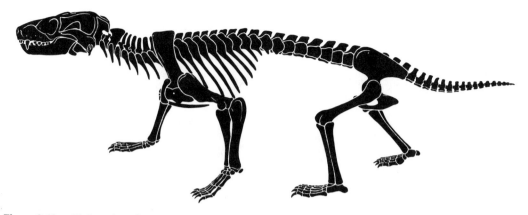

Figure 9-10. *Thrinaxodon,* a lower Triassic cynodont, found in South Africa and Antarctica. About one-third natural size.

early mammal. In recent years rather numerous, complete, and widespread fossils of tritylodonts have been found—additional *Tritylodon* in Africa, *Bienotherium* in western China, *Oligokyphus* (of Jurassic age) in England, complete skeletons of *Kayentatherium* from Arizona, and fragmentary but definitive bones from Argentina. Evidently these therapsids spread across the globe during late Triassic history.

The tritylodonts were small animals, in which the skull had a high sagittal crest and huge zygomatic arches for the accommodation of very strong jaw muscles. There was a well-developed secondary palate. The dentition was peculiar. A pair of enlarged anterior incisor teeth above and below was separated by a gap or diastema from the cheek teeth. The cheek teeth (seven on each side) were rather square, and each tooth bore rows of cusps, arranged longitudinally. The upper teeth had three such rows to each tooth; the lower teeth had two. The two rows of cusps in the lower teeth fitted in the grooves between the three rows of upper cusps. Evidently, the lower jaw moved forward and backward when the jaws were closed, thus grinding food between the cusped teeth in somewhat the same fashion as some modern rodents grind their food.

In many respects the tritylodont skull was very mammalian in its features. Certainly, because of the advanced nature of the zygomatic arches, the secondary palate, and the

specialized teeth, these animals had feeding habits that were close to those of some mammals. They obviously ground their food into small bits; thus, they made it available for quick transformation into energy, as might be expected in very active, and perhaps warm-blooded animals. Yet, in spite of these advances, the tritylodonts still retained the reptilian joint between the quadrate bone of the skull and the articular bone of the lower jaw. It is true that these bones were very much reduced, so that the squamosal bone of the skull and the dentary bone of the lower jaw (the two bones involved in the mammalian jaw articulation) were on the point of touching each other. Nevertheless, the old reptilian bones were still participants in the articulation and, therefore, the tritylodonts technically may be regarded as reptiles.

Finally, during late Triassic times appeared a group of theriodont reptiles that essentially bridge the gap between the advanced theriodonts and the primitive mammals. These were the ictidosaurs or chiniquodonts, of Triassic and early to middle Jurassic age. Perhaps, they are best exemplified by the unusual genus, *Diarthrognathus,* from the Triassic beds of South Africa.

In the ictidosaurs many of the characters that in other therapsids had reached a high state of perfection were carried far toward the mammalian condition. The temporal opening was very

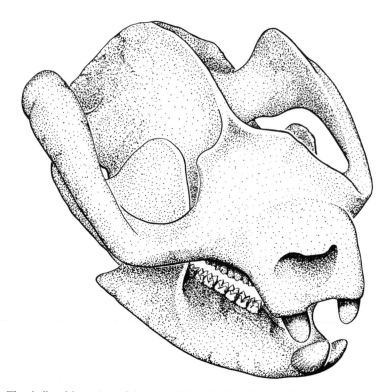

Fig. 9-11. The skull and lower jaw of the therapsid reptile, *Bienotherium,* about natural size. This illustrates the strongly mammal-like nature of the skull, jaw, and dentition in this advanced mammal-like reptile.

large indeed, and was confluent with the orbit. There was a great emphasis of certain skull bones and a complete suppression of others. The teeth were in an advanced stage of development.

But, the most interesting and fascinating point in the morphology of the ictidosaurians (at least, as seen in *Diarthrognathus*) was the double jaw articulation. In this animal, not only was the ancient reptilian joint between a reduced quadrate and articular still present but also the new mammalian joint between the squamosal and dentary bones had come into functional being. Thus, *Diarthrognathus* was truly at the dividing line between reptile and mammal in so far as this important diagnostic feature is concerned. In the mammals, the quadrate and articular bones have migrated from the articular region to the middle ear where, as we shall learn, they have been transformed into two of the bones concerned with the transmission of vibrations from the eardrum to the inner ear.

Because, in the ictidosaurs, the transformation of the quadrate and articular bones had not taken place, these animals can be placed arbitrarily within the reptiles. All of which indicates how academic is the question of where the reptiles end and the mammals begin.

The ictidosaurs were almost across the threshold that separated reptiles from mammals, and in so evolving they had brought about the doom of the synapsids, and eventually of the reptiles, as dominant vertebrates. They established the foundation for the subsequent rise and triumph of the mammals, but this was not to take place for many millions of years. During a time lapse of perhaps one hundred million years after the rise of the first mammals from ictidosaurian ancestors the reptiles were to be supreme. They dominated the Mesozoic era completely, and it is with the long and involved story of reptilian glory that we shall now concern ourselves, deferring a consideration of mammalian beginnings to a later chapter.

A Coal Forest

Conquest of the Land

10

EARLY LAND-LIVING VERTEBRATES

To go back a little in our story, we recall that the Devonian period was a crucial time in the evolutionary history of the backboned animals, for it was then that the several classes of Pisces, of fishes in the very broad sense of the word, became well oriented along the lines of evolutionary development that they were to follow during subsequent geologic ages. It was the period of the first flowering of the vertebrates.

The Devonian period was an important time for other forms of life as well as for the vertebrates, for this was the age during which the plants established themselves on the lands of the earth. And although it is probable that invertebrates first ventured on the land in Silurian times, it was not until the Devonian period that animals without backbones became numerous and well adapted for land life. By late Devonian times primitive forests covered the lands, forming havens for various scorpions and spiders and ancient insects that scurried about over the ground or climbed into the green shelter of the plant life. We might say that the stage was set for the egress of vertebrates from their ancestral watery home.

As we have seen, the transition among vertebrates from life in the water to life on the land, the transition from fish to amphibian, took place at the very end of the Devonian period. With the opening of Carboniferous times the early land-living vertebrates were feeling their way into a new environment, where plants and insects and other forms of life offered new food supplies and the opportunities for new ways of living.

During the Mississippian period the amphibians enjoyed an initial dominance on the land that was the natural result of their position as the first and for the time being the only land-living vertebrates. There were no other animals to challenge them. But their complete dominance was short lived, because in Pennsylvanian times the reptiles arose to share the land with the amphibians. Although the reptiles, even in their most primitive manifestations, were generally more efficient and better adapted as land-living animals than the amphibians, they by no means suppressed their primitive but persistent predecessors. For a long time amphibians and reptiles lived side by side, and it is probable that many of the large labyrinthodonts were able to compete on more or less equal terms with their reptilian contemporaries. It was not until well after the beginning of the age of dinosaurs, when reptiles had advanced to comparatively high stages of development, that the large labyrinthodont amphibians finally vanished from the land faunas.

As might be expected, the proportion of amphibians is high in the early faunas of land-living vertebrates, but it decreases through Carboniferous and Permian times, in the progression from earlier to later vertebrate associations. For instance, the Pennsylvanian faunas of Linton, Ohio, of Mazon Creek, Illinois, of Joggins, Nova Scotia, and of Kounová in Czechoslovakia are composed almost entirely of amphibians or of fishes and amphibians, with reptiles constituting insignificant proportions of the assemblages. It is probable that the preponderance of amphibians in these faunas does not reflect a true picture of the times, because these were associations of swamp-dwelling animals, in which amphibians would naturally be predominant. Nevertheless it would appear that amphibians were very abundant in many of the Carboniferous land faunas. In this connection it should be mentioned that the lower Permian faunas of Texas, although rich in amphibians, are even richer in reptiles. In the later Permian faunas, for instance those of Europe, of Russia, and of South Africa, the reptiles are overwhelmingly predominant, whereas amphibians are comparatively scarce. Some of these differences may be the result of differences in environments, but, allowing for this, it is probable that they do reflect to a considerable degree the expansion of the reptiles during late Paleozoic times.

ENVIRONMENTS OF EARLY LAND-LIVING VERTEBRATES

What were the conditions that led to the divergent evolutionary trends among amphibians and reptiles at the end of the Paleozoic era? Why did the amphibians predominate until about the beginning of the Permian period, after which the reptiles became the ruling land vertebrates?

Probably in Mississippian and Pennsylvanian times lands were low all over the earth. These were periods of uniformity in topography and in climates. Dense forests of primitive plants covered the lands, and it would appear that tropical environments extended from the equatorial regions to high latitudes in both the northern and the southern hemispheres. Swamps were abundant. This was the great age of coal formation.

Such environments would have been favorable to the adaptive radiation of amphibians. It is significant that during early Carboniferous times the transition from ancestral crossopterygian fishes to land-living amphibians was completed.

With the close of the Pennsylvanian period and the advent of the Permian period a change of environments took place. There were uplifts of the continental areas. The topography of the land became more varied than it had been previously, and with this development there were correlative changes in climate. The uniformity so characteristic of the Carboniferous age gave way to varied climatic conditions and consequently to varied environments. Local environments ranged from streams, ponds, and swamps to dry uplands. There was probably an alternation of wet and dry seasons taking place each year, and there was even extensive glaciation in the southern hemisphere at the beginning of Permian time. In these varied environments of the Permian pe-

riod the reptiles became the dominant land-living vertebrates.

THE RELATIONSHIPS OF UPPER PALEOZOIC CONTINENTS

On several preceding pages it is mentioned that the occurrences of certain vertebrates lend support to the concept of an ancient world in which the continents were conjoined to form two great supercontinents, Laurasia in the northern hemisphere and Gondwana in the southern hemisphere. It should be added that this concept also envisages the two supercontinents initially joined to form one great immense parent landmass, Pangaea.

At this place it is pertinent to digress briefly from our discussion of early land-living vertebrates to consider the comprehensive theory of Plate Tectonics and Continental Drift, a theory that truly marks a revolution in geological thought. Our view as to the distributions of terrestrial vertebrates through geologic time, including the early amphibians and reptiles that have been described in Chapters 7, 8, and 9, is closely correlated with the Plate Tectonic theory. Therefore it is necessary to know and appreciate the probable arrangements of ancient continents, which are most logically explained by Plate Tectonics, if we are to understand the relationships of past vertebrate faunas.

The Plate Tectonic theory may be summarized as follows:

1. The crust of the earth—the lithosphere and the underlying mantle down to a depth of about 100 kilometers—forms a hard integument.
2. Beneath this crust is the asthenosphere, hotter and weaker than the lithosphere, and forming a plastic substratum to the latter.
3. The lithosphere is divided into at least eight major and several minor plates that have been and are moving in relation to each other.
4. Tectonic activity within the lithosphere occurs particularly along the boundaries of these plates.
5. The plate margins are:
   a. Separating along ridges, where hot magmas are upwelling to add new materials to the lithosphere.

b. Colliding along trenches, where subduction is taking place and lithospheric materials are being returned to the lower parts of the mantle.

c. Sliding past each other horizontally, along transform faults.

6. The amount of crust created by the upwelling of magmas along ridges is equaled by the amount destroyed by subduction along trenches. Thus the earth remains a closed system.

In middle Paleozoic times the landmass of the earth was concentrated in the parent entity known as Pangaea, which in turn was composed of the conjoined northern and southern supercontinents, Laurasia and Gondwana, respectively. With the transition from Paleozoic to Mesozoic times rifting began within Laurasia and Gondwana and the several continents as we know them began to take form. The continental masses, carried on tectonic plates like blocks of wood frozen in slabs of ice, drifted apart during Mesozoic and Cenozoic times to their eventual present positions.

Rifting, with the resultant separation of North America from Europe, formed the North Atlantic Ocean. Similarly the separation of South America from Africa formed the South Atlantic Ocean. Thus these ocean basins are younger than the continents they separate, and the same is true for other ocean basins as well. Australia and Antarctica drifted from their original connections with Africa and subsequently broke apart to drift separately to their present positions. Peninsular India, originally occupying a position between the eastern border of Africa and Antarctica, drifted to the northeast to an eventual collision with the greater Asiatic landmass.

The collision of plates often caused the wrinkling of the crust to form mountain ranges, the most notable of which is the Himalaya uplift.

Geological, geophysical, paleontological, and biological evidence supports in concordant details the chain of events outlined above.

Figure 10-1. The postulated relationships of land masses during Permian time, according to the concepts of Plate Tectonics and Continental Drift, and the evidence of the distributions of land-living tetrapods. At this stage of geologic history there presumably was a single supercontinent, Pangaea, composed of a northern hemispheric part, Laurasia (including North America and Eurasia lacking peninsular India), and of a southern hemispheric part, Gondwana (including South America, Africa, peninsular India, Antarctica, and Australia, with New Zealand).

THE DISTRIBUTION OF UPPER PALEOZOIC VERTEBRATE FAUNAS

Although the invasion of the land by vertebrates took place with the transition from Devonian to Mississippian times, there is virtually no record of what happened immediately after this event. The Mississippian was seemingly a time of extensive marine inundations, and very few continental deposits have survived from that age to give us a record of what life was like on the land. The marine limestones of the Mississippian period preserve an extensive record of invertebrate faunas, with here and there the remains of fishes, but except for a few scattered localities there are no fossils of land-living vertebrates from this portion of the geologic record.

With the advent of the Pennsylvanian period conditions changed, and there were broad continental areas where extensive swamps and dense tropical forests of primitive plants covered the land. Consequently in the continental sediments of the Pennsylvanian period there are various fossil localities at which early land-living vertebrates have been found in North America and central Europe. These occurrences were not random, but were along the Carboniferous equator, which traversed Laurasia through North America—from the central states across Nova Scotia, across southern England, and into central Europe. Along this equatorial zone of a closely joined North America and Europe there were low-lying swamps that provided tropical havens for the early amphibians and reptiles.

Among the earliest of the Pennsylvanian vertebrate faunas are those already mentioned—the faunas of Joggins, Nova Scotia, of early Pennsylvanian age, and those of Linton, Ohio, and Mazon Creek, Illinois, of middle Pennsylvanian age. At Linton the fossils are found in a cannel coal, at Mazon Creek, in concretionary nodules. The preservation of the fossils at Joggins is particularly interesting. Here are many trunks of ancient trees, standing upright in the sediments, just as they stood in life. It would appear that these trees died and decayed while they were still standing. Mud and sand washed into the hollow tree trunks, and in many of these natural coffins were buried the skeletons of early amphibians, often beautifully and completely preserved.

In some of the English coal fields are also found early fossil amphibians.

In the upper Pennsylvanian sediments the record of early land-living vertebrates becomes even more complete than it was in the earlier sediments. The Pittsburgh faunas of Pennsylvania, West Virginia and Ohio give us a glimpse of varied amphibians and reptiles directly antecedent to the Dunkard faunas that occupied this area during early Permian times. Similar faunas occur at Nýřany and Kounová, Czechoslovakia, and the many resemblances of these faunas to the late Pennsylvanian faunas of North America indicate that there must have been a free interchange of land-living vertebrates between these areas.

The upper Pennsylvanian land faunas, valuable as they are as a record of early terrestrial vertebrates, are restricted as compared with the faunas of the Permian. This was the period in geologic history when land-living vertebrates enjoyed their first great evolutionary expansion, when the varied land surfaces, the diverse climates, and probably the alternation of seasons gave scope for a wide adaptive radiation of the amphibians and the reptiles, especially the latter, into many differing environments. This was the period when the reptiles became the truly dominant land animals, even though the labyrinthodont amphibians remained as important constituents of the faunas. In truth, this was the beginning of the Age of Reptiles.

The finest lower Permian faunas are those of the red beds of northern Texas, Oklahoma, New Mexico, and Utah. In Texas a sequence of richly fossiliferous formations extends up from the Pennsylvanian-Permian boundary. The lower part of this sequence is designated as the Wichita series, the upper part as the Clear Fork, and within the successive horizons of these series it is possible to trace the progressive evolution of various lower Permian amphibians and reptiles. It is in this sequence that much of the history of labyrinthodont evolution is revealed. The record of the captorhinidan reptiles is also extensive, and that of the pelycosaurs outlines virtually the entire evolutionary development of these reptiles. The New Mexico sediments, known as the Abo and the Cutler formations (the latter extending into Utah, where as a result of recent work it has been elevated to the status of a group that contains several constituent formations), are generally correlative with the Texas sediments.

It has been thought for many years that a seaway, extending from the south, separated Texas from western New Mexico in early Permian time, thus isolating the tetrapod faunas of the two regions from each other. Recent studies indicate that this seaway did not reach sufficiently far north to isolate completely the faunas of Texas, Utah, and New Mexico. Instead it would seem that north-central Texas and southeastern Utah were then large deltas bordering the northern shore of the seaway. Faunal differences thus may be due to differing environments of lowland and upland rather than to the presence of a marine barrier.

The lower Permian faunas of other continents are less well documented than the faunas of Texas, but they give some evidence of what land life was like in other regions. In Autun, France, a fauna is found that is in many ways similar to the fauna of the Texas red beds, and similar lower Permian fossils are found in Germany, especially around Dresden. Some fossils are found in southern Brazil and some in northern India.

The history of tetrapod life in North America, beyond the Clear Fork Beds of Texas, is continued in the San Angelo, the Flower Pot, and the Hennessy formations of Texas and Oklahoma—the first two being constituents of the Pease River series. Fossils from these horizons, although generally not so abundant or complete as the ones of the earlier Wichita and Clear Fork series, are of particular importance because they demonstrate a close relationship between the middle Permian vertebrates of North America and those of the Old World, especially the ones found in the Permian zones I to III of Russia. Therefore, for a continuation of the story of Permian vertebrates, particularly of the tetrapods, we must study continental areas other than North America.

In two regions of the Old World, middle and upper Permian sediments are exposed in continuous and unparalleled series from which extensive collections of vertebrates have been made. One of these regions is in northern Russia, where sediments of the Dvina series are exposed, and the other is in the Great Karroo Basin of South Africa, the locale of the famous Karroo series. The Dvina sediments form a continuous sequence ranging through the middle and upper Permian period and into the lower portion of the Triassic period. The series has been divided into five zones (possibly six), of which one is of early Permian age, the next three are of middle Permian age, one of late Permian age, and one of early Triassic age. The Karroo series, one of the classic stratigraphic sequences of earth history, also ranges through the Permian and Triassic periods, in this case covering the complete extent of both these geologic ages. It begins with the Dwyka and Ecca beds of early Permian age, passes upward into the Lower Beaufort beds of middle and late Permian age, and continues in the Middle and Upper Beaufort and Stormberg beds of Triassic and early Jurassic age. These sediments succeed each other without many appreciable breaks, and show a graduation from the Permian into the Triassic period—one of the few places in the world where such a passage between two major geologic systems can be seen.

The faunas of the Permian of Russia and South Africa show many close similarities that indicate an intimate relationship between these regions during the final stages of Paleozoic history. Evidently there were continuous land connections between the two areas, and similar ecological and climatic conditions, so that land animals could migrate back and forth. The faunas were predominantly of the "upland" type, consisting for the most part of animals that lived on open plains. These are the faunas that show the greatest development of the therapsids, the mammal-like reptiles, particularly in the South African area. In the Karroo there is an almost overwhelming array of therapsids—dinocephalians, dicynodonts, and theriodonts—through thousands of feet of thickness of the Beaufort beds. Although the therapsids make up the bulk of the Karroo and Dvina faunas, there are also other prominent constituents, notably the large pareiasaurs, and labyrinthodont amphibians.

It is from these two regions that much of our knowledge of middle and late Permian vertebrates has been derived. Other upper Permian localities are scattered, and for the most part have yielded but fragmentary faunas. In central Europe are the classic Kupferschiefer and Zechstein of Germany, predominantly marine beds coming above the lower Permian Rothliegende. In southern Scotland is the Cutties Hillock locality, from which a small fauna has been recovered.

Other localities that might be mentioned, all of late Permian age, are the Ruhuhu and Tanga beds of East Africa, the Chiweta beds of Nyassaland, the Bijori beds of central India, and the upper Newcastle coals of Australia.

The correlative relationships of the Permo-Carboniferous vertebrate-bearing sediments, so briefly described in the foregoing paragraphs, are indicated by the accompanying chart.

VERTEBRATES AT THE CLOSE OF PALEOZOIC TIMES

The broad uplifts of continental land masses in Pennsylvanian and Permian times resulted in the firm establishment of the vertebrates upon the land, their differentiation into varied lines of evolution, and their wide distribution throughout the world. So it was that by the close of the Permian period the land was dominated by cold-blooded tetrapods, by amphibians, and particularly by reptiles. The labyrinthodonts had passed the zenith of their evolutionary history and were on the decline toward a secondary return to the water. They were numerous and still successful as land-living animals, but they were gradually giving way to the more aggressive reptiles. The primitive reptiles, the captorhinidans, had virtually completed their evolutionary development and, for one small remnant that was to continue into and through the Triassic period, were at the point of extinction. The pelycosaurs, too, had gone through their evolutionary history, but were succeeded by their descendants, the therapsids. The mammal-like reptiles were, at the close of the Permian period, in the full flush of their expansion, and they were destined to continue and to reach great heights of progressive adaptation in the subsequent Triassic period.

Finally there were various other reptiles, of small consequence in Permian history, but that were to be very important and numerous in later periods of the earth's story. These were the protorosaurs and their relatives; the eunotosaurs, doubtfully the ancestors of the turtles; and the eosuchians; and the thecodonts, crocodilians, the flying reptiles, and the dinosaurs that were to be the rulers of the Mesozoic era.

CORRELATION OF UPPER PALEOZOIC VERTEBRATE-BEARING SEDIMENTS

		North America	England and Europe	Russia	S. Africa Karroo	E. Africa, etc.	Other Continents
Permian	Upper	Pease River, Flower Pot, San Angelo, Hennessey	Cutties Hillock — Zechstein	Dvina IV	Lower Beaufort — Balfour *Daptocephalus* zone	Ruhuhu, Tanga, Chiweta, Mangwa	Upper Newcastle coals, Australia; Bijori, India
	Middle	Clark Fork, Wichita; Abo-Cutler, Garber, Wellington, Dunkard	Magnesian limestone — Kupfer-schiefer	III · II	Middleton *Cistecephalus* zone; Koonap *Tapinocephalus* zone		
	Lower		Autun, Branau — Rothlie-gende	I	Ecca		Itarare, Brazil
Pennsylvanian	Upper	Pittsburgh, Danville	Nýřany, Kounová	O	Dwyka		
	Middle	Linton, Mazon Creek	English coal fields				
	Lower	Joggins	English coal fields				
Mississippian		Mauch Chunk, Albert Mines (N. B.), Mississippi Valley	Edinburgh coal field, Scotland; Bristol, England				

Coelophysis

Early Ruling Reptiles

11

THE ADVENT OF A NEW AGE

Up to this point the history of the back-boned animals has been outlined from its early stages, from Cambrian through Devonian times, to a climax at the end of the Permian period. Then, highly varied fishes inhabited the streams and the ponds of the continents, and the oceans that surrounded the lands, numerous and diverse representatives of the earliest land-living vertebrates, the amphibians, lived in the watery and swampy environments of the land areas, and the reptiles, in those times the most advanced of the terrestrial vertebrates, were establishing the pattern of dominance that was to characterize their history for many millions of years to come. Some excursions have been made into the later history of certain vertebrates during the telling of the story, but on the whole the account has been largely concerned with the animals that lived before the close of the Paleozoic era.

We now come to a new phase in the history of vertebrate life, the development of back-boned animals during the Triassic period, the earliest subdivision of the Mesozoic era. The Triassic period was important as a time of transition between the old life of the Paleozoic era and the progressive and highly varied new life of the Mesozoic.

Many Triassic fishes, amphibians, and reptiles were developing in the directions of future vertebrate evolution; at the same time certain elements in the Triassic faunas represented the holdovers of persistent forms from Paleozoic times.

Among the land-living vertebrates, the holdovers from the Permian that continued into and beyond the Triassic period were the labyrinthodont amphibians. The labyrinthodonts, in the form of advanced temnospondyls, enjoyed a final burst of prolific evolution during the Triassic period before they became extinct. The progressive therapsids, particularly the very mammal-like theriodonts, reached advanced stages of structural development before they too disappeared, toward the close of

Triassic times. The Triassic procolophonids represent on the other hand an interesting remnant of little evolutionary importance. Finally, the Triassic protorosaurs to be discussed below, and the eosuchians represent comparatively small groups of reptiles, continuing from equally restricted Permian ancestors.

As contrasted with these holdovers from the Permian period, there was a great array of new land-living vertebrates, many of them the ancestors of flourishing evolutionary lines that were destined to continue through the Mesozoic era, and sometimes on through the remainder of geologic time right up to the present day. Thus ancestors of the frogs appeared during Triassic times, the progenitors of a group of vertebrates that has been and is singularly successful. Likewise, the first representatives of another very successful group, the turtles, made their appearance in the Triassic period. Furthermore, the predecessors of the lizards, today the most numerous of reptiles, arose in late Triassic time. Rhynchosaurs were distributed across scattered continental regions of the earth during the Triassic period.

The Triassic period was the time when various highly specialized marine reptiles were entering upon their long histories. For instance, the first ichthyosaurs are found in rocks of Triassic age. The placodonts and the nothosaurs also lived during Triassic times, and from the nothosaurs, during the middle part of the Triassic period, the plesiosaurs arose.

The reptilian orders known as the Protorosauria (a small but diverse group), the Trilophosauria (known from a very few genera), and the Rhynchosauria (bizarre reptiles that were especially characteristic of ancient Gondwana) were here and there occupants of Triassic landscapes. They may be regarded perhaps as "experiments" in reptilian evolution that did not survive beyond the upper limits of Triassic time.

Especially notable among the new animals that appeared during the Triassic period were the thecodonts, the crocodilians, the dinosaurs, the pterosaurs, the ictidosaurs, and the first mammals. The thecodonts were the first representatives of those reptiles known as archosaurs, and they were the direct ancestors of the reptiles that were to be so completely dominant during the later periods of Mesozoic history, namely, the crocodilians, the flying reptiles, and particularly the numerous dinosaurs. The thecodonts set the archosaurian pattern that was extraordinarily successful for more than one hundred million years.

Although the mammals were destined to play a very small evolutionary role while the dinosaurs were dominant during the remainder of the Mesozoic era, they came into their own with the advent of Cenozoic times.

THE THECODONTS

The diapsid reptiles, defined and listed on page 108, are divisible into three large groups, perhaps at the level of infraclasses, each composed of several orders. One of these, the Lepidosauromorpha, contains the most primitive of the diapsids, the Eosuchia, and in addition the Squamata, or lizards and snakes, and the Sphenodonta or tuataras. Another, the Archosauromorpha, contains the Thecodontia, the Crocodilia or crocodiles, the Pterosauria or flying reptiles, and two dinosaurian orders, the Saurischia and the Ornithischia, as well as the Pterosauria, the Trilophosauria, the Rhynchosauria, and the Thalattosauria and Protorosauria. Of these two groups, the latter are especially important to the student of evolution, and among the first five archosauromorph orders mentioned above, the thecodonts are of particular concern to us at this place.

The thecodont reptiles were comparatively limited in their adaptive radiation and in the numbers of their genera and species, and they were very definitely limited in their geologic history. These reptiles appeared just prior to the beginning of the Triassic period and they became extinct at the close of Triassic times. Yet brief as was the history of the thecodonts, it was an evolutionary development of great consequence.

The thecodonts may be divided into five suborders; the proterosuchians—these being the most primitive members of the order, largely of early Triassic age but adumbrated by a single genus of late Permian age; the ornithosuchians—these comprising a central group of the order and showing varied adaptations as small and large carnivores; the rauisuchians—these being very large middle and late Triassic carnivores that paralleled in some respects the large carnivorous dinosaurs of Jurassic and Cretaceous age; the aetosaurians—these being highly specialized, armored, herbivores; and the phytosaurs—these also being highly specialized reptiles, but adapted for life as large predators, remarkably like the subsequent crocodilians in form and function.

The proterosuchians, of which the genus *Erythrosuchus* from the lower Triassic of South Africa is characteristic, were clumsy reptiles, with stout bodies and limbs, generally short tails, and frequently with relatively large skulls. The proterosuchian skull retained various primitive features such as the frequent presence of palatal teeth, a pineal foramen (which was lost at an early stage during the evolution of the archosaurs), and a backward-sloping occipital region with no emargination for the eardrum. These thecodonts were fully quadrupedal, showing little if any of the trend toward bipedalism that was to be so marked in archosaurian evolution. The pelvis was generally rather primitive; its only distinct archosaurian feature was a slight elongation and a marked downturning of the pubis.

The proterosuchians represent an early and seemingly a sterile evolutionary branch among the thecodonts. Of more consequence in the history of the order of reptiles were the ornithosuchians, among which many of the trends that were to characterize archosaurian evolution made their appearance. *Euparkeria*, another thecodont from the lower Triassic of South Africa, represents very nicely the basic ornithosuchian structure.

Euparkeria, a small carnivorous reptile, perhaps two feet in length, was lightly built. The

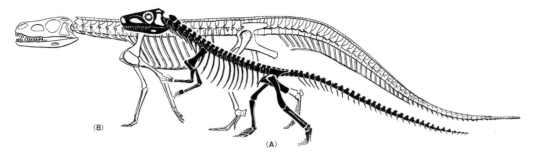

Figure 11-1.   Skeletons of thecodonts. **(A)** *Euparkeria*, an ornithosuchian from the Lower Triassic of South Africa. Skeleton about one meter in length. **(B)** *Ticinosuchus*, a rauisuchian from the Middle Triassic of Europe. Skeleton about three meters in length.

bones were delicately constructed and many of them were hollow. This little reptile seems to have been partially bipedal. When it was resting or moving about slowly, it probably walked on all four limbs; but it probably ran on the hind limbs alone, with the forepart of the body lifted high above the ground and with the forelimbs swinging freely. At such times the body was pivoted at the hips, and the rather long tail served as a counterbalance to the weight of the body. Very possibly the hands were used part of the time for grasping—perhaps as an aid of feeding. The pelvis was of rather primitive form as in the proterosuchians, but the ilium was somewhat expanded and the pubis was strongly downturned as a first indication of the various specializations that in time were to characterize the archosaurian pelvis. In the shoulder girdle, the scapula and the coracoid were the dominant bones, but the other pectoral elements were still present, although greatly reduced.

The skull, like the rest of the skeleton, was lightly constructed. It was narrow and deep, with a large eye and two large temporal openings on each side. It was further lightened by the development of a very large opening in front of the eye, the antorbital opening. Such fenestrations were typical of almost all of the archosaurians. Finally, there was a large opening in the side of the lower jaw, again a character that was carried through archosaurian phylogeny. There were small teeth on the pal-

ate, and there was a full array of sharp teeth around the margins of the jaws, in the premaxillary and maxillary bones of the skull, and in the dentary bone of the lower jaw. These teeth were set in sockets—the thecodont type of tooth implantation so characteristic of the archosaurs.

Some of the ornithosuchians retained this primitive archosaurian plan. Thus *Lagosuchus*, from the Middle Triassic of South America and *Ornithosuchus*, from the Upper Triassic of Europe, adhered to the pattern established in *Euparkeria*. *Ornithosuchus* was characterized by two rows of armor plates down the middle of the back, a common character among the ornithosuchians.

Within the suborder of the carnivorous rauisuchians there was a strong trend toward giantism. Such genera as *Rauisuchus* from South America and *Postosuchus* from North America were moderately large giants, four meters or more in length, with enlarged skulls armed with dagger-like teeth. Indeed, the adaptations of the whole animal, in which the skull was carried on a short, powerful neck, the body was balanced at the hips so that locomotion was largely effected by the strong, bird-like hind limbs, while the fore limbs, although robust, were probably of restricted use for locomotion, were remarkably similar to those of the gigantic carnivorous dinosaurs that were to dominate the terrestrial faunas of late Jurassic and Cretaceous times. One would suppose that such highly specialized thecodont carnivores might have continued into later Mesozoic years, yet with the close of Triassic time the rauisuchians became extinct, to yield their function as dominant predators to the theropod dinosaurs. This surrender of dominance during the transition from Triassic to Jurassic time may have been owing to such esoteric factors as the details of locomotor specializations, whereby the form of the pelvis, of the femur and of the ankle joint in the dinosaurs were more nicely adjusted for rapid running than was the case in the rauisuchians.

The development of armor reached an extreme in some of the later thecodonts, for example, in *Aetosaurus* and *Stagonolepis* of Europe and in *Typothorax* and *Desmatosuchus* of North America, all of late Triassic age. These reptiles were literally encased with heavy bony armor plates (covered in life with horny plates),

making them well-nigh impregnable to attack. Such a complete suit of armor added a great deal of weight to the animal; hence, these reptiles were quadrupedal, but obviously secondarily so. In them the forelimbs are notably smaller than the hind limbs, indicating their derivation from bipedal ancestors. They all had small skulls and weak teeth and obviously were noncarnivorous. *Stagonolepis* was characterized by a sort of piglike snout; perhaps, it rooted in the ground for tubers and other underground plants. It is evident that aetosaurs were harmless vegetarians, well protected from attack by aggressive predaceous reptiles, which no doubt included their thecodont cousins, the rauisuchians and phytosaurs.

The Triassic phytosaurs, named from the characteristic genus *Phytosaurus*, trended toward large size at an early stage in their history, and correlatively they returned secondarily to a

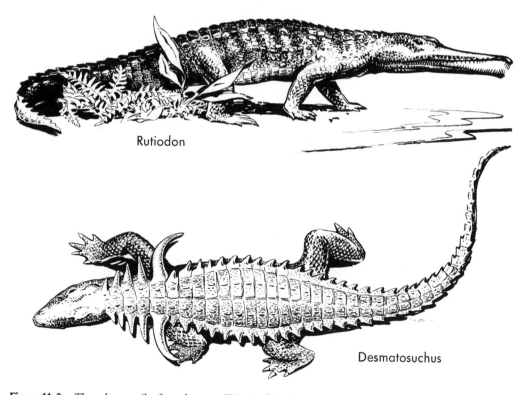

Figure 11-2. Thecodont reptiles from the upper Triassic of North America. *Rutiodon* was a phytosaur, *Desmatosuchus* was an aetosaur. These were large reptiles, often ten feet or more in length. Prepared by Lois M. Darling.

four-footed pose, like some of the armored aetosaurs. The skull and the body became elongated and very crocodilian-like or, as it is better to say, the crocodiles subsequently became phytosaur-like, since the phytosaurs developed special adaptations that were later imitated by the crocodiles.

The phytosaurs, such as *Phytosaurus* and *Rutiodon*, were highly predaceous, aggressive reptiles that lived in streams and lakes, and preyed upon fishes or any animals that they could catch. The front of the skull and the lower jaws were elongated and studded with sharp teeth. Unlike the crocodiles, however, the nostrils in the phytosaurs were set far back, just in front of the eyes, and in many phytosaurs were raised upon a volcanolike eminence that protruded above the level of the skull roof. Consequently these reptiles could float down stream with only the nostrils exposed, which was most advantageous to a water-living predator.

The legs and feet were strong, for progression on land, and although the front legs were large they were smaller than the hind legs, thus revealing the ultimate bipedal ancestry of these reptiles. The body was protected by heavy bony scutes, which in life must have been covered with horny epidermis. All in all, the phytosaurs were strikingly similar in appearance to modern crocodilians, and it seems rea-

sonable to think that their habits were likewise crocodilian-like. They were among the dominant reptiles of late Triassic times, and some of them reached gigantic proportions.

The extraordinary resemblances between the phytosaurs and their crocodilian successors, as well as between certain rauisuchians and the large theropod dinosaurs, constitute prime examples of parallel evolution through time. It must be realized that the phytosaurs were not ancestral to the crocodilians, nor were the rauisuchians ancestral to the theropod dinosaurs; in both instances there was a repetition of form and function for essentially identical modes of life. But in both cases the earlier protagonists of their particular life styles became extinct, to be succeeded by their highly successful imitators.

THE PROTOROSAURS

The protorosaurs, mentioned at the beginning of this chapter, had their origins in the Permian period probably as derivatives from early eosuchians, and continued through Triassic times. The Permian forms were rather small, lizard-like reptiles (again this oft-repeated comparison with lizards) that probably lived in the undergrowth of that period, feeding upon small reptiles and insects. The pelvis was a platelike structure, and the skull was characterized by an upper temporal opening on ei-

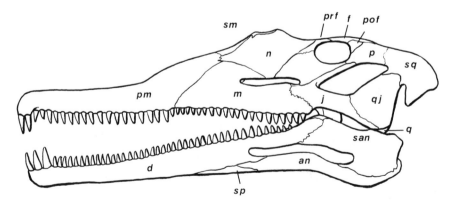

Figure 11-3. The phytosaur, *Rutiodon*. Lateral view of skull and jaw. About one-tenth natural size. For abbreviations, see pages 447–449.

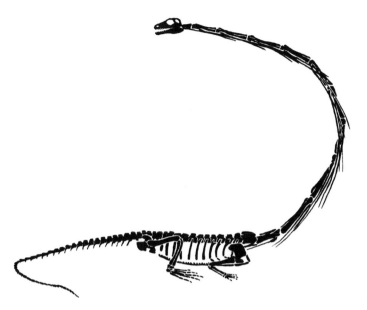

Figure 11-4. *Tanystrophaeus*, a remarkably long-necked protorosaurian from the Middle Triassic of Europe. Skeleton about three meters in length.

ther side, bounded below by a very deep squamosal bone.

From their Permian beginnings the protorosaurs enjoyed a limited but in one case a most bizarre range of adaptive radiation during the course of Triassic history. The one case involves the genus *Tanystrophaeus* from the Triassic of central Europe, in which the amazingly long neck, each constituent vertebra being lengthened—giraffe fashion, was so elongated as to exceed by three times the length of the body. Otherwise, in the structure of the skull and the proportions of the body, this rather large reptile, as much as ten feet or more in length, is not particularly unusual. One family of protorosaurs, the prolacertids, is made up of small, very lizard-like reptiles that for many years were considered to be progenitors of the lizards; it is now realized that their relationships may be with the archosaurs. *Prolacerta*, a typical genus from southern Africa, is also found in the Triassic sediments of Antarctica, thus adding one of numerous paleontological resemblances, or even identities, to indicate that Antarctica was once closely connected with Africa in the Triassic supercontinent of Gondwana.

THE TRILOPHOSAURS

This group of Triassic reptiles is known principally from the North American genus, *Trilophosaurus*, in which the skull was very deep and the jaws were provided with teeth that were transversely broadened into chisel-like blades. We can only speculate as to the food of this reptile; probably it was a plant-eater, and the bladelike teeth were used for chopping vegetation.

THE RHYNCHOSAURS

The rhynchosaurs, once considered as related to the sphenodonts as exemplified by the modern genus, *Sphenodon*, are now thought to constitute a separate reptilian order. These strange reptiles enjoyed a brief period of ascendancy during middle and late Triassic times, as indicated by abundant fossils found particularly in Africa, India, and South America. Some of the rhynchosaurs attained a con-

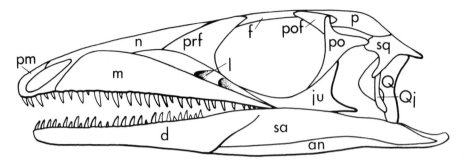

Figure 11-5.   Skull and lower jaw of the lower Triassic protorosaur, *Prolacerta*, found in South Africa and Antarctica. Note the lack of a connection between the jugal, and the quadratojugal and quadrate bones, thus providing a movable articulation between the skull and lower jaw. Slightly larger than natural size. For abbreviations, see pages 447–449.

siderable size—*Scaphonyx*, from the Triassic of Brazil, being an animal that frequently stood more than two feet high at the shoulder. Since these reptiles were sturdily built, with stocky bodies, *Scaphonyx* may have had a live weight of almost two hundred pounds. The skull in the rhynchosaurs was broad and deep, with the premaxillary bones elongated and pointed, evidently to perform the function of anterior teeth. The maxillary bones of the skull were provided with numerous small teeth, crowded together on either side of a longitudinal groove. The edges of the lower jaws were bladelike, to fit into this groove when the mouth was closed. It may be assumed that this strange apparatus perhaps was an adaptation for eating husked fruits; certainly the rhynchosaurs lived a very special kind of life and fed on a very special diet. Evidently the environmental conditions that favored these highly adapted reptiles may have persisted for a geologically short time, because with the close of Triassic history the rhynchosaurs became extinct.

END OF THE TRIASSIC

The association of Paleozoic holdovers and of new progressive forms in the Triassic faunas gave to these assemblages a distinctive, heterogeneous appearance that sets them off from the faunas preceding and following them. Indeed, the mixture of labyrinthodont amphibians and procolophons with progressive mammal-like reptiles and archosaurs,

including the early dinosaurs exemplified by the genus *Coelophysis*—to be described in the next chapter—was a meeting of the old and the new that involved various crosscurrents of struggle and competition. This was truly a time of transition.

In the course of the competition between these varied groups of tetrapods some were bound to fall by the wayside, so that during and at the end of Triassic times there were rather extensive extinctions among the amphibians and reptiles. In general the progressive types prevailed, and the archaic forms disappeared under the pressure of competition from highly developed animals with which they were unable to contend.

As will be shown later on, the end of the Cretaceous period was a time of broad extinctions, when four orders of reptiles, containing numerous genera and species, were completely effaced. It is not so generally realized that the close of the Triassic period was also a time of great extinctions, when at least seven orders of reptiles disappeared, while two other tetrapod orders, the labyrinthodont amphibians and the mammal-like therapsid reptiles, vanished *after* the close of Triassic time. Mention should also be made of the marine nothosaurs and placodonts, discussed in another chapter, that also disappeared with the close of Triassic history. Perhaps the extinction of genera and species was not so great in late Triassic times as at the end of the Cretaceous period, but the suppression of several orders of tetrapods must

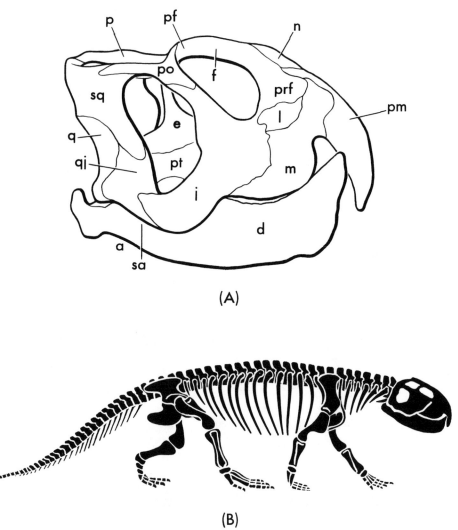

(A)

(B)

Figure 11-6.   The Triassic rhynchosaur, *Scaphonyx*, from South America. **(A)** Lateral view of skull. Note the pointed premaxillary bone, which would seem to have functioned as a large tooth; note also the beak-like lower jaw. About one-fifth natural size. **(B)** The skeleton may be six feet or more in length. For abbreviations, see pages 447–449.

be viewed as a series of evolutionary events of considerable significance.

Thus the close of the Triassic period saw the virtual but not complete disappearance of the widely distributed and numerous temnospondyl labyrinthodonts, certainly a successful group of Triassic tetrapods. The procolophonian reptiles of the Triassic, unlike the labyrinthodonts, were a mere remnant of their Permian forebears, so that their passing

at the end of Triassic times was a matter of small consequence. Nevertheless it marked the extinction of a group of reptiles that had been upon the earth for many millions of years. The protorosaurs, trilophosaurs, and rhynchosaurs likewise, were small groups in Triassic faunas, but again their extinction at the close of the Triassic period removed still other orders of reptiles from the earth. The thecodonts can be regarded in a somewhat different light. These were progressive reptiles that disappeared at the close of the Triassic period, because of competition from the new reptiles of which they were the ancestors. In a similar fashion, most of the mammal-like reptiles, which disappeared even before the close of the Triassic period, vanished because of the highly progressive nature of their descendants; they evolved themselves into oblivion. One group of therapsids, the dicynodonts, continued to the end of the Triassic, when they seemingly were caught in the wave of extinction that affected various other tetrapod groups.

With the extinction of the procolophons, protorosaurs, therapsids (except for a few stray lingerers), nothosaurs, placodonts and thecodonts at the end of Triassic times the stage was set for a new chapter of Mesozoic history. Tetrapod evolution was oriented along new and progressive lines of great variety. For the next hundred million years there were to be giants on the earth.

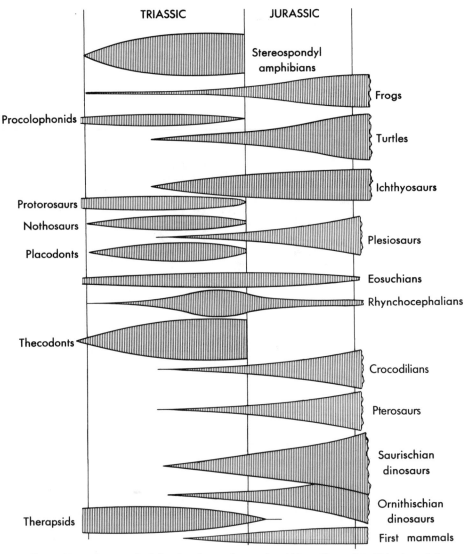

Figure 11-7.   Range and relative abundance of tetrapods within and beyond the Triassic period.

Apatosaurus

12

Triumph of the Dinosaurs

INTRODUCTION

We come now to a consideration of the dinosaurs, the great reptiles that were dominant on land through one hundred million years of Mesozoic history. What were the dinosaurs?

The word "dinosaur" was invented by Sir Richard Owen more than a century ago to designate certain large fossil reptiles that were then being recognized and described for the first time. The word is a combination of Greek roots meaning "terrible lizard," a purely descriptive term, which, like so many scientific names, must not be taken literally. Many of the dinosaurs undoubtedly were terrible animals when they were alive, but they were not lizards, nor were they related to lizards except in a most general way. In the early days of paleontological science the Dinosauria were regarded as a natural group of reptiles, but as knowledge of these long extinct animals was expanded most authorities concluded that the term embraces two distinct reptilian orders. Consequently the word dinosaur is now a convenient name, but not necessarily a systematic term.

The two orders of dinosaurs are designated as the *Saurischia* and the *Ornithischia*, these names being based upon the form of the pelvis—a basic character in the evolution of the dinosaurs. In the Saurischia the pubic bone of the pelvis extends down and forward from its juncture with the ilium and the ischium, the dorsal and the postero-ventral bones of the pelvis, respectively. In the Ornithischia the pubis has rotated so that it occupies a position ventral and parallel to the backwardly extending ischium. Of course there are numerous other characters by means of which the two orders of dinosaurs are distinguished, each from the other.

Furthermore the two orders of dinosaurs may be divided into suborders that indicate the evolutionary divergence that took place in these reptiles. At this point it is useful to list the suborders of dinosaurs.

Order Saurischia
  Suborder Staurikosauria
    Probably the earliest and most primitive saurischians. These were carnivores with relatively large skulls, a primitive saurischian pelvis and a bipedal mode of locomotion. Lower part of Upper Triassic.
  Suborder Theropoda
    Persistent carnivorous saurischians. These dinosaurs retained the bipedal mode of locomotion through their evolutionary history. Upper Triassic through Cretaceous.
  Suborder Sauropodomorpha
    Herbivorous saurischians, for the most part gigantic and quadrupedal. The largest of the dinosaurs. Upper Triassic through Cretaceous.
Order Ornithischia
  Suborder Ornithopoda
    Among these were the most primitive of the ornithischians, the heterodontosaurs and fabro-

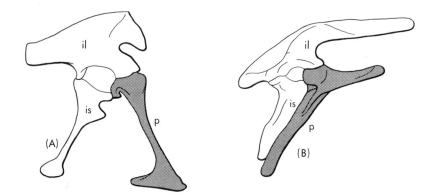

Figure 12-1.   The pelvis in the two dinosaurian orders. **(A)** Saurischian pelvis, with a forwardly directed pubis. **(B)** Ornithischian pelvis, with the pubis parallel to the ischium. For abbreviations, see pages 447–449.

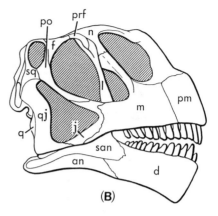
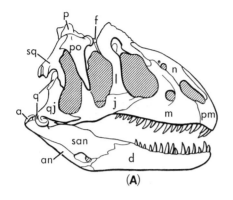
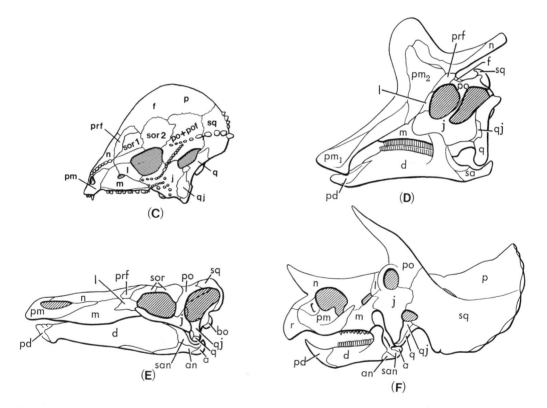

Figure 12-2. Skulls of dinosaurs. Saurischian skulls. **(A)** *Allosaurus*, a large, carnivorous theropod, one-sixteenth natural size. **(B)** *Camarasaurus*, a gigantic, herbivorous sauropod, one-ninth natural size. Ornithischian skulls. **(C)** *Prenocephale*, a dome-headed pachycephalosaur, one-sixth natural size. **(D)** *Lambeosaurus*, an aquatic ornithopod, one-fourteenth natural size. **(E)** *Stegosaurus*, an herbivorous stegosaurian, one-seventh natural size. **(F)** *Triceratops*, an herbivorous, horned ceratopsian, one-seventeenth natural size. For abbreviations, see pages 447–449.

saurs, and the highly specialized trachodonts or duck-billed dinosaurs, which were large, semi-aquatic herbivores, with the front of the skull and jaws broadened into a flat bill. The ornithopods were dominantly bipedal, but many of them had strong forelimbs, capable of being used for quadrupedal locomotion. Upper Triassic through Cretaceous.

Suborder Pachycephalosauria

Small to large bipedal herbivores in which the skull roof often was inordinately thickened to form a heavy, bony dome. The dentition was simple. Cretaceous.

Suborder Stegosauria

Heavy, quadrupedal herbivores, perhaps adapted for life on "uplands" away from marshes and rivers. The hind legs were always much longer and heavier than the forelegs; the skull was small. Plates, spikes, and scutes formed dermal armor and protuberances on the body and tail. Principally Jurassic, extending into the lower Cretaceous.

Suborder Ankylosauria

Heavy, quadrupedal herbivores, similar in adaptations to the stegosaurs, but very strongly armored with thick bony plates that completely encased the back and sides of the body and tail. Cretaceous.

Suborder Ceratopsia

Quadrupedal herbivores, with subequal development of forelimbs and hind limbs. The skull was greatly enlarged, particularly by an extension of some of the posterior elements to form a large frill that extended back over the shoulders. The front of the skull was narrow and deep like a parrot's beak, and there were generally horns on the nose, above the eyes, or in both locations. Upper Cretaceous.

THE EARLIEST SAURISCHIANS

It would appear that the earliest known dinosaurs are primitive saurischians found in Brazil and Argentina, in sediments that are variously interpreted as either of late middle or of early late Triassic age. Such dinosaurs, here accommodated within the suborder Staurikosauria, are exemplified by the genus *Staurikosaurus* from Brazil in which the skull evidently was very large with relation to the size of the body (only the lower jaw is known and is as long as the femur) and the jaws are armed with sharp, piercing teeth. The tibia is longer than the femur in this relatively small saurischian dinosaur, and the hollow limb

bones indicate a fully bipedal pose. Evidently *Staurikosaurus* was an active predator, perhaps two meters in length, establishing a pattern of life that was to be followed through Mesozoic history by the varied theropod saurischians.

Herrerasaurus, from Triassic sediments in Argentina that are essentially correlative with the rocks in which *Staurikosaurus* is found, is comparable in many ways with the Brazilian dinosaur. *Herrerasaurus* is the larger of the two, about four meters in length, and consequently has a relatively robust skeleton. It was a bipedal dinosaur with a large skull, and with strong hind limbs in which the two proximal bones of the ankle, the astragalus and calcaneum, were integrated with the lower hind limb bones, the tibia and fibula respectively—this being a very characteristic dinosaurian feature. Furthermore, the pelvis in *Herrerasaurus* shows many resemblances to the same very important bony complex in *Staurikosaurus*. In both the ilium was short and deep, an inheritance from thecodont ancestors, while the plate-like pubic bone was expanded distally. The forelimbs were relatively small. Perhaps *Herrerasaurus* shows a very early trend toward large size in the saurischian dinosaurs, a trend that was to be especially strong during the course of dinosaurian evolution.

THE CARNIVOROUS THEROPODS

The earliest carnivorous theropod dinosaurs are well represented by the North American genus *Coelophysis*, in recent years made known by extraordinarily complete and beautifully preserved fossil skeletons, found in the upper Triassic sediments of northern New Mexico. *Coelophysis* was an animal about six feet in length, so lightly built (the bones were hollow) that in life it probably did not weigh more than forty or fifty pounds. It was strictly a bipedal reptile, with the hind legs very strong and birdlike and well adapted for walking, and with the short front limbs bearing mobile hands that must have been useful for grasping and tearing food. The body was bal-

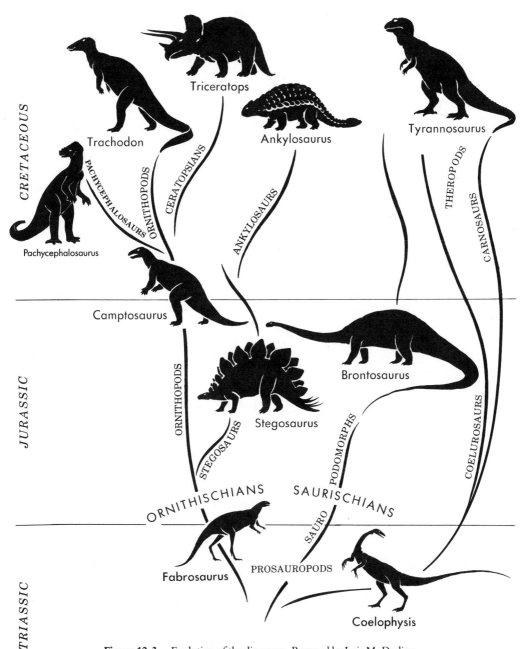

Figure 12-3. Evolution of the dinosaurs. Prepared by Lois M. Darling.

anced at the hips, and the tail was long and slender. The neck was rather long, and it carried at its end a delicately constructed skull equipped with sharp teeth.

From this account it is apparent that *Coelophysis* inherited the basic archosaurian features that had been established during early Triassic times by the primitive thecodont reptiles. Beyond these characters, however, were the structural modifications that made *Coelophysis* a theropod dinosaur. The skull was long and narrow with large temporal and preorbital fenestrae, in which respects it had developed features that were to be particularly characteristic of the later theropod dinosaurs. The jaws were provided with sharp, serrated, laterally compressed teeth, set in deep sockets. These teeth indicate that *Coelophysis* was strongly carnivorous, probably preying upon small or medium-sized reptiles.

The pelvis, a basic key to dinosaurian relationships, is more advanced in *Coelophysis* than in the staurikosaurian dinosaurs, described above. In this dinosaur the ilium was expanded fore and aft and there was a long sacral attachment involving several vertebrae. From the ilium on each side the pubis extended forward and down, and the ischium extended backward and down. Both these bones were long, the pubis especially so, and their juncture with the ilium was effected by bony processes rather than by a solid attachment, so that the socket or acetabulum for reception of the ball-like head of the femur was open or perforate, not closed as in the more primitive thecodonts.

Coelophysis probably represents the basic adaptations of the theropod dinosaurs, of reptiles adjusted to a life on the dry uplands, where the ability to run fast and move quickly was of prime importance, not only for the capture of food in the form of small animals but also for escape from enemies. From such a beginning many later saurischian dinosaurs had their phylogenetic growth.

From an ancestor more or less like this the theropods evolved along four general lines of adaptation. In the first place, certain theropods, the coelurosaurians, remained small, and retained for the most part the primitive aspect of the ancestral types. Such, of course, was *Coelophysis*, and such was *Ornitholestes*, a little coelurosaurian about five or six feet in length, that lived during late Jurassic times. The description that has been given above for *Coelophysis* applies in a general way to *Ornitholestes*. Perhaps the most striking specialization to be seen in this persistently small, primitive carnivore, as compared with its Triassic predecessors, was in a certain degree of elongation of the forelimbs, particularly the hands, not for a secondary return to four-footed locomotion as in many of the other dinosaurs, but rather as an adaptation for increased efficiency of grasping. In addition to elongation, there was suppression of the fourth and fifth digits, so that the hand consisted of three very long, flexible fingers, terminating in sharp claws. A hand like this was obviously very useful for catching small animals, probably reptiles for the most part, but perhaps insects too. Certainly the teeth of *Ornitholestes* indicate that this active little dinosaur ate all sorts of small game. It probably frequented the undergrowth of the late Jurassic forests, where it was able to hunt successfully by reason of its agility.

A variation of this line of theropod adaptation, taking place during the Cretaceous period, was marked by an increase to moderate size and high specialization of the skull. This branch of theropod evolution may be designated as the ornithomimosaurians. The late Cretaceous genus, *Ornithomimus*, a dinosaur about the size of a large ostrich, exemplifies this adaptational trend. *Ornithomimus* had very long, slender hind limbs and very birdlike feet, which indicate that it must have been a rapid runner, much like the modern ostriches. The enlarged forelimbs were relatively even more elongated than were those of its coelurosaurian relative, *Ornitholestes*, and it too had the hand limited to three long, grasping fingers. The most remarkable specializations of *Ornithomimus* were in the skull which, borne at the end of a long and sinuous neck, was very small and completely toothless. The jaws were specialized as a beak, very similar to the beak of an ostrich. Indeed, the comparisons of this dinosaur in various aspects of its anatomy to an ostrich are so striking that it is often called the "ostrich-dinosaur" which, of course, does not imply any direct relationships but does indicate that *Ornithomimus*, during late Cretaceous times, lived a life that is paralleled by the large flightless birds of today. It

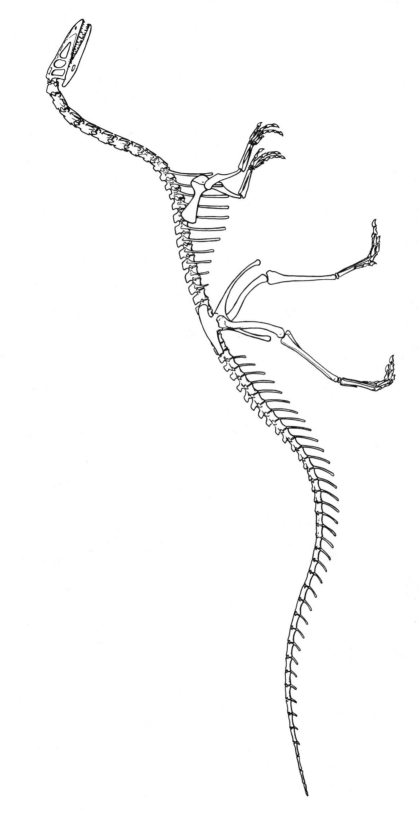

Figure 12-4. Skeleton of the North American upper Triassic theropod. *Coelophysis,* illustrating the structure in an early and generally primitive coelurosaurian dinosaur. About six feet in length. Prepared by Lois M. Darling.

probably ate many things—small reptiles, insects, fruits, and any other food that it could catch and handle. At the slightest sign of danger, this dinosaur was able to run away—to live and feed another day; hence, it survived until the close of the Cretaceous period.

The forelimbs of a gigantic ornithomimid were found a few years ago in the Cretaceous sediments of Mongolia.

A third line of theropod adaptation, only recently recognized, is that of the deinonychosaurs, which evolved during the Cretaceous period. They were small- to medium-sized theropods, characterized by remarkable specializations. The deinonychosaurs were bipedal, like the other theropods, but the forelimbs and the hands were remarkably large. The skull, too, was large, and the margins of the jaws were set with saberlike, serrated teeth. In some of these dinosaurs, as exemplified by the genus *Dromaeosaurus*, the braincase of the skull was unusually large, so that one may wonder if such dinosaurs were more "intelligent" than their contemporaries. Perhaps the most interesting specializations of these dinosaurs are found in the hind limbs, the feet, and the tail. In *Deinonychus* the feet became specialized away from the typical birdlike form, characteristic of most of the theropods, and became rather asymmetrical. The second digit (in other theropods, a lateral toe) was in these dinosaurs much enlarged, and was furnished with a huge, scimitarlike claw. The third and fourth digits were of approximately equal size, and it appears that the axis of the foot passed between them, instead of through the third digit as in other theropods. The large second toe with its bladelike claw seems to have been carried in an elevated position when the animal was walking or running; apparently, it was a highly specialized offensive and defensive weapon—not a part of the locomotor apparatus.

An additional bizarre feature was added to the appearance of *Deinonychus* by the structure of the tail, which throughout most of its length (starting not far behind its junction with the sacrum) was stiffened by long, bony rods extending longitudinally from the zygapophyses of the vertebrae. Each rod covered a span of eight or ten vertebrae; the cumulative effect was a bundle of eight or more closely packed bony rods on the two sides of each vertebra. Consequently, the tail must have functioned like a long board or girder, with the joint between it and the body just behind the pelvis. It has been suggested that this stiff and peculiar tail may have served as a very effective balancing appendage.

It is becoming evident that the deinonychosaurs were rather widely distributed in Cretaceous times; their remains have been found in North America and Mongolia. The discovery of these strange dinosaurs within very recent years is one more example of the fact that there are always new things to be found in the field of paleontology.

The fourth line of theropod evolution, the carnosaurians, a group separate from the coelurosaurians and deinonychosaurs, attained its culmination in late Jurassic and Cretaceous times. Here the trend was toward gigantic size. The carnosaurs, typified by the genera *Allosaurus* of the upper Jurassic and by *Albertosaurus* and *Tyrannosaurus* of the Cretaceous period, grew to be the largest land-living meat-eaters of all time. In so doing they retained the bipedal pose that had typified the ancestral theropods; in fact, bipedalism was intensified among the carnosaurs, because in these dinosaurs the hind limbs became tremendously strong and heavy, and the forelimbs and hands were inordinately reduced. In *Allosaurus* the arms and the clawed hands, though relatively small, were still capable of being used as an aid to feeding; but in *Tyrannosaurus*, a giant, some forty feet in length, that stood about twenty feet high to the top of its head and that had a probable weight during life of six or eight tons, the forelimbs were reduced to such a degree that they must not have been very useful. Since there was this reduction in the forelimbs, the activities of predation, of killing and feeding, were concentrated in the skull and jaws. Consequently the skull in the carnosaurs became tremendously en-

Allosaurus

Ornithomimus

Ornitholestes

Figure 12-5. Theropod dinosaurs, drawn to the same scale. *Allosaurus*, a giant carnosaur, was about thirty feet long; *Ornitholestes* and *Ornithomimus* were light, running dinosaurs. Prepared by Lois M. Darling.

larged, and the long jaws were armed with huge, daggerlike teeth. Long jaws and large teeth gave an effective bite for dealing with large prey, which constituted the food of these dinosaurs. There were greatly enlarged openings in the skull that cut down weight, so that it became a series of bony arches—a frame for muscle attachments—in which the braincase was comparatively small. Even with such adaptations, the skull with its large, strong muscles to operate the jaws was obviously heavy, and because of this weight the neck in

the carnosaurs was shortened, which served to avoid adverse leverages. All in all the giant carnosaurs of late Mesozoic times were remarkably well adapted for hunting and killing other large reptiles, especially other dinosaurs.

THE TRIASSIC PROSAUROPODS

A third suborder of saurischian dinosaurs is the Sauropodomorpha, containing two infraorders, the Prosauropoda and Sauropoda. The prosauropods, restricted to the upper

Triassic, and lower Jurassic, were among the earliest of the saurischians, and from their appearance at the base of the upper Triassic, or perhaps even in the upper reaches of the middle Triassic, they experienced a considerable breadth of adaptive radiation. Like most animals at the beginning of a line of evolutionary radiation, the earliest members of the group were of relatively small size. However, there was a strong trend to giantism among the prosauropods; hence, they were the largest of the Triassic dinosaurs, with some genera attaining lengths of twenty feet or more.

The genus *Plateosaurus* from the Triassic beds of southern Germany was a very characteristic member of the Prosauropods. *Plateosaurus* was a saurischian of considerable size; it had very heavy hind limbs to support the weight of the body; the forelimbs were relatively large and, perhaps, were used in part for locomotion, although this dinosaur was predominantly a bipedal form; the skull was small; and the teeth were flattened, evidently for cutting plant food rather than for eating meat. It might be added that both the neck and the tail in *Plateosaurus* were long; that the ilium in the pelvis was of primitive form and was short and deep (quite in contrast to the ilium of the theropods); the pubic bones were platelike; and the hind foot was not birdlike, but instead was heavy, with four functional toes. Probably, most of the prosauropods were (like *Plateosaurus*) rather clumsy herbivores, but there is the possibility that some of them may have been ponderous carnivores. These dinosaurs seem to have constituted a short and sterile branch of saurischian evolution, although there is evidence that some of them were, in effect, ancestral to the gigantic sauropods of Jurassic and Cretaceous age.

THE GIANT SAUROPODS

The trend in saurischian evolution that was foreshadowed by some of the prosauropods culminated in Jurassic times in the large sauropod dinosaurs, exemplified by such genera

as *Apatosaurus, Diplodocus,* and *Brachiosaurus,* and was carried through the Cretaceous period by similar sauropods. Here we see the dinosaurs reaching their greatest size, attaining lengths of sixty to eighty feet and more and probable body weights during life of as much as eighty tons. These were the largest land-living animals ever to have lived, although it must be remembered that they have been greatly exceeded in size by some of the large modern whales. It is very likely that the giant sauropod dinosaurs represent the maximum size that can be reached by land-living animals, owing to the physical limitations of bone, muscle, and ligament.

The sauropods were essentially similar to each other, differences being those of size and proportion and of details in the skeleton, especially in the skull. Since they had become giants they were completely quadrupedal, for their bulk was too great to permit a bipedal pose, although there is reason to think that some of them could stand on their hind legs, using the heavy tail as a prop, in order to feed upon the high foliage of trees. Even so, the forelimbs were considerably smaller in most of the sauropods than the hind limbs, thereby indicating the possible bipedal ancestry of these animals. The limbs were extraordinarily heavy, and the bones were dense and solid, to give strong support to the great weight of the body. The feet were very broad and, as revealed by fossil footprints, had large foot pads, like those in the feet of elephants, making a strong, elastic cushion to take up the impact of each step and to give support and traction. The toes were accordingly short, and not all of them bore claws. In *Apatosaurus,* for instance, there was a single large claw on the inner toe of each front foot, and claws on the inner three digits of the hind foot. Some brontosaur footprints discovered in lower Cretaceous rocks in Texas give graphic and dramatic evidence of the size of the feet, the length of the stride, and the manner of walking in these big sauropods.

The pelvis was a huge structure for strong

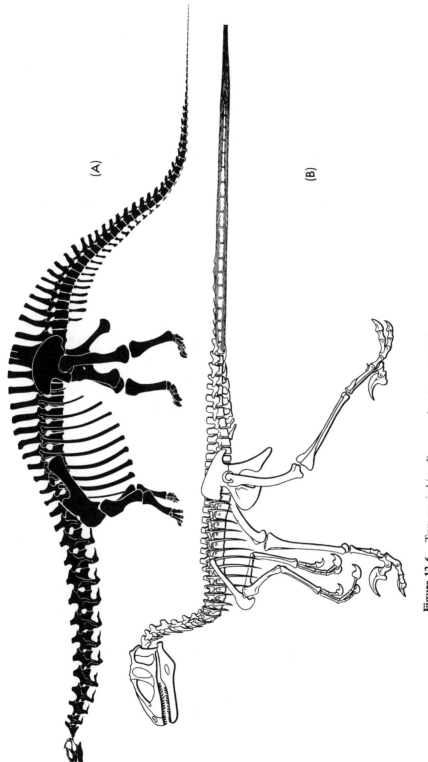

Figure 12-6. Two saurischian dinosaurs showing widely divergent adaptations. **(A)** *Apatosaurus*, a gigantic sauropod from the upper Jurassic beds of North America. The skeleton is about seventy feet in length. **(B)** *Deinonychus*, a carnivorous theropod of early Cretaceous age. The skeleton is about eight to ten feet in length. Note the enlarged hands, the specialized feet in which the second digit with its gigantic claw (evidently a powerful weapon) was retracted when the animal was walking or running and the straight tail, posteriorly stiffened by long, bony rods.

support and muscle attachments, as might be expected in dinosaurs weighing many tons, and the shoulder blades were likewise very long and heavy. The vertebrae were large, particularly in the neck and back, and, as an adaptation to achieve strength and size without being inordinately heavy they contained deep hollows in the sides of the centra and neural arches, thus eliminating bone where it was not needed yet at the same time allowing for expanded articular surfaces. These vertebrae were distinguished by accessory articulations, the hyposphene and hypantrum, medial to the zygapophyses, that strengthened the vertebral column. Both the neck and the tail were very long, and the vertebrae had strong spines that provided large areas for the attachment of strong muscles.

The skull was comparatively small. In many sauropods the nostrils were raised to the top of the skull, as an adaptation for breathing with only the top of the head protruding above the water, whereas teeth were generally limited to the front portion of the jaws. In some sauropods such as *Diplodocus* the teeth seem surprisingly weak for so large an animal, being pegs no greater in diameter than pencils. In other sauropods, *Camarasaurus*, for example, the teeth were leaf shaped or spatulate.

This description, which applies to almost any of the sauropods, indicates herbivorous dinosaurs of great size that probably fed on soft, lush vegetation. Certainly their relatively small jaws and weak teeth would not permit them to feed upon very tough plants, although certain evidence indicates that some of them may at times have eaten freshwater mollusks. We wonder how such a small mouth could take in enough food to keep such a large animal alive, but it must be remembered that these reptiles may have had comparatively low rates of metabolism; consequently their food requirements were possibly small as contrasted with those of the large mammals we know today, elephants, for instance. It seems likely that the sauropods frequented swamps and rivers and lakes, and fed largely on the vegetation in these habitats, either along the shore or under water. And these waters were not only places for the sauropods to feed; they also afforded protection from their enemies. If attacked or threatened by the giant theropods, like *Allosaurus* or his relatives, the sauropods might retreat into the water, where they could wade or swim, with only the head exposed above the surface, and where the big carnivores could not follow them. The sauropod adaptations were strikingly successful, for these dinosaurs persisted until the end of Mesozoic times and became widely distributed over all the continental regions of the earth.

THE EARLIEST ORNITHISCHIANS

Although the saurischian dinosaurs had become well established throughout the world during late Triassic times, it would appear that the ornithischian dinosaurs did not similarly expand until about the beginning of Jurassic history. There are some scattered evidences of ornithischians in Upper Triassic sediments, but on the basis of recent discoveries it is apparent that the earliest well documented record of ornithischian beginnings is to be found in rocks that now are generally regarded as of very early Jurassic age. In such sediments there have been found several genera, notably *Fabrosaurus* and *Heterodontosaurus* in Africa and *Scutellosaurus* in North America that, upon a foundation of well-preserved skeletons, afford a rather clear view of ornithischian beginnings.

These particular genera show quite clearly that by this stage of geologic history the basic characters so typical of the Ornithischia were well established. In the ornithischian dinosaurs the pubic bone of the pelvis had rotated backward so that it occupied a position parallel to the ischium. This contiguity of the pubis and ischium is similar to the arrangement of these bones in birds, hence the name Ornithischia or "birdlike pelvis." In many of the ornithischians the ilium was greatly extended both front and rear, and there was a large anterior process of the pubis, the prepubis, reaching

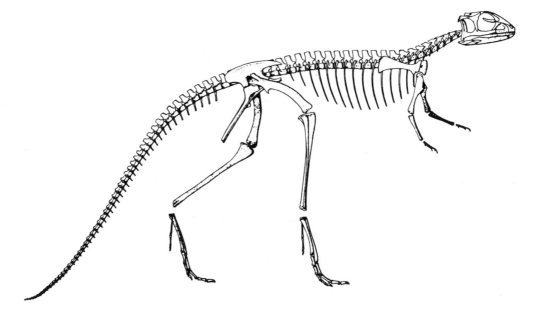

Figure 12-7. Skeleton of the South African upper Triassic ornithopod *Fabrosaurus*, illustrating the structure in an early, primitive ornithischian dinosaur. About one-seventh natural size.

forward beneath the front of the ilium. Consequently the pelvis as seen from the side was a tetraradiate structure, its four processes or prongs being formed by the anterior and posterior extensions of the ilium, by the prepubis, and by the closely appressed ischium and pubis. The ornithischians usually had lost the teeth in the front of the mouth in both upper and lower jaws, and generally this portion of the skull and jaw took the shape of a beak of some sort; in most of these dinosaurs there was a single new bony element on the front of the lower jaw, the *predentary*, that formed the lower portion of this cutting beak. The teeth, limited to the sides of the jaws, were often highly modified for cutting or chewing vegetation, because all of the ornithischians were herbivorous. Generally, the ornithischians were never as completely bipedal as were many of the saurischians—most of these dinosaurs showing a secondary return to a four-footed pose at an early stage in their history. The ends of the toes usually bore flat nails or

hooves, rather than claws. The known history of the ornithischians is much more varied, and the range of adaptations in these dinosaurs was wider than in the saurischians.

The earliest ornithischian dinosaurs belong to the suborder Ornithopoda, a group that spans the entire range of ornithischian history, while within the ornithopods these ancestral genera may be placed within a lesser category known as the Hypsilophodontia. As for these first hypsilophodonts, all of them were of small size as might be expected in prototypical or ancestral genera. *Fabrosaurus* is a bipedal dinosaur with long, slender hind limbs and relatively small forelimbs and hands. The skull was short and rather deep and the cheek teeth were flattened and triangular with serrated edges—the kind of teeth that would lead to the highly evolved dentitions in the later ornithischian dinosaurs. The antorbital opening, which in the saurischian dinosaurs was typically very large, was in *Fabrosaurus* quite small, anticipating the small or even obliterated fenestra

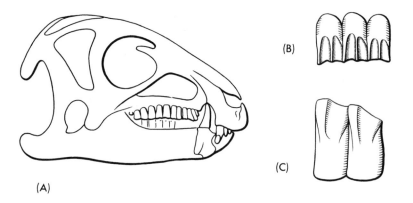

Figure 12-8. *Heterodontosaurus*, from the upper Triassic of South Africa, one of the earliest known ornithischian dinosaurs. **(A)** Lateral view of the skull and jaw, about one-half natural size. Note the premaxillary teeth (absent in most ornithischians), the predentary bone at the front of the lower jaw, a large lower "canine" tooth, and the contiguous, flat-crowned cheek teeth, adapted to an herbivorous diet. **(B)** Enlarged view of three upper cheek teeth, in lateral aspect. **(C)** Enlarged view of two lower cheek teeth, in lateral aspect.

that characterizes the ornithischian dinosaurs. All in all this early ornithischian dinosaur serves as a generalized form from which later members of the order might have evolved.

Heterodontosaurus is known from two nicely articulated skeletons, which afford excellent knowledge of this early ornithopod. In this dinosaur the skull, about four inches in length, was characteristically ornithischian, with a depressed jaw articulation, with a separate, edentulous predentary bone on the front of the lower jaw (which is one of the most striking features among the ornithischian dinosaurs), and with small, rather specialized teeth in the sides of the jaws, obviously adapted for cutting or chopping plant food. Quite surprisingly, there was a large "canine" tooth in the front of the lower jaw. The postcranial skeleton was characterized, among other things, by a typical ornithischian type of pelvis, with the pubic bone paralleling and contiguous to the ischium. This dinosaur was strongly bipedal, with elongated hind limbs. The forelimbs were robust and the clawed hands were large, evidently adapted for grasping. This dinosaur seemingly represents an early side branch of ornithischian evolution and was not ancestral to any of the later ornithopod dinosaurs.

Scutellosaurus, also one of the very early ornithischian dinosaurs, is particularly noteworthy because of the abundant armor plates that covered the body, some of which look very much like miniature stegosaurian plates—*Stegosaurus* being a Jurassic ornithischian. For this reason it has been suggested that *Scutellosaurus* may be an early stegosaur. Other plates of *Scutellosaurus* resemble on a small scale the plates of the Cretaceous armored dinosaurs—the ankylosaurs. It was less bipedal in its locomotor adaptations than *Fabrosaurus*. Its forelimbs were relatively large and the tail was very long, evidently as a counter-balance for a body made heavy by the weight of the armor. It would appear that *Scutellosaurus* was becoming secondarily quadrupedal, thus anticipating the method of locomotion that was universal among the armored dinosaurs of later Mesozoic times.

The presence of such varied ornithischian dinosaurs in the early Mesozoic world adumbrates the wide range of adaptive radiation that was to mark the evolutionary history of the ornithischians during the years when these great reptiles were among the rulers of the Jurassic and Cretaceous continents.

ORNITHOPOD EVOLUTION

Hypsilophodon, a small ornithopod from the Lower Cretaceous sediments of Europe, would appear to be a conservative descendant from fabrosaur ancestors. This small dinosaur was lightly built for an ornithischian, with a long tail and rather long, supple toes in the hind feet, for which reason it has been suggested that *Hypsilophodon* may have a tree-climbing reptile, a supposition that may be doubted. There were still teeth in the premaxillary bones, a particularly primitive character found among only a few of the ornithischians.

The later Jurassic and Cretaceous ornithopod dinosaurs may be divided into two categories, the iguanodonts and the hadrosaurs. The iguanodonts are represented in Upper Jurassic sediments by the camptosaurs, which embody in their morphology rather generalized types from which more specialized ornithopods may have evolved. The skeleton of *Camptosaurus* is well known; it will be described as representative of a generalized ornithischian dinosaur.

This was a small to medium-sized dinosaur, sometimes no more than six or seven feet in length, sometimes as much as twenty feet long. It was a predominantly bipedal animal, with strong hind limbs, but the front legs were sufficiently robust so that this dinosaur could, if it so desired, walk around on all four feet. Perhaps the quadrupedal method of locomotion was utilized when *Camptosaurus* was moving about slowly to feed, whereas reliance was placed on the long hind legs alone when this animal needed to run away—to escape attacks from large, carnivorous dinosaurs and other predatory reptiles. In its general build *Camptosaurus* was a heavier animal than the theropod dinosaurs of similar size, and it seems likely that this primitive ornithopod was not a particularly rapid runner. The hind feet were rather broad and the four functional toes (the fifth toe was much reduced) pointed forward. The fingers of the hands were short and terminated in blunt nails, which makes it seem likely that the hands were not used for grasping, as they were in the meat-eating dinosaurs. The pelvis was of the ornithischian type already described; the tail was heavy.

Camptosaurus was an inoffensive plant-eater, and this is reflected in the various adaptations of the skull and dentition. The skull was comparatively low, and rather long. The temporal openings were large, but the opening in front of the eye was relatively small, thereby instituting a trend of development among the ornithischians, already mentioned, that commonly led to the reduction or even the complete suppression of this particular fenestra. The lower jaw was somewhat shorter than the length of the skull, because of the forward and downward extension of the quadrate bone on which the lower jaw articulated. Because of the ventral extension of the quadrate, the articulation of the lower jaw was placed at a level below the line of the teeth, so that the articulation of the jaw on the skull was offset. This had the mechanical advantage of bringing the cheek teeth together at approximately the same time, like a crushing mill, when the mouth was closed, rather than shearing them past each other like scissors, as was the jaw action in the theropod dinosaurs. This adaptation has proved very useful to plant-eating vertebrates through the ages, since it allows a maximum amount of dental surface to be applied to the food in a minimum amount of time and with a minimum amount of motion, a factor of importance to animals that must ingest a great quantity of green vegetable food.

Correlated with the offset jaw articulation in *Camptosaurus* was an elevated coronoid process on the jaw, allowing for the attachment of strong temporal muscles. There were broad, leaflike teeth in the sides of the jaws, serving admirably as choppers to cut the food, but in the front of the jaws there were no teeth at all. The premaxillary bones of the skull were somewhat broadened around a very large opening for the nostrils to form a flat beak, and there was the typical predentary bone on the front of the lower jaw, shaped to bite against the upper beak. Evidently leaves and stems were cropped with this beak, which in life had a horny covering, and then passed back by the tongue to the cutting teeth in the sides of the jaws, to be chopped into compar-

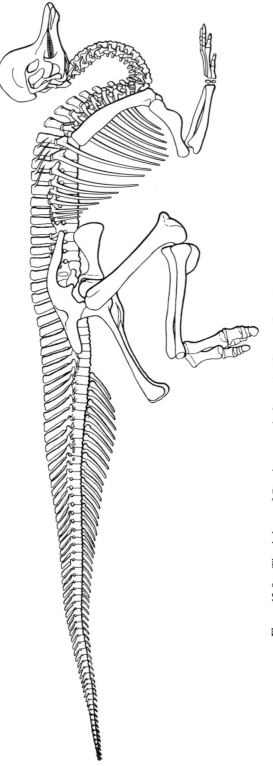

Figure 12-9. The skeleton of *Corythosaurus*, a hadrosaurian dinosaur from the upper Cretaceous of North America. The skeleton is about thirty feet in length. Note the characteristic ornithischian pelvis, and the great crest on the top of the skull, formed by the premaxillary and nasal bones.

atively small masses that could be easily swallowed.

The more specialized iguanodonts are typified by the genus *Iguanodon*, found in lower Cretaceous sediments in Europe. *Iguanodon*, incidentally the first dinosaur to be scientifically described, was essentially an enlarged camptosaur that reached a length of thirty feet or more. In this reptile the thumb was enlarged into a sharp spike that may have been used as a weapon for defense.

The most spectacular and successful ornithopod dinosaurs were the trachodonts or hadrosaurs of late Cretaceous age. Most of the trachodonts, like *Anatosaurus*, grew to great size, attaining lengths of thirty or forty feet and weights during life of several tons. In spite of their size, these dinosaurs like the other ornithopods were predominantly bipedal, the hind limbs being very heavy and the feet broad. But the most characteristic adaptations in these dinosaurs were in the head.

The skull in the hadrosaurs was elongated, and in front both skull and lower jaw were very broad and flat, in form rather similar to the bill of a duck. Consequently, they are often known as the "duck-billed dinosaurs." The lozenge-shaped teeth in the sides of the jaws were enormously multiplied in number and closely appressed, forming in each jaw a solid "pavement" that was evidently used like a grinding mill to crush the food. Often there were five hundred or more teeth, above and below—a remarkable adaptation for the comminution of food!

In addition to the extraordinary specializations of the jaws and teeth, many of the hadrosaurs were distinguished by unusual adaptations in the nasal region of the skull. In these genera the premaxillary and nasal bones were pulled back over the top of the skull to form a hollow crest. In *Corythosaurus*, for instance, the premaxillaries and nasals formed a high, helmet-shaped crest on top of the skull. In *Lambeosaurus* they formed a sort of hatchet-shaped crest, and in *Parasaurolophus* they formed a long tubular crest, extending far back behind the occipital region. Dissections show that the nasal passage, running

from the external to the internal nares, traversed the crest in a loop.

What was the purpose of this long nasal passage? Numerous suggestions have been advanced to explain this interesting adaptation in the crested hadrosaurs. For example, it has been suggested that the elongated nasal passage served as an accessory air-storage chamber that was utilized by these reptiles when the head was submerged under water. This explanation is not very convincing, particularly because the volume of air that might have been held in the nasal loop would have been relatively small. Again, the crest has been explained as an air lock, to prevent water from entering the nasal passage. But there is good reason to think that the duck-billed dinosaurs had sphincter muscles around the nostrils to close them, as do modern crocodiles. Perhaps the most plausible theories to explain the crests and their contained nasal loops are (1) that they were expansions to give large areas for mucous membranes, thereby increasing the sense of smell, or (2) that they were resonating chambers to increase the power of the voice. It seems reasonable to believe that these structures had a functional significance.

Certainly there is abundant evidence that the hadrosaurs were semiaquatic dinosaurs. They are generally found in sediments that were deposited in rivers or lakes, or even in the shallows along marine coastlines. Because of the manner in which the hadrosaurs were buried it is not uncommon to find the impression of the skin fossilized, as well as the bones; indeed several fossilized "mummies" have been preserved by a combination of fortunate circumstances, to show, except for details of color, just what the trachodonts looked like. These excellent fossils reveal the fact that the hadrosaurs had webs of skin between the toes in the hand. The skin was leathery and roughly pebbled. In the backbone there were strong, ossified tendons, adding to the strength of the vertebral column, and indicating, along with other features, that these animals swam by powerful sculling movements of the tail.

From the available evidence it is possible to reconstruct in detail the appearance and the mode of life of the duck-billed dinosaurs. They probably fed by wading in shallow waters near the shores, grubbing along the bottom for water plants. The remarkable development of dental batteries in these dinosaurs, as described above, has led some authorities to suppose that the hadrosaurs may have fed, at least in part, upon tough, fibrous terrestrial vegetation. If danger threatened when these dinosaurs were feeding on land, they probably would dash into deep water and swim away. Such adaptations enabled them to survive in a world inhabited by giant, aggressive theropod dinosaurs. Perhaps the success of the hadrosaurs, as exemplified by the remarkable variety among late Cretaceous genera, was their ability to live in different environments.

One factor that may have been of particular importance to the success of the hadrosaurs would appear to have been their highly evolved reproductive behavior. Recent discoveries of the Upper Cretaceous hadrosaur, *Maiasaura*, in Montana have revealed not only the carefully constructed nests of these dinosaurs but also aggregations of skeletons of very small individuals that have been interpreted as representing possible "nurseries", where newly hatched hadrosaurs were guarded and protected by adults.

Discoveries such as this, together with detailed studies of dinosaurian footprints and trackways, are yielding interesting interpretations of many aspects of behavior in these ancient reptiles. Such studies have indicated that the dinosaurs had indeed evolved intricate patterns of behavior that almost certainly contributed to their evolutionary success.

THE PACHYCEPHALOSAURS

The pachycephalosaurs or "dome-headed" dinosaurs, often included among the ornithopods, are perhaps better considered as belonging to a separate ornithischian suborder, the Pachycephalosauria, derived from ornithopod ancestors. They are well exemplified by the genera *Stegoceras* and *Pachycephalosaurus* of late Cre-

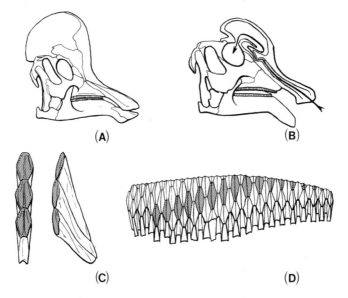

Figure 12-10.   Hadrosaurian skulls, jaws and teeth. **(A)** *Corythosaurus*, from the Upper Cretaceous of North America, showing the expansion of the nasal and premaxillary bones to form a hollow crest on the top of the skull. **(B)** *Procheneosaurus*, from the Upper Cretaceous of North America, showing the passage of air (arrow) through the nostrils and the hollow crest. **(C)** Three hadrosaurian lower teeth in lateral view (left) and anterior view (right). **(D)** The lower dental battery of a hadrosaurian dinosaur, showing the succession of teeth. This illustrates the manner in which there was a constant replacement of individual teeth as they were worn down by the mastication of abrasive vegetation.

Corythosaurus

Pachycephalosaurus

Camptosaurus

Figure 12-11. Ornithopod dinosaurs, drawn to the same scale. *Corythosaurus* was about thirty feet long. Prepared by Lois M. Darling.

taceous age, as indeed is the case for all of the members of this suborder. In these dinosaurs, the first rather small, the second growing to great size, the skull roof became remarkably thickened, to form a tremendous boss of dense, heavy bone above the brain. As a result of this thickening of the skull the upper temporal opening was completely obliterated and the lateral one was reduced almost to the point of suppression, and in the larger genus numerous knobs decorated the margins and the front of the skull. It is now generally believed that the thick, dome-like skull of the pachycephalosaurs was used as a battering ram, probably in sexual combat— similar to the manner in which male Bighorn Sheep use their thick, coiled horns.

THE STEGOSAURS

The stegosaurs or plated dinosaurs were, we might say, camptosaurs that grew large, reverted secondarily to a four-footed mode of locomotion, and developed peculiar plates on the back and spikes on the tail. They are among the earliest of the known ornithischians, appearing in rocks of early Jurassic age. *Stegosaurus*, from upper Jurassic sediments, was a dinosaur twenty feet or more in length, with legs that were heavy and feet that were short and broad. Even though this reptile was permanently quadrupedal the hind legs were much larger than the forelegs, making the pelvic region the highest point of the body. The shoulders were low. The skull was very camptosaurid-like and remarkably small. *Stegosaurus* is famous for having a brain much smaller than the enlargement of the spinal nerve cord in the sacrum, where the nerves of the hind limbs and the tail came together, which has given rise to the erroneous popular belief that this animal had two "brains."

There was a double row of possibly alternately arranged, bony, triangular-shaped plates down the middle of the back. The edges of these plates were thin; the bases were thickened and obviously were imbedded in the back, so that the plates stood vertical. In life these plates were probably covered by a horny layer. On the tail there were four long, bony spikes. The purpose of the tail spikes is obvious, for *Stegosaurus* could swing them against an adversary with damaging results; but the function of the plates on the back has been variously interpreted. At one time it was thought that they might have served some protective function; if so they left the flanks of the animal

completely exposed. Alternatively, the plates were supposed to have served as displays—for species or sexual recognition. Some recent wind-tunnel experiments, combined with studies of plate morphology, indicate that they may have been thermoregulators, acting as fins to dissipate body heat. The strongly vascular structure of the plates probably accommodated large blood vessels, allowing an abundant and rapid flow of blood, thus making the plates very effective for temperature control.

The stegosaurs, primarily a Jurassic group, continued into the beginning of Cretaceous times and then became extinct—the second large category of dinosaurs to disappear.

A few words may be added at this place concerning the ornithischian genus, *Scelidosaurus*. This dinosaur, found in the Lower Jurassic sediments of southern England, was for many years the oldest known ornithischian dinosaur, its position in this respect being superseded only recently by the discovery of the very early ornithischians discussed above. *Scelidosaurus* may be a very early stegosaur, or it may be a predecessor of the ankylosaurs or armored dinosaurs; its relationships await future studies for clarification. It is a rather large ornithischian, being some twelve feet or more in length. The skeleton was robust and it would appear that this was a quadrupedal dinosaur. The skull was relatively small. In life *Scelidosaurus* was distinguished by numerous bony armor plates, imbedded in the skin.

THE ANKYLOSAURS OR ARMORED DINOSAURS

Although some of the adaptations in the ornithopod and stegosaurian dinosaurs are difficult to interpret, those in the ankylosaurs are self-evident. The ankylosaurs were armored ornithischian dinosaurs of middle Jurassic and Cretaceous age, showing specializations for defense that were imitated by some of the edentates among the mammals, millions of years later.

Two families constitute this dinosaurian suborder, the more primitive nodosaurids and the advanced ankylosaurids. It would appear that the ankylosaurs, in the large sense of the word, had their origins in Europe during Jurassic time and then migrated into North America and Asia during the Cretaceous period.

Ankylosaurus was a bulky, quadrupedal reptile, some twenty feet in length, not very high at the back, and very broad. Its legs were heavy; and, like *Stegosaurus*, it had hind limbs that were much longer than its forelimbs, and its feet were short. Its skull was very broad as compared with its length. The top of the head and the entire back were completely covered by a continuous armor of heavy polygonal, bony scutes, and along the sides of the body there were long, bony spikes. The armored tail terminated in a great mass of bone that obviously formed a club or bludgeon. When danger approached, these dinosaurs needed only to go down on the belly to become a sort of pillbox that must have been hard to penetrate. And woe to the daring predator that approached too near; he stood in dire peril of bone-cracking blows from the heavy club on the end of the tail. The teeth were singularly small and weak, indicating that the ankylosaurs must have fed upon soft plants.

THE CERATOPSIANS OR HORNED DINOSAURS

The last of the dinosaurs to evolve were the ceratopsians, appearing in late Cretaceous times. Their evolutionary history, compared to the history of other dinosaurs, was brief, but in the span of the upper Cretaceous these reptiles went through a remarkably varied range of adaptive radiation, to produce some of the most spectacular and interesting of all dinosaurs. The horned dinosaurs were, indeed, successful, but the present evidence seems to indicate that their success was geographically limited; their fossils are found only in North America and northeastern Asia. It seems probable that in late Cretaceous time the ceratopsians were confined to these areas by marine barriers—notably the great inland sea of middle North America to the east and the Turgai Straits of Central Asia to the west.

The beginning of the ceratopsians probably can be seen in a small ornithopod genus, *Psittacosaurus*, from the Cretaceous sediments of Mongolia. This was a generally primitive, bipedal ornithischian, in many respects not far removed from some of the smaller ornithopods. It was specialized, however, in that the skull was rather short and deep, with its front portion narrow and hooklike, resembling in some respects the beak of a parrot. This dinosaur may not actually have been ancestral to the ceratopsians, but the modifications in the skull puts it in a position approximating a ceratopsian progenitor.

The first true ceratopsians are represented by *Leptoceratops* of North America and *Protoceratops* of Mongolia. These were small dinosaurs, but they had developed most of the specializations that were to characterize the horned ceratopsians through the extent of their phylogenetic history. Even though small they were definitely quadrupedal, with the forelimbs secondarily enlarged and with the feet rather broad, and quite certainly adapted to the single function of walking.

It is in the skull, however, that these dinosaurs show the greatest specializations, for here there was an adaptational trend quite unlike that seen in any of the other dinosaurs. Briefly, the skull was deep and narrow in its front portion, forming a beak like that of *Psittacosaurus*, and it was enlarged to such a degree as to be about a third or a fourth the total length of the animal. The enlargement of the skull was brought about in part by an actual increase in size of the structure as a whole and in part (particularly in *Protoceratops*) by a backward growth of the parietal and squamosal bones to form a large perforated "frill" at the back of the skull. Because of the great development of the frill, the skull was balanced on the occipital condyle, that portion behind the condylar joint being about equal to the region of the skull in front of the condyle.

What was the function of the frill in the skull of these early ceratopsians? A careful analysis of the anatomy indicates clearly that the frill

Stegosaurus

Ankylosaurus

Figure 12-12. Plated and armored dinosaurs, drawn to approximately the same scale. Each of these dinosaurs was about twenty feet long. Prepared by Lois M. Darling.

in all ceratopsians was primarily an enlarged area for the origin of strong temporal muscles, running from this region in the skull to the lower jaw. These muscles provided enormous force for closure of the jaws, such action being in all but the earliest ceratopsians strongly vertical, thereby making the teeth highly efficient shearing devices. The teeth, individually similar to the teeth of hadrosaurs, are arranged in single rows, and there obviously was constant and rapid replacement of the teeth to maintain their efficient shearing function. Also the frill afforded attachments for strong neck muscles to control the movement of the head. Secondarily it may have had some

protective function, since it projected back over the vital neck and shoulder region. A third possible function of the frill was for display, correlated with intraspecific combat among males for the establishment of dominance hierarchies. Again, it is quite possible that the frill was in part a thermoregulatory structure—providing a large area for the absorption of heat when the animal was in the sun or for the dissipation of heat when the animal was in the shade. Thus it would appear likely that the frill was a multi-purpose specialization in the ceratopsians, thereby being one of the prime factors in the evolutionary success of these dinosaurs.

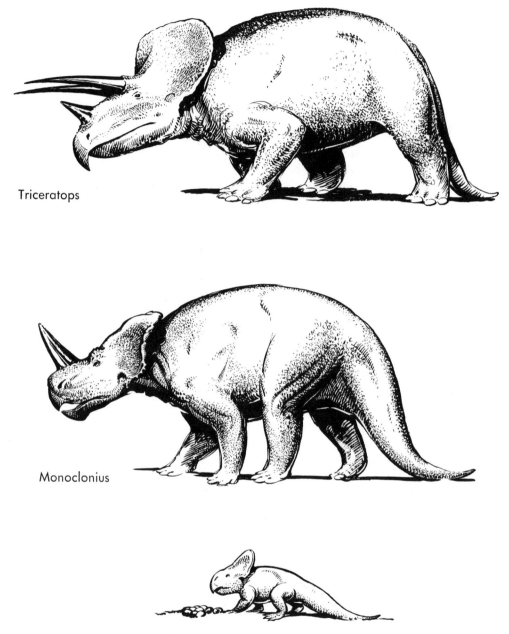

Triceratops

Monoclonius

Protoceratops

Figure 12-13.   Three ceratopsians or horned dinosaurs, drawn to the same scale. *Triceratops*, from the Lance formation of North America, was about twenty-five feet long. *Monoclonius*, also from North America, is found in the somewhat older Oldman formation of the Belly River series, and *Protoceratops*, one of the most primitive of the ceratopsians, is found in the Bain Dzak (or Djadochta) formation of Mongolia. Prepared by Lois M. Darling.

These early ceratopsians were essentially hornless, although in the largest known individuals of *Protoceratops* there was the beginning of a small horn on the nose. Except for the lack of horns, *Protoceratops* and *Leptoceratops* were small prototypes of their giant descendants.

Protoceratops is one of the most completely known of all the dinosaurs, because our information on this dinosaur is based not only on a large series of individuals showing all stages of growth from the newly born infant to the adult, but also on several nests of fossil eggs, some of them containing portions of embryos. These dinosaur eggs are similar in shape to the eggs of lizards, being elongated, and somewhat larger at one end that at the other. The egg shell was evidently calcareous and tough, and its surface was decorated by a pattern of small, sinuous striations. The preserved clusters of eggs show that the female *Protoceratops* scooped out a nest in the sand, in which she deposited the eggs in several concentric rings, much as do present-day turtles. Evidently the eggs were then covered with sand, and were incubated and hatched by the heat of the sun. Luckily for twentieth-century paleontologists, some of the eggs laid so many millions of years ago by this little dinosaur failed to hatch.

From a protoceratopsian ancestry the evolution of the horned dinosaurs was marked first by an increase in size, to such a degree that many of the later members of the group attained lengths of twenty-five feet and body weights of six or eight tons, and second by the development of horns on the skull. In addition, there was in the genus *Triceratops* a secondary closing of the fenestrae in the frill.

The development of horns in the ceratopsians is interesting by reason of the diversity shown in this adaptation. Among some of the horned dinosaurs like *Monoclonius* there was a single large nasal horn on the front of the skull. In another genus, *Styracosaurus*, there was not only a large nasal horn but a series of long, bristling spikes around the borders of the frill. In still other genera, like *Chasmosaurus, Penta-*

ceratops, and *Triceratops*, the nasal horn was augmented by two large brow horns, one above each eye.

The horned dinosaurs evidently were something like rhinoceroses among modern mammals. They were large, upland-dwelling herbivores that defended themselves by fighting effectively, using the horns as powerful weapons. The large ceratopsians were certainly formidable; the huge skull, provided with long, sharp horns, was thrust forward with tremendous force by the powerful neck muscles and by the momentum of the heavy body, borne upon strong, sturdy limbs. Any dinosaur that attacked one of the large ceratopsians had a fight in store for him.

We are tempted to wonder why there was so much variety in the development of horns in these dinosaurs. Why was not a single pattern of horns sufficient? Here we may compare the horned dinosaurs with the modern antelopes of Africa, which exist today in such a very bewildering array. Perhaps there is an analogy to be drawn between the antelopes and the horned dinosaurs. We know that the horns in antelopes are utilized not only for defense against enemies but also for display, and for combat among males of the same species to establish their positions of relative dominance within the herd. It is noteworthy that the various species of antelopes show patterns of intraspecific fighting based upon the shapes of their horns. Therefore it seems reasonable to suppose that the same considerations may apply to the horned dinosaurs, which evidently inhabited certain Cretaceous landscapes in large herds.

At the end of the Cretaceous period the ceratopsians were among the most numerous of the dinosaurs, and it would appear that *Triceratops*, the last of these reptiles in North America, roamed in great numbers across the face of the continent. But with the transition to Tertiary times these numerous and highly successful dinosaurs died out, as did all the other members of the two great orders of reptiles that dominated middle and late Mesozoic lands.

DINOSAUR GIANTS

The dinosaurs evolved in great variety along widely divergent lines, as we have seen, and became adapted to many modes of existence, so that there were dinosaurs of various forms and sizes living during the Mesozoic era, some carnivorous, some herbivorous, some large and some small. Size range among the dinosaurs ran the gamut from reptiles no larger than big lizards or medium-sized crocodiles to huge animals eighty feet in length and weighing during life up to forty or fifty tons and more. Although mere size cannot be cited as a universal dinosaurian character, since there were many small dinosaurs, these reptiles nevertheless were typified by a preponderance of giants. Of all the dinosaurs certainly the great majority were giants, if we define giants as animals more than twenty feet in length and weighing several tons.

Why should there have been so many giants among the dinosaurs? This is a difficult question. It is probable, however, that giantism in the dinosaurs is to be correlated in part with Mesozoic environments. The herbivorous dinosaurs lived at a time when there was an abundance of tropical or subtropical vegetation over many of the continental areas of the earth. There was an adequate food supply, and this favored evolution to great size among the herbivores. As the herbivores grew into giants, the carnivores that preyed upon them likewise became giants.

For any animals, and especially reptiles, there are certain advantages in being a giant. Large size is a measure of protection. Moreover, the ratio of surface area to mass decreases as size increases; consequently the large animal has proportionately less surface to absorb heat and radiate heat. This means that the large animal requires less food in proportion to its size than the small animal, which is important for a reptile, with a low rate of metabolism and a varying body temperature. Moreover we know from experiments with living crocodilians of various sizes that body temperature in an individual reptile fluctuates more slowly than it does in a small reptile, because a large mass requires more time to heat up and cool off than does a small mass. It is therefore possible that their very size, which perhaps caused body temperatures to remain fairly constant, may have given the giant dinosaurs some of the attributes of active, endothermic or "warm-blooded" tetrapods, such as birds and mammals.

In recent years some students of the dinosaurs have argued (often with great vehemence) that the dinosaurs were truly endothermic, manufacturing their own body heat, rather than ectothermic or "cold-blooded" as are modern reptiles, which derive their body heat from the temperature of the environment surrounding them. Some cogent arguments have been advanced to support the concept of dinosaurian endothermy, notably the structure of the limbs in these reptiles giving to them an "upright" stance (a pose corroborated by the evidence of dinosaur footprints) and the histology or microstructure of the bones, which is similar in many respects to bone histology in mammals. Other authorities have questioned these arguments and adhere to the more conservative and traditional concept of the dinosaurs as reptilian ectotherms.

Perhaps this is not strictly an "either-or" question. It may be that the smaller dinosaurs, notably the theropods, which very probably were ancestral to the birds, were endothermic and such endothermy may have persisted in their gigantic Jurassic and Cretaceous descendants. But there is good reason to suppose that the great majority of dinosaurs were ectothermic with many of the attributes of "warm-blooded" tetrapods, as suggested above.

This very lively debate is much too complex to summarize here. Suffice it to say that the concept of endothermic dinosaurs is interesting and appealing, but at the present time is neither proven nor disproven.

However we may view dinosaurian physiology, it is clear that the Mesozoic era was an age of giants, when things were on a big scale. The ecological relationships between hunted and hunter were similar to those we see at the present day, but magnified. Instead of antelopes and lions, or bison and wolves, there were herbivorous and carnivorous dinosaurs.

Then near the close of Mesozoic times environmental changes began to take place. There were changes in topography and climates and consequently in vegetation. These changes, though fairly rapid in geologic time, were very slow in terms of years, and we would suppose that the dinosaurs might have adjusted themselves to such changes. Yet they failed to do so, and they became extinct. A new pattern of land life became established, where the advantage of size no longer was of such importance. In this new world, where agility and the power of sustained activity were paramount, the dinosaurs vanished, and the mammals became supreme.

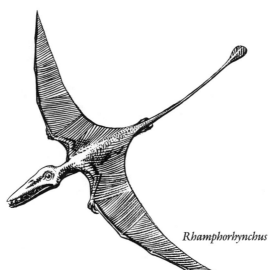

Rhamphorhynchus

Aerial Reptiles

13

GLIDING REPTILES

During late Permian times, the first tetrapods with adaptations for locomotion through the air appeared. These were the coelurosauravids, small primitive diapsids known from specimens found in Madagascar and Europe. *Coelurosauravus*, a typical member of the group, had a generally lizard-like body but with greatly elongated ribs that extended outward from the reptile's flanks. A membrane of skin probably covered the ribs, thus forming a broad, surface that acted as an airfoil. When coelurosauravids dropped from a height, they could glide to a lower spot in much the same way that the unrelated gliding lizard *Draco* does today.

The early experiment in aerial locomotion demonstrated by the coelurosauravids lasted for a short time, since members of the group are known only from upper Permian deposits. By late Triassic times, however, two other independent experiments in gliding appeared among tetrapods. One was in early lepidosauromorph diapsids of the family Kuehnosauridae, named after the genus *Kuehneosaurus* from upper Triassic deposits of Europe. These small reptiles show a development of the ribs into a gliding surface very similar to that of the coelurosauravids and *Draco*, although they are not closely related to either. The genus *Icarosaurus*, from upper Triassic rocks of New Jersey, is characteristic of the kuehneosaruids. This group became extinct at the end of the Jurassic period.

The other type of gliding adaptation that evolved by late Triassic times is represented by *Sharovipteryx* (formerly *Podopteryx*) from upper Triassic sediments of the central Asian part of the Soviet Union. In this small primitive diapsid, the gliding membrane was not associated with the ribs, which were of normal lizardlike proportions, but instead extended backwards from the hind legs to the base of the tail like a cape. There also may have been a small membrane associated with the relatively tiny front limbs. This is the only known example in tetrapods of a gliding surface primarily associated with the hind legs. Other types of gliding and flying adaptations mainly involve elongated ribs, both front and hind limbs ("flying" frogs and geckos), the torso as a whole ("flying" snakes), or just the front limbs (birds, bats, and pterosaurs).

PROBLEMS OF FLIGHT

During Triassic times the first widespread invasion of the oceans by air-breathing tetrapods took place. That this event did not occur before the Triassic period is probably owing to the fact that until then the tetrapods had not attained a degree of evolutionary development sufficiently advanced to become adapted to the rigorous requirements for complete marine life. Similarly it is interesting to see that true flight (as opposed to gliding) among the vertebrates was not established until late Triassic times, probably because until then tetrapod anatomy was not sufficiently advanced to permit adaptations in these animals that could satisfy the extremely rigorous requirements for flying. Then two groups of tetrapods, the pterosaurs or flying reptiles and later the birds, quite independently took to the air.

The requirements for flight in vertebrates are indeed restrictive, and only a few groups of tetrapods have been able to meet them during the last two hundred twenty-five million years of evolutionary history. In the first place, flying animals must overcome the downward pull of gravity. The problem is not particularly serious for small invertebrates like insects, but for vertebrates, the smallest of which exceed all but insect giants, gravity has been and is a problem of the first importance.

For instance, flying vertebrates must be comparatively light, particularly in relation to the strength of the muscles that move the wings. Because of the physical limitations of muscle strength, bone strength, the relationships of wing surface to body weight, and the like, there are definite upper limits in size beyond which the flying vertebrate cannot go,

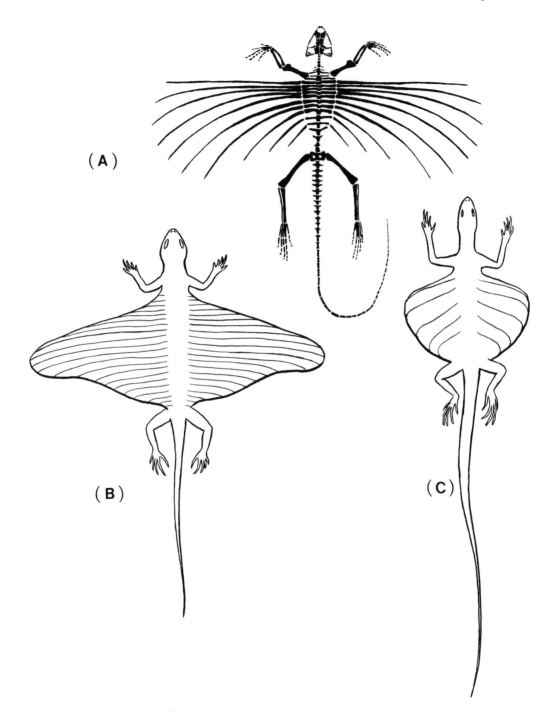

Figure 13-1. Skeleton of the early lepidosauromorph diapsid **(A)** *Icarosaurus*, from the late Triassic of eastern North America, about one-half natural size, and body outlines of **(B)** *Coelurosauravus*, a primitive diapsid from the late Permian of Madagascar and Europe, about one-fourth natural size, and **(C)** the living gliding lizard *Draco*, about one-half natural size.

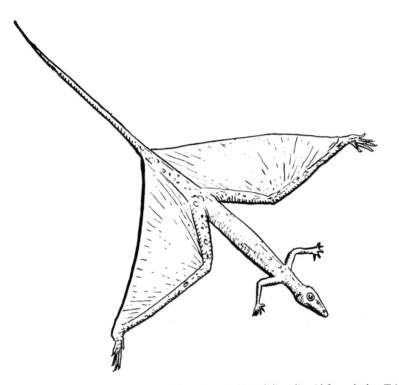

Figure 13-2. Reconstruction of *Sharovipteryx (Podopteryx)*, a primitive gliding diapsid from the late Triassic of central Asia, USSR. About one-half natural size.

and in this respect it is rather different from the man-made airplane, which continues to increase in size as the powers of motors increase. To achieve lightness, the bones of flying vertebrates are commonly hollow, with thin outer walls. Up to a certain point this type of skeletal structure is strong, and it gives sufficient areas for the attachments of large muscles, so necessary to flying vertebrates.

Flying vertebrates must have wings, and throughout vertebrate history these wings have been developed by transformations in the front limbs. Another requirement is a strong backbone as a girder for the support of the flying machine. Powerful arm muscles are necessary to move the wings up and down, and this means that some of the areas for attachment of the muscles, generally the breast bones, must be tremendously enlarged. There must be a landing gear of some sort, usually supplied by modifications of the hind limbs.

Of course other parts of the anatomy are involved in flight beside the skeleton and the muscular system. For instance, flying vertebrates must have a flight surface, either a membrane as was developed in the flying reptiles and as in the present-day bats, or modifications of the body cover, like the feathers of birds. Flight involves the senses, too. Flying vertebrates either have very powerful vision, as do the birds, and as probably did the pterosaurs, or they have some other method of guiding their flight, such as the sonar system of the bats. Flight presupposes a delicate sense of balance and nervous control, with consequent high specializations of the cerebellum in the brain, and of the entire nervous system.

Finally, flight, so far as we know it in the living tetrapods, requires a high rate of body metabolism. Whether this stipulation can be applied to the extinct flying reptiles is a moot point.

Adaptations for flight carry certain disadvantages, particularly those that go along with comparatively small size. On the other hand, flight frees tetrapods from the trammels of earthbound locomotion; flying vertebrates are free to move about over areas of considerable size and to cross many barriers that limit the movements of land-living animals. These advantages of flight need no elaboration; the distribution and success of the modern flying birds are indications of the benefits of being able to fly.

PTEROSAURS

The pterosaurs were archosaurian reptiles that became adapted for flight by the end of the Triassic period. They evolved in considerable variety during Jurassic times, and some of them continued through the Cretaceous period, at the end of which they became extinct.

Perhaps the pterosaurs can be described by outlining the principal characters of the late Jurassic form, *Rhamphorhynchus*. This was a reptile about two feet in length, with a characteristic archosaurian skull. Thus there were two temporal openings behind the large eye, and in addition a large preorbital fenestra. The front of the skull and the jaws were elongated and were supplied with long, pointed teeth that projected forward to a considerable degree, probably as an adaptation for fish-catching.

The skull was borne upon a fairly long and flexible neck. The back behind the neck was short and solid, and there was a continuous series of ribs between the pectoral and pelvic girdles. In *Rhamphorhynchus* there was a very long tail, perhaps twice the length of the backbone in front of the pelvis, and impressions in the rock indicate that it was supplied with a rudder-shaped membrane at its end.

In the forelimbs the humerus was strong, the radius and ulna rather elongated, and the fourth finger tremendously elongated to form the principal support for a wing membrane, the presence of which is amply proved by impressions in the rocks. The fingers in front of the fourth digit were reduced to small hooks that may have served the pterosaur as hangers for roosting in trees or on cliffs, while the fifth finger was lost. Projecting forward from the wrist was a spikelike bone, the pteroid bone, that helped to support the wing membrane. The scapula and coracoid were strong, and the coracoid was attached ventrally to an enlarged sternum, which served as an area for the origin of the large pectoral muscles that moved the wings. In some advanced pterosaurs (but not *Rhamphorhynchus*) the upper end of the shoulder blade was attached to the backbone by a special bony element, the notarium, thus giving added strength to the shoulder girdle.

The hind limbs in *Rhamphorhynchus* were comparatively small and slender, as they were in all the pterosaurs, and it is probable that the wing membranes were attached to them.

Such is the picture of a typical pterosaur. It was evidently a reptile capable of sustained flight, an aerial carnivore that probably fed by swooping down to the water of lakes or lagoons to catch fish swimming at the surface. Like the modern bats, it may have been exceedingly awkward and comparatively helpless on the ground, standing and moving in a quadrupedal posture. Alternatively, like modern birds, it may have walked and run on the ground rather well on its two hind legs. It is currently a matter of controversy as to whether pterosaurs were inefficient, slow-moving quadrupeds or swift bipeds when on the ground.

During Jurassic times there were various pterosaurs similar to *Rhamphorhynchus*—the rhamphorhynchoids. However, in late Jurassic times another group of pterosaurs, the pterodactyloids, appeared as derivatives from the rhamphorhynchoids. In these pterosaurs the tail was reduced to the point where it was almost suppressed. Likewise the dentition was reduced so that in the most advanced members of the group teeth were completely absent and the jaws took the form of a birdlike beak.

The pterodactyloids continued through much of the Cretaceous period, reaching the culmination of their evolution in the genera

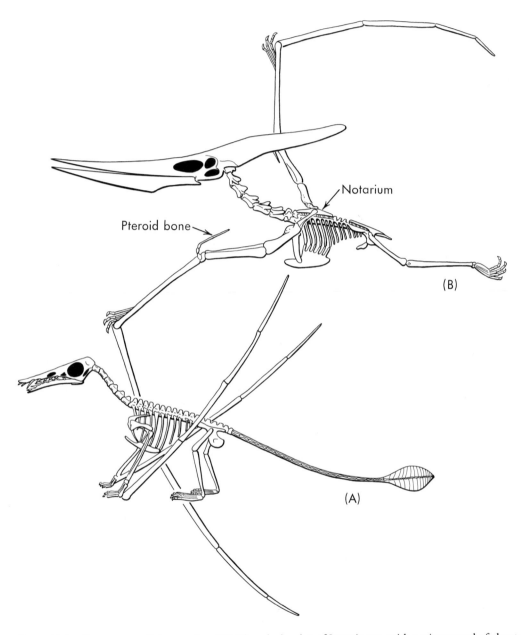

Figure 13-3.   Pterosaurs or flying reptiles. **(A)** *Rhamphorhynchus* of Jurassic age, with a wing spread of about two feet. **(B)** *Pteranodon* of Cretaceous age, with a wing spread of about twenty feet. The pteroid bone in the wrist of *Pteranodon* was a support for the wing membrane; the notarium served to attach the upper edge of the scapula to the backbone, thus making a strong base for the great wing. Note the large fenestra in front of the eye of this reptile.

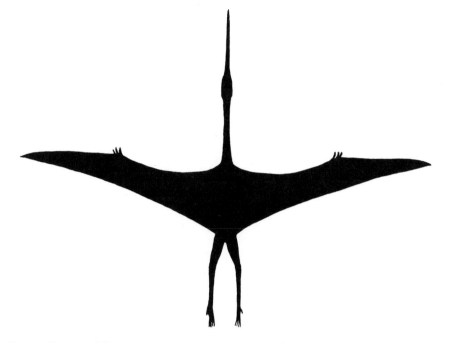

Figure 13-4. Silhouette of the giant pterosaur *Quetzalcoatlus*, from the late Cretaceous of Texas. The wingspread was approximately 40 feet.

Pteranodon, found in the Niobrara formation of Kansas, and *Quetzalcoatlus,* from the Tornillo beds of Texas. *Pteranodon* had a wing spread of more than twenty-five feet! Even so, the body was comparatively small, bulking no larger than the body of a turkey. The jaws formed a long, toothless beak, and the back of the skull was extended posteriorly into a crest, the purpose of which is a matter of conjecture. It seems probable that *Pteranodon* was a gigantic fish catcher, swooping down to scoop its prey from near the surface of the inland sea. The presence of fossil fish remains in the body cavities of some specimens of *Pteranodon* is the basis for this supposition.

Large as it was, *Pteranodon* was dwarfed by *Quetzalcoatlus* with a probable wing spread of forty feet! This pterosaur had a long neck, and it has been suggested that it was a sort of huge, reptilian "vulture," feeding on the carcasses of giant dinosaurs.

Certain interesting questions arise in connection with the pterosaurs. Since they were reptiles, and since reptiles as we know them have a comparatively low rate of metabolism, how could the pterosaurs maintain flight for any appreciable length of time? Perhaps much of their flight was of a soaring type, which did not require the expenditure of much energy. But perhaps pterosaurs had a high metabolism that could provide sufficient energy for actively flapping flight.

In this connection it should be mentioned that some years ago wind tunnel experiments were performed on models of *Pteranodon* to test its aerodynamic potentialities, and especially to determine how so large a flyer could take off from the ground or from the surface of the water. It was found that if this giant pterosaur were to spread its wings and face into the wind, even a light breeze would take it up like a kite. The same probably was true for *Quetzalcoatlus.*

Aside from their soaring abilities it is difficult to imagine the pterosaurs, especially the smaller ones, functioning without sources of energy beyond those known or supposed for other reptiles. Some years ago the fossils of a pterosaur, *Sordes pilosus*, were discovered in central Asia in sediments of late Jurassic age. The specimens clearly showed the imprints of hair-like strands densely covering the body. So it would appear that this pterosaur may have been "warm-blooded," thereby requiring an insulating coat to maintain its high body temperature. Perhaps this may have been true for all members of the order.

Why did the pterosaurs become extinct? Perhaps because they were in competition with the birds, which were becoming modernized in late Cretaceous times. The birds were evidently much more efficient than the pterosaurs, and it is quite likely that their perfection contributed to the end of the flying reptiles.

Archaeopteryx
and Gull

Birds

14

BIRDS

It may seem incongruous to introduce the birds into our discussion at this place, but actually it is not. The birds have been called "glorified reptiles"; as a matter of fact, we can be more specific and designate the birds as "glorified archosaurians." With such a concept in mind, it is quite pertinent to consider the birds at this time.

If we were to do justice to the marvelous adaptations and the interesting facts of distribution that characterize modern birds, an entire volume would be required. Such is not possible nor is it suitable in a discussion like this one; consequently our attention will be limited to the major characteristics of the birds and to a short résumé of their evolutionary history.

Birds have solved the problems of flight differently from pterosaurs or flying reptiles and bats. In the birds there are feathers that not only form the flight surfaces but also serve to insulate the body from its environment. The hind legs are very strong in most birds, giving to these vertebrates the double advantage of being able to run or walk on the ground as well as to fly in the air. Indeed, some of the water birds, such as ducks and geese, have the distinction of being able to move around proficiently in the water, on land, and in the air, a range of natural locomotor ability that has never been attained by any other vertebrates.

As we know from modern birds, these vertebrates are highly organized animals, with a constant body temperature and a very high rate of metabolism. In addition, they are remarkable in having evolved extraordinarily complex behavior patterns, such as those of nesting and song, and the habit among many species of making long migrations from one continent to another and back each year.

THE EARLIEST BIRDS

The first birds appeared during late Jurassic times. These birds are known from four very good skeletons, two incomplete skeletons, and an isolated feather, all from the Solnhofen limestone of Bavaria, Germany. This fine-grained rock, which is extensively quarried for lithographic stone, was evidently deposited in a shallow coral lagoon of a tropical sea, and fortunately for us flying vertebrates occasionally fell into the water and were buried by the fine limy mud, to be preserved with remarkable detail. In this way the late Jurassic bird skeletons, which have been named *Archaeopteryx*, were fossilized. And not only were the bones preserved in these skeletons, but also imprints of the feathers.

If the indications of feathers had not been preserved in association with *Archaeopteryx* it is likely that these fossils would have been classified among the dinosaurs, for they show numerous theropod characters. These were animals about the size of a crow, with an archosaurian type of skull, a long neck, a compact body balanced on a pair of strong hind limbs, and a long tail. The forelimbs were enlarged and obviously functioned as wings.

The skull had the two posterior temporal openings so characteristic of the archosaurians, but they were reduced by enlargement of the bones surrounding the brain. There was a very large eye opening, containing a ring of sclerotic plates, and in front of the orbit there was a large antorbital opening. The front of the skull and the lower jaws were elongated and narrowed into a beak, in which were well-developed teeth.

The neck was long and flexible, and the back was comparatively short and compact. A short, strong back is, of course, essential to a flying animal. The sacrum was long, giving a strong attachment between the elongated ilium of the pelvis and the backbone. The two other bones of the pelvis on each side, the pubis and the ischium, were transformed into rods, and the pubis had rotated to a ventral or posterior position. The hind limbs were strong and very bird-like, with three clawed toes pointing forward on each foot and one short toe directed to the rear. This type of hind limb and foot was typical also of the theropod dinosaurs. Evidently *Archaeopteryx* was able to walk and run in much the same fashion as a chicken. The bony tail was characteristically reptilian and about as long as the rest of the vertebral column. In the forelimb the scapula was slender, the bones of the arm were also long and slender, and the hand, composed of the first three digits, was much elongated. All the bones of the skeleton were delicately constructed.

As preserved, the skeletons of *Archaeopteryx* show long flight feathers extending out from the hand and the lower arm bones, whereas other feathers are indicated on the body. The tail was unique in that it had a row of feathers down each side.

Here was a truly intermediate form between the archosaurian reptiles and the birds. The skeleton alone was essentially theropodan, but with some characters trending strongly toward the birds. The feathers, on the other hand, were typical bird feathers, and because of them *Archaeopteryx* is classified as a bird—the earliest and most primitive member of the class. The wing feathers, closely comparable to the primary feathers of modern birds, would indicate not only that this animal was able to fly, or at least to glide, but also that it was warm-blooded. The expanded braincase would indicate that it had already evolved a comparatively complex central nervous system, which is so important to a flying animal.

It should be noted that there is a recently discovered collection of fragmentary bones from the upper Triassic Dockum beds of Texas that are said to represent the oldest bird. These scattered specimens can be reconstructed to form a bipedal, crow-sized animal, which has been given the name *Protoavis*. If these fossils are truly avian, then they would represent the oldest known bird, older than *Archaeopteryx* by approximately 75 million years.

There is much controversy, however, concerning the Dockum specimens. Many workers think the bones are not of a bird, but instead represent a new kind of theropod dinosaur with some birdlike features. Because

no indications of feathers have been found associated with the bones of *Protoavis*, and because the specimens are fragmentary and isolated, it will be difficult to determine whether the genus is a theropod-like bird or a bird-like theropod. The expected differences in bone structure between these two options may not be distinct enough to allow a definite decision. Unless better, more complete specimens of *Protoavis* are found, it is unlikely that this controversy will be solved soon.

ORIGIN OF BIRDS AND AVIAN FLIGHT

It has long been evident that birds are descended from archosaurian reptiles, and for many years it was thought that they had a thecodont ancestry. Recent studies and new fossil discoveries, however, present convincing evidence that birds are direct descendants of small theropod dinosaurs. Thus, in one sense, dinosaurs did not become completely extinct because one line of theropods evolved into all the birds alive today.

Although the vast majority of workers today hold the view that birds are descendants of theropods, there is a small group who think the similarities between birds, especially *Archaeopteryx*, and small theropods are the result of convergent evolution in the two lineages from a common ancestor that was an advanced ornithosuchian thecodont. In this minority view, theropods and birds are not ancestor and descendant, respectively, but rather are sister groups that evolved from the same ancestor group.

Another controversy in bird evolution is the determination of the intermediate steps between the first birds and their archosaurian ancestors, especially with regard to the origin of flight. Two main theories have been put forward to account for the structural changes necessary for avian flight. One argues that bird ancestors were bipedal reptiles that ran swiftly on the ground. They originally developed a cover of small, primitive feathers probably as

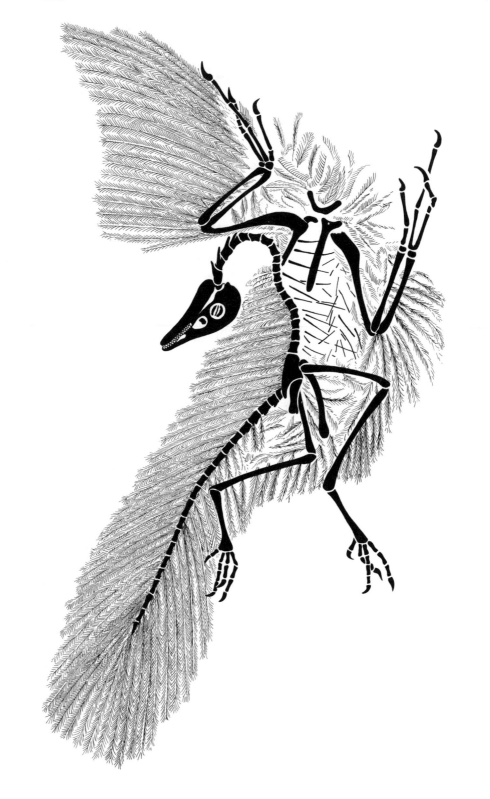

insulation to conserve body heat. Then feathers on the front limbs became elongate either to help the reptiles leap after and swat down insect prey from the air, or to allow them to jump greater distances while running by providing a gliding surface. The other theory postulates that the first birds climbed trees, from which they glided to the ground or to other trees, as do modern flying squirrels. At first the feathered forelimb was too small and ill-adapted to allow for more than gliding, but as time went on it enlarged and eventually was able to support the animal in flight. Thus flight was an outgrowth of climbing and gliding. In both scenarios, flapping of the feathered front limbs would not only improve their original function, but would also lead to true flapping flight.

Archaeopteryx was still in a transitional stage of adaptation between its archosaurian, probably theropod, ancestors and its fully developed avian descendants. It probably spent most of its time in the air gliding, but occasionally it may have flapped its wings to gain some added thrust forward. Such is the evidence of Jurassic birds. At the time, they faced strong competition from pterosaurs or flying reptiles, which were probably more adept on the wing than were the first birds.

CRETACEOUS BIRDS

By Cretaceous times the birds had progressed far along the evolutionary road that was to lead to modern birds. In most respects the skeleton of the Cretaceous birds had become "modernized." There was a coalescence of the skull bones so characteristic of modern birds, with a further reduction of the temporal openings. There was the development of a high degree of pneumaticity in the bones of the skeleton. The pelvis and sacrum were strongly coalesced to make a single structure, serving as an anchor between the strong hind legs and the body. The bones of the hand were also fused as in modern birds, not free from each other as they were in *Archaeopteryx*. The long, bony tail was suppressed. In some of the Cretaceous birds there was a great enlargement of the sternum for the origin of powerful pectoral muscles, sure proof of the perfection of flight. However, the Cretaceous birds were primitive in that they retained some teeth in the jaws. One of the best-known of the Cretaceous birds is the genus *Hesperornis* from the Niobrara chalk of Kansas. This was a bird specialized for swimming and diving, very much in the fashion of the modern loons and grebes. The body was somewhat elongated, the jaws were long, the feet were adapted for paddling, and the wings were suppressed. Another well known Cretaceous bird is *Ichthyornis*, from the same marine chalk in Kansas. It was a gull-like shore bird with a strongly keeled sternum, indicating that it was a good flyer.

CENOZOIC BIRDS

By the beginning of Cenozoic times the birds were completely modernized. They had reached their present high stage of skeletal structure, and so far as we can tell there has been little major structural evolution among the birds during the last fifty to seventy million years. However, there have been certain events in the history of birds during Cenozoic times that are worth noting. Perhaps the most striking fact about early Cenozoic avian history was the rapid radiation of large, ground-living birds such as *Diatryma*, from early Tertiary rocks of Europe and North America, and *Phororhacos*, from Miocene deposits of South America. It has been suggested that for a time there was active competition between the large terrestrial birds and the early mammals. If so, it was a passing phase, and the mammals soon became the dominant land animals. From then on birds developed for

Figure 14-1. A skeleton with impressions of feathers of the earliest known bird, *Archaeopteryx*, as it was found. About one-half natural size.

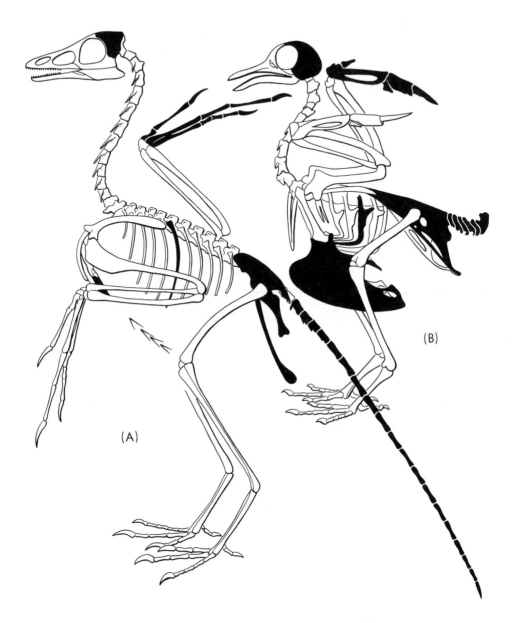

Figure 14-2.   A comparison of the skeletons in the earliest known bird, *Archaeopteryx* of Jurassic age (**A**), and in a modern pigeon (**B**). Comparable regions of the skeleton (brain case, hand, sternum, rib, pelvis, tail) are shaded black. Not to scale. In the modern bird the brain case is expanded, the bones of the wing are coalesced, the pelvis is fused into a single, solid structure, the bony tail is reduced, the ribs are expanded, and the sternum or breast bone is greatly enlarged for the attachment of strong wing muscles. All these adaptations, and many others, make the modern bird an efficient flying animal.

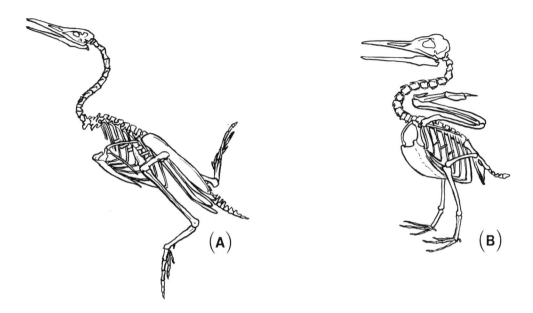

Figure 14-3.   Skeletons of upper Cretaceous birds from the Niobrara Chalk of Kansas. **(A)** *Ichthyornis* and **(B)** *Hesperornis*. (A) is about one meter long; (B) about one-fourth natural size.

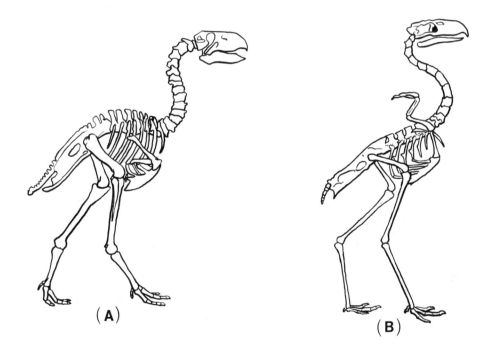

Figure 14-4.   Skeletons of early, large, flightless birds. **(A)** *Diatryma*, from the early Tertiary of Europe and North America and **(B)** *Phororhacos*, from the Miocene of South America. (A) is about two meters tall; (B) about 160 centimeters tall.

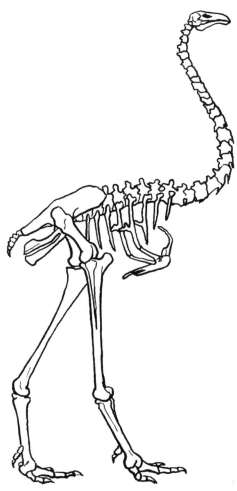

the most part as flying vertebrates, although a few large, flightless birds, the so-called ratites, have continued to the present day in most of the continental regions. The ostriches lived in Eurasia during late Cenozoic times and still inhabit Africa, the rheas are found in South America, the emus and cassowaries in the Australian region, and within the last few thousand years the so-called elephant bird inhabited Madagascar and the moas New Zealand. All these bird have been marked by growth to a great size, a secondary reduction of the wings accompanied by a flattening of the sternum and a strengthening of the legs.

Most of the modern birds are remarkably similar to each other in structure. Their great variety has been brought about by adaptations to many different types of life, which in turn has entailed changes in proportion, differences in coloration, and a wide range of differentiation in behavior patterns. Certainly the birds are at the present time highly successful vertebrates, sharing with the teleost fishes and the placental mammals the fruits of evolutionary achievement, as indicated by numerous species showing wide ranges of adaptation, by great numbers of individuals, and by a wide distribution throughout the world. It has been estimated, quite conservatively, that there are perhaps one hundred billion birds in our modern world.

Because of the structural uniformity of Ce-

Figure 14-5. Skeleton of the moa, *Dinornis*, from the Pleistocene and Recent of New Zealand. More than two meters tall.

SONG BIRDS
Robin

FOREST BIRDS
Owl

OCEANIC BIRDS
Albatross

UPLAND BIRDS
Pheasant

ADAPTATIONS TO VARIED HABITATS

WADING AND SHORE BIRDS
Plover

FLIGHTLESS GROUND BIRDS
Ostrich

Archaeopteryx

AQUATIC BIRDS
Penguin

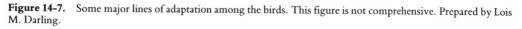

Figure 14-7. Some major lines of adaptation among the birds. This figure is not comprehensive. Prepared by Lois M. Darling.

Figure 14-6. Silhouette of the giant, condor-like bird *Argentavis*, from the Miocene of Argentina. The wing-spread was approximately 25 feet.

nozoic and modern birds, their classification is a difficult subject. Modern birds are classified to a large degree upon external characters, which of course cannot be utilized for fossil birds. Moreover, characters that in other vertebrates would be regarded as of comparatively minor importance are accorded ordinal rank in the classification of birds. Yet in spite of the relatively small differences that separate many of the major groups of birds, these vertebrates nonetheless show a wide range of adaptive radiation.

For instance, Cenozoic and modern birds range in size from some hummingbirds that are among the smallest of land-living vertebrates to the giant moas and elephant birds, now extinct. The largest known flying bird is *Argentavis*, a condorlike form from Miocene sediments of Argentina. It had a wingspan of approximately twenty-five feet! Birds range in their powers of locomotion from flightless, ground-living birds to the masters of the air, like the albatross or the swallows. They range in their adaptations to diet from the strictly plant-eating and seed-eating birds, like some of the fowls and songbirds, to aggressive birds of prey like the owls and hawks. Their colors range from very plain birds to the brilliant birds of paradise that live in the tropics. Their songs range from modest chirps to the remarkably complex songs of some of the perching birds. Their nests range from simple affairs to very elaborate structures. In their distribution they range from sedentary birds to migrants that fly back and forth across almost half the circumference of the globe.

These statements, so well known as to need no elaboration at this place, merely point up the fact that there was a great deal of evolution among the birds during Cenozoic times, in spite of the general structural similarity of these vertebrates. It must be remembered that the basic avian structure was determined at an early stage in the evolutionary history of birds because of the rigorous limitations placed upon a flying vertebrate. Consequently adaptations in the birds have been along lines that are not always indicated by the details of anatomy, a fact that makes these vertebrates highly interesting to the student of recent animals but difficult subjects for the paleontologist.

The classification of birds by orders is presented at the end of this book.

BIRDS IN THE FOSSIL RECORD

Of all the classes of vertebrates, the birds are least known from their fossil record. As we have seen, knowledge of Jurassic birds is based largely on six skeletons, whereas our information about Cretaceous birds, though more extensive, is nonetheless very scanty. Fairly numerous fossils of Cenozoic birds are known, but for the most part they consist of fragmentary bones. Only in unusual fossil deposits, such as the tar pits of the Pleistocene, are Cenozoic fossil birds found with any degree of completeness. These conditions restrict our knowledge of the evolution of birds through time. The study of modern birds is followed in great detail by many professional zoologists, and amateur bird watchers can be numbered by the millions. The study of fossil birds is one of the most restricted aspects of paleontology.

Elasmosaurus

Aquatic Reptiles

15

TETRAPOD ADAPTATIONS TO LIFE IN WATER

During their long evolutionary history the tetrapods had freed themselves from any dependence on water to become, as reptiles, animals that throughout their entire life history were completely terrestrial. Some of them went back to the water, and all the various adaptations that had made reptiles efficient and independent land-living animals needed to be modified. These newly secondarily aquatic (marine and/ or freshwater) animals no longer had to contend with the problems of gravity or desiccation; no longer did they have to propel themselves across a dry land surface. Rather, we might say, they assumed the old problems with which their fish ancestors had contended many millions of years past—the problems of buoyancy, of propulsion through the water, and of reproduction away from the land.

The reptiles had efficient lungs. These lungs were not abandoned when reptiles went back to the water, but rather were used for breathing, in place of the gills that had long since been lost. The reptiles had legs and feet. These were transformed into paddles, similar in function to the fins of the fish. The old fish tail had long since disappeared, so that some of the aquatic reptiles evolved substitute tails for propulsion, tails that imitated the fish tail to an astonishing degree. The old fish method of reproduction by eggs viable in water had been abandoned, so the aquatic reptiles that were unable to come out on the land developed substitute methods, such as live birth of their young, for continuing their kind in an environment where there was no place to lay an egg in a protecting nest. It can thus be seen that among these reptiles there was a reversal in the trend of evolution—from distant aquatic fish ancestors, through intermediate land-living amphibian and reptile ancestors, to aquatic reptilian descendants.

EARLY AQUATIC REPTILES

From early Permian to middle Triassic times, there appeared several independent reptilian experiments in adaptations to an aquatic environment. The earliest known example is the early Permian captorhinidan *Mesosaurus*, mentioned previously (Figure 8-7). It was a small form, about one meter long, that lived in the marine to brackish waters of shallow basins, using swimming as its main means of locomotion. Other early aquatic reptiles include a side branch of eosuchian diapsids that lived during late Permian and early Triassic times. *Hovasaurus* and *Tangasaurus* from upper Permian beds of Madagascar are characteristic of this group. They were small reptiles similar in overall proportions to mesosaurs, except that their tails were much longer and deeper, and the head was not as elongate. As in mesosaurs, the limbs were still relatively large and the feet were not changed into paddles. The aquatic eosuchians are often found with aggregations of small stones within their rib cages. These probably acted as ballast to help keep these small reptiles under the water surface.

A later group of early aquatic reptiles were the thallatosaurs, an obscure group of primitive diapsids known only from middle Triassic rocks. *Askeptosaurus* from marine deposits of Switzerland is a typical member. It had a longer and more streamlined body than mesosaurs or the aquatic eosuchians, and its limbs were smaller too. In contrast to the condition in mesosaurs and aquatic eosuchians, the nostrils of thallatosaurs were placed well back along the snout, a very common vertebrate adaptation to an aquatic life.

All of these early aquatic reptiles were only moderately adapted to a life in water, and none was completely successful because none left descendants. During the Mesozoic era, however, several groups of reptiles, especially the euryapsids, evolved adaptations to an aquatic environment that went well beyond those shown by these early experiments.

THE ICHTHYOSAURS

The Triassic period marked the first time during the evolution of the vertebrates that land-living tetrapods turned in any appreciable numbers to a life lived in marine waters.

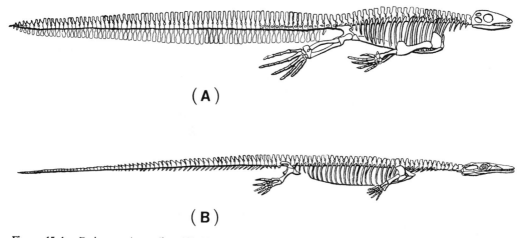

(A)

(B)

Figure 15-1. Early aquatic reptiles. **(A)** *Hovasaurus*, an eosuchian diapsid from the late Permian of Madagascar, about one-half meter long. **(B)** *Askeptosaurus*, a thallatosaurian diapsid from the middle Triassic of Switzerland, about two meters long.

Since then this trend has been several times repeated, but in the Triassic period it was a major step in vertebrate history. Of course many amphibians of late Paleozoic and Triassic times were aquatic, but few of the amphibians were marine; the first widely developed marine tetrapods were reptiles.

The ichthyosaurs, in many respects the most highly specialized of the marine reptiles, appeared in early Triassic times. Their advent into the geologic history of the reptiles was sudden and dramatic; there are no clues in pre-Triassic sediments as to the possible ancestors of the ichthyosaurs. It is only through an interpretation of the anatomical structures of these highly specialized reptiles that we are able to make some estimations as to their origin, which probably was from early diapsid ancestors. The basic problem of ichthyosaur relationships is that no conclusive evidence can be found for linking these reptiles with any other reptilian order. However, the presence of an upper temporal opening in ichthyosaurs suggests affinities with euryapsid reptiles of the subclass Diapsida. Therefore, ichthyosaurs are here included in the infraclass Euryapsida, although they could also be put in their own subclass.

The ichthyosaurs adhered closely throughout their history to a single pattern of adaptation so that the description of one genus of ichthyosaur applies in most respects to a majority of the other genera. The genus *Ichthyosaurus* from the Jurassic sediments of many widely separated localities around the world serves very well as a characteristic member of the group. Its remains are frequently found in black shales, in which not only the bones but also the body outline are occasionally preserved; therefore we have definite information as to the body shape and some of the soft parts in this interesting reptile. From such fossils we know that *Ichthyosaurus* was a fishlike reptile, ranging up to ten feet or more in length. It had a streamlined body, increasing in size from the head to a point not far behind the shoulder region and then decreasing uniformly from here back to the tail. There was no neck in the true sense of the word, and the back of the elongated head merged into the body without a break of the streamlining, as is essential in a fast-swimming vertebrate. The four legs were modified into paddles, and at the posterior end of the body there was a caudal fin, strikingly similar in shape to the caudal fin of many fishes. In addition, as shown by imprints in the rock, there was a fleshy dorsal fin, a new structure to take the place of the bony dorsal fish fin.

From this it is apparent that *Ichthyosaurus*

Ichthyosaurus

Figure 15-2. Ichthyosaurs like this, similar to modern porpoises in size and habits, lived in the oceans of the world during Jurassic and Cretaceous times. Prepared by Lois M. Darling.

was essentially like a big fish in its mode of life, and the various adaptations that were described in a previous chapter as advantageous for the fast-swimming fishes can be cited at this place as being applicable to the ichthyosaurs. In short, these animals moved through the water by the rhythmic oscillation of the body, the muscular waves going from the front to the back and then being transmitted to the tail to scull the animal through the water. The four paddles were used as balancers, to control movements up and down through the water, to aid in steering, and to assist in braking. The dorsal fin was a stabilizer to prevent rolling and side slip. In the shape of the body and the caudal fin, the ichthyosaurs showed many similarities to some of the very speedy fishes, such as modern mackerel or tuna, so it would seem obvious that these reptiles were fast swimmers, pursuing their prey through the wide expanses of Mesozoic seas.

It may be useful to note some of the details of various tetrapod structures that were modified in *Ichthyosaurus*. Since this reptile was buoyed up by the water, the vertebrae had lost their interlocking joints and were simplified as flattened discs, bearing neural spines. The double-headed ribs articulated to these discs and flared out to form a body that was deeply rounded in cross-section. Posteriorly the backbone turned down suddenly, to occupy the lower lobe of the fleshy tail fin; hence the tail in the ichthyosaurs may be designated as a reversed heterocercal type. The bones of the limbs were shortened, while the wrist and ankle bones and the bones of the fingers and toes were modified into flattened hexagonal elements, closely appressed to each other. The bones of the fingers were increased in number (hyperphalangy) and often there was an increase in the rows of phalanges (hyperdactyly), thereby adding to the length and width of the paddle. In life the whole structure was enclosed within a continuous sheath of skin.

The skull was elongated, principally by a lengthening of the jaws. The jaws were provided with numerous conical teeth with highly infolded enamel.

The enormous eye is evidence that the ichthyosaurs depended very largely on the sense of vision. The nostrils, as is so common in aquatic tetrapods, were set far back on the top of the skull, thereby facilitating breathing when the ichthyosaur came to the surface. The back of the skull was compressed, and there was a single upper temporal opening, typical of the euryapsids.

How did the ichthyosaurs reproduce? It is probable that these reptiles were unable to come out on land, just as modern porpoises and whales are unable to leave the water. Therefore they could not lay their eggs in the sand or in nests, as did their reptilian cousins. Fortunately for us some fossils from Germany show unborn embryos within the body cavity of the adult, and in one specimen the skull of an embryo is located in the pelvic region, as if the little ichthyosaur were in the process of being born when death overtook the mother. So it is evident that these reptiles were ovoviviparous—that they retained the egg within the body until it was hatched, as do some modern lizards and snakes.

This description indicates the high degree of specialization for marine life that was attained by the ichthyosaurs in general. What has been said applies with but few modifications to all the Jurassic and Cretaceous ichthyosaurs.

There was a dichotomous evolution of the ichthyosaurs, beginning in middle Triassic time. The two lines of adaptation have been called the latipinnate, or "wide-paddle" and longipinnate, or "long-paddle" branches, the names quite obviously referring to the shapes of the appendages. The latipinnate ichthyosaurs, having their origins in a small middle Triassic genus, *Mixosaurus*, were abundant during early Jurassic times, but became extinct at the end of that period. The longipinnate

ichthyosaurs, first represented by the middle Triassic genus *Cymbospondylus* continued into late Cretaceous time.

The Triassic ichthyosaurs were somewhat more primitive than their descendants in that the skull was relatively shorter, the paddles smaller, and the tail less downcurved. Interestingly, the largest ichthyosaurs are from late Triassic deposits. *Shonisaurus* from Nevada was a giant, attaining a length of up to 15 meters.

PLACODONTS AND SAUROPTERYGIANS

Whether or not the ichthyosaurs are included within the Euryapsida, this category does constitute a useful group for the Mesozoic marine reptiles often designated as the Sauropterygia— the Nothosauria and Plesiosauria—and for the Placodontia. The nothosaurs and placodonts were confined to the Triassic period; the plesiosaurs range in age from late Triassic time to the end of the Cretaceous period.

Adaptations among these reptiles for life in the ocean were quite different from those characteristic of the ichthyosaurs. Whereas the ichthyosaurs were fast-swimming, fish-shaped reptiles that swam by a sculling movement of the body and tail, using the paddles for balance and control, the placodonts and sauropterygians were comparatively slow-swimming animals that rowed through the water with large, strong paddles.

PLACODONTS

In early Triassic times the placodonts, a group limited in age to this geologic period, became specialized for life in shallow, marine waters, where they fed upon mollusks that they picked off the sea bottom. These reptiles were rather massively constructed, with a stout body, a short neck and tail, and paddle-like limbs. In the genus *Placodus* there was a strong, ventral "rib basket" of bony rods that helped to support the viscera as well as to provide protection for the lower surface. On the back there was a row of bony nodules above the vertebrae, indicating that the upper surface of the body in this reptile was armored. In the pectoral and pelvic girdles the ventral bones were comparatively strong, but the bones in the upper portions of the girdles were reduced. The limb bones were moderately long and the feet were flattened to form rather small paddles.

The skull was short, with the squamosal bone and other bones beneath the upper temporal opening very deep. The nostrils were not terminal, but had retreated to a position immediately in front of the eyes. In the lower jaw there was a high coronoid process, for the attachment of strong masticating muscles that originated on the powerful temporal arch of the skull. The teeth of *Placodus* were specialized in a most interesting fashion. The front teeth in the premaxillary bones and in the front part of the dentary protruded almost horizontally, and it is evident that they formed efficient nippers. Behind them the teeth of the maxillaries and the palatine bones in the skull and of the back portion of the lower jaw were reduced in numbers but broadened to form huge, blunt grinding mills, which when brought together by the strong jaw muscles must have been capable of crushing tough sea shells. Evidently *Placodus* swam along slowly, plucking various mussels and other shells off the sea floor and crushing them with the strong jaws and teeth.

Placodus was one of the more generalized members of this group of reptiles. In late Triassic times the shallow seas of Europe and the Middle East were inhabited by some remarkably specialized placodonts, among which the genera *Placochelys* and *Henodus* are of particular interest. In these genera the body was broad and flat, and was heavily armored by dorsal and ventral scutes that were coalesced to form a heavy shell. In *Henodus* the teeth were reduced to a single pair on each side, and there was a broad, transverse beak, obviously covered in life by horny plates. In short, these reptiles displayed a series of structural adaptations that were to some degree analogous to the ones characteristic of the large sea turtles of later geologic ages.

NOTHOSAURS

Contemporaneous with the placodonts were the nothosaurs, which reached the height of their development in late Triassic times. These were small to medium-sized, elongated reptiles, with very long, sinuous necks. In them the ventral portions of the girdles were heavy, as in the placodonts, and, similarly, there was a well-developed rib basket. The limbs were somewhat elongated and the feet were modified as short paddles. Limbs and feet were rather strong in the nothosaurs, and it is quite probable that these animals could come out on land, as do modern seals and sea lions. The skull was comparatively small and flat, and instead of a deep squamosal beneath the temporal opening the arch was emarginated on its lower surface to form a rather slender rod. The external nares were slightly retreated, and the internal nares, piercing the front of an almost solid palate, were beneath the outer nostrils. The margins of the long jaws were set with numerous sharp teeth.

The nothosaurs, of which the genus *Nothosaurus* is typical, paddled through the water as did the placodonts. The jaws and teeth in these reptiles were clearly fish traps, so it is apparent that as the nothosaurs swarm along they darted the long, flexible neck from side to side or forward, thus catching fish that came within their range.

PLESIOSAURS

This method of living and feeding was obviously successful, for the Triassic nothosaurs were followed by the plesiosaurs, which became numerous and of worldwide extent during Jurassic and Cretaceous times. In essence, the plesiosaurs followed the nothosaurian pattern, but became specialized through an increase in body size, a great increase in the size and efficiency of the paddles, and improvements of the fish-catching jaws.

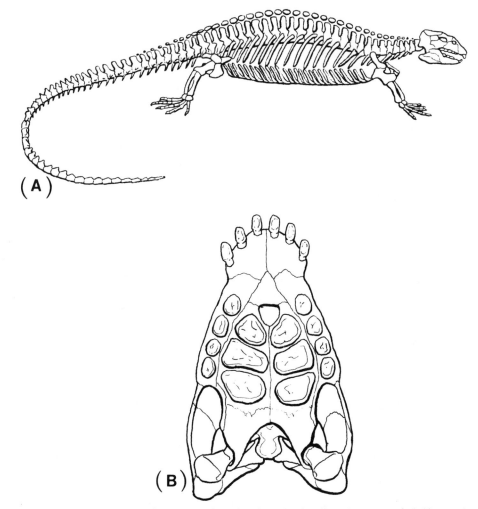

Figure 15-3. The **(A)** skeleton and **(B)** palate of the placodont *Placodus*. (A) is about one and a half meters long; (B) about one-third natural size.

The increase in size among the plesiosaurs began during early Jurassic times and continued through the remainder of the Mesozoic era, reaching its culmination in late Cretaceous times. Many of the Jurassic plesiosaurs were ten to twenty feet in length, whereas some of the late Cretaceous forms reached lengths of forty feet or more. In a great many plesiosaurs much of this length was represented by a very long neck, so that the body remained comparatively short. The body was broadened by the lateral extension of the ribs, each of which had a single articulation with the vertebrae in-

stead of the two articulations that are general in reptiles. The vertebrae had flattened ends on the centra, weak zygapophyses and tall spines, making for a flexible backbone with adequate surfaces for the attachment of strong back muscles. There was a strong ventral rib basket.

The ventral portions of the girdles were very large in the plesiosaurs, whereas the upper elements, the scapula in the pectoral girdle and the ilium in the pelvis, were much reduced. The strong ventral portions of the girdles extended back and forward of the ar-

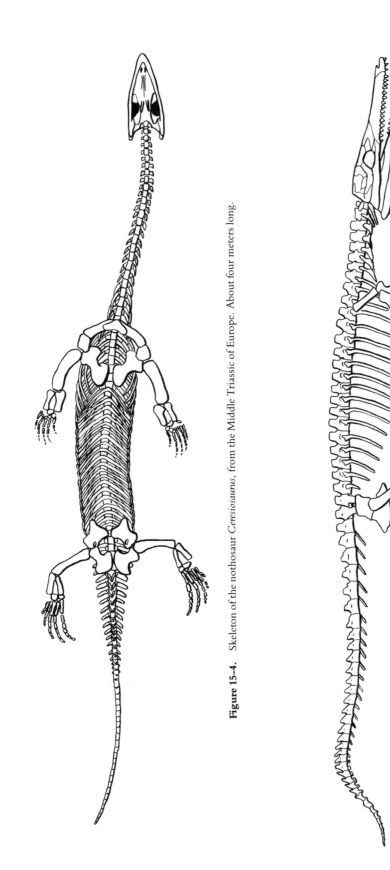

Figure 15-4. Skeleton of the nothosaur *Ceresiosaurus*, from the Middle Triassic of Europe. About four meters long.

Figure 15-5. A typical geosaur crocodilian of the Early Jurassic. About three meters long.

ticulations for the limbs, giving origins for powerful muscles that not only pulled the paddles back with great force, but also moved them forward with almost equal force. Consequently the plesiosaurs could row forward or backward, or they could combine these motions among different paddles, to rotate quickly. We might compare swimming in a plesiosaur with a rowboat manned by two skillful oarsmen.

The paddles in the plesiosaurs were very large. The upper limb bones were heavy and elongated, for attachment of the strong rowing muscles, but the lower limb bones and the bones of the wrist and ankle were short. The phalanges of the digits were multiplied in number, thus adding to the length of the paddles, but were always cylindrical, never flattened into discs as in the ichthyosaurs.

The skull was in effect a derivative of the nothosaur skull, with large eyes and temporal openings, an emarginated squamosal bone, nostrils in front of the eyes, an almost solid palate, and jaws that were provided with very long sharp teeth for catching and holding slippery fish.

From the beginning of their history the plesiosaurs followed two lines of evolutionary development. In one line, that of the pliosaurs or short-necked plesiosaurs, the neck was comparatively short, but the skull became much elongated, especially through elongation of the jaws. Such plesiosaurs are typified by *Pliosaurus* of the Jurassic and *Trinacromerum* of the Cretaceous. One of the pliosaurs, *Kronosaurus*, from the Cretaceous of Australia, reached huge dimensions with a skull about twelve feet in length!

In the other line, the long-necked plesiosaurs, the trend in evolution, apart from increase in size, was toward great elongation of the neck. These plesiosaurs evidently continued the nothosaur habit of rowing through shoals of fish, to catch their prey by darting the head this way and that. In Jurassic genera, such as *Muraenosaurus*, the paddles were very

large and the neck was elongated so that it equaled the body in length. However, the culmination of this branch of plesiosaurian evolution was reached by the elasmosaurs of upper Cretaceous times, characterized by the genus *Elasmosaurus*, in which reptiles there was a prodigious increase in the length of the neck, so that it was frequently as much as twice the length of the body and contained no fewer than sixty vertebrae!

The plesiosaurs, both the short-necked and the long-necked types, continued with unabated vigor to the end of Cretaceous times. Then they became extinct, as did so many of the dominant upper Cretaceous reptiles.

MARINE CROCODILIANS, LIZARDS, AND TURTLES

Three other groups of Mesozoic marine reptiles will be briefly considered at this place, even though the reptilian orders to which they belong will not be discussed until a later chapter. The geosaurs or thalattosuchians were members of two related families of early Jurassic to early Cretaceous crocodilians that went to sea. In these crocodiles a reversed heterocercal tail evolved, by a sharp downturning of the caudal vertebrae, and the limbs were modified to form paddles. With these specializations added to the characteristic crocodilian structures the geosaurs were well suited for life in the sea, but their existence was a short one, confined mainly to the Jurassic period.

The dolichosaurs, aigialosaurs, and mosasaurs were nothing more than varanid-type lizards that become dwellers in a marine habitat. The first two of these three groups, which appeared in early Cretaceous times, were generally small forms, only partially adapted for life in the sea. However, the aigialosaurs appear to have been ancestral to the late Cretaceous mosasaurs, which evolved as highly specialized marine reptiles. In so doing the mosasaurs followed a trend toward giantism, as have so many ocean-living vertebrates, so that some of the last and most specialized of these marine lizards, such as *Tylosaurus*, were

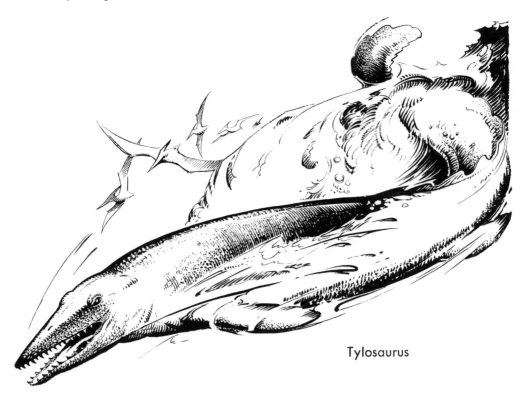

Tylosaurus

Figure 15-6.   A mosasaur, a Cretaceous marine lizard, twenty feet or more in length. Prepared by Lois M. Darling.

animals thirty feet or more in length. Adaptations in the mosasaurs for life in the oceans were brought about by a retreat of the nostrils to a rather posterior position on the top of the otherwise varanid-like skull, by the transformation of the limbs into flat paddles, as might be expected, and by a deepening of the long tail, to form a scull. However, there was no bending of the backbone either up or down, and it would appear that the tail in these reptiles extended back in a straight line from the pelvic region to the tip. The mosasaurs were obviously efficient swimmers, propelling themselves through the water by lateral undulations of the body and the powerful tail, using the paddlelike limbs for balancing and steering.

The marine lizards appeared quite suddenly in Cretaceous times, and like the geosaurs their history was a short one. But while it lasted it was a highly successful line of adaptation, for these reptiles became world wide in their distribution during the later phases of Cretaceous history. They died out near the close of Cretaceous times, as did so many of the large, dominant Cretaceous reptiles, but their close relatives, the varanid lizards, have continued to the present day and are now widely distributed in the Old World.

Although a majority of chelonians, (turtles and tortoises) live either in freshwater or on dry land, the history of this group includes a number of marine forms. In fact, the largest chelonian ever known to have existed, *Archelon*, was a sea turtle that lived during the Cretaceous period and had a shell more than eight feet long. *Protostega*, also of Cretaceous age, was another giant marine turtle, although not quite as big as its relative *Archelon*. Thus modern

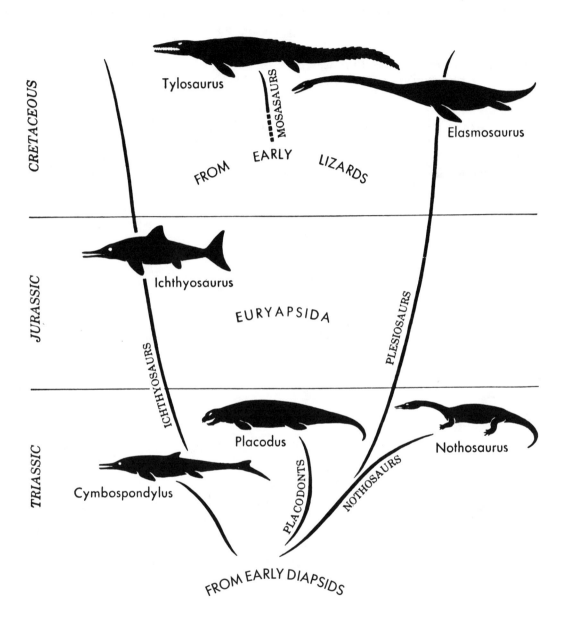

Figure 15-7. Generalized relationships of Mesozoic marine reptiles.

marine or sea turtles, which are among the largest living chelonians, have an ancestry that includes true giants.

CHAMPOSAURS AND PLEUROSAURS

Finally, two small groups of diapsids should be mentioned in any discussion of aquatic reptiles. Champosaurs were small to medium-sized, freshwater reptiles that had superficial resemblences to crocodilians but, which were distinct from the order Crododylia and are placed in their own order, Choristodera. Champosaurs ranged in age from the late Cretaceous to the Eocene and have been found in North America, Europe, and East Asia. Thus they were one of the few Mesozoic reptile groups that survived the mass extinction and the end of the Cretaceous period. Pleurosaurs were a side branch of sphenodontan diapsids that became specialized for an aquatic mode of life. They have been found only in Europe in rocks of early Jurassic to early Cretaceous age. They had greatly elongated bodies with small fore and hind limbs and a long tail, and their nostrils were placed well back of the snout tip. Although they left no descendants and thus were ultimately unsuccessful, the pleurosaurs provide us with evidence of yet another reptilian experiment in adaptations to an aquatic environment.

Jurassic Scene

Years of the Dinosaurs

THE RELATIONSHIPS OF
MESOZOIC CONTINENTS

As has been briefly outlined in Chapter 10, the great Pangaean supercontinent seemingly was intact at the close of Paleozoic time, its northern moiety, Laurasia, being composed of North America and Eurasia except for peninsular India, and its southern part being composed of Africa, South America, Antarctica, Australia, and the Indian peninsula. During the Permian–Triassic transition there may have been the beginnings of rifting within Pangaea, but it probably did not progress to any marked extent through the remainder of Triassic time. By the end of the Triassic period rifting probably had developed to the extent that South America was beginning to break away from Africa by a clockwise rotation, thus initiating the opening of a South Atlantic Ocean between the southeastern border of South America and the southwestern border of Africa. Similarly, by a clockwise rotation of Laurasia the North Atlantic Ocean was beginning to appear between eastern North America and the Mauritanian bulge of Africa. At the bottom of the world, however, southern Gondwana was still intact, the conjoined Antarctic-Australian landmass being in contact with the southern tips of South America and Africa and peninsular India being wedged between the southeastern coast of Africa and Antarctica.

This disposition of Africa, Antarctica, Australia, and peninsular India probably persisted through Jurassic time and well into the Cretaceous period. In late Cretaceous time the Indian peninsula finally broke away from Gondwana to drift rapidly to the northeast, eventually to collide with mainland Asia. The South Atlantic and North Atlantic openings increased by continental rotations, but northeastern South America retained a junction with northwestern Africa, while North America was attached to northwestern Europe.

Thus even though the breakup of Pangaea steadily progressed during Mesozoic history, there still remained continental connections whereby active tetrapods could wander from one part of the world to another.

It must be remembered that this discussion involves the relationships and movements of the great continental blocks with respect to each other. In addition to such large-scale events there were vertical oscillations within the several continental regions, causing mountain uplifts and depressions, these latter leading to the invasions of the continental blocks by relatively shallow seaways. These seaways, acting as barriers, could affect the distributions of land-living tetrapods. For example, during parts of Cretaceous time North America was bisected by a north to south seaway; Asia was similarly divided by a seaway known as the Turgai Strait.

VARIED FAUNAS OF THE
TRIASSIC PERIOD

It has already been shown that the Triassic was a period of transition, a time when various groups of tetrapods, persisting from the Permian period, lived side by side with numerous new reptiles, evolving along lines that were to be characteristic and dominant during the middle and later portions of the Mesozoic era. Consequently the Triassic vertebrate faunas were marked by the variety of animals composing them, retaining on the one hand certain aspects of the late Paleozoic tetrapod assemblages and containing on the other many progressive Mesozoic elements. In addition to the zoological variety of early Mesozoic faunas there was a great range of ecological adaptations, brought about by the many different land environments of Triassic times.

It will be remembered that the Permian period was a time of varied climates and environments, which favored the development of the reptiles so that they became the truly dominant land animals. The evidence seems to indicate that this was true for the Triassic period to an even greater degree than it had been for

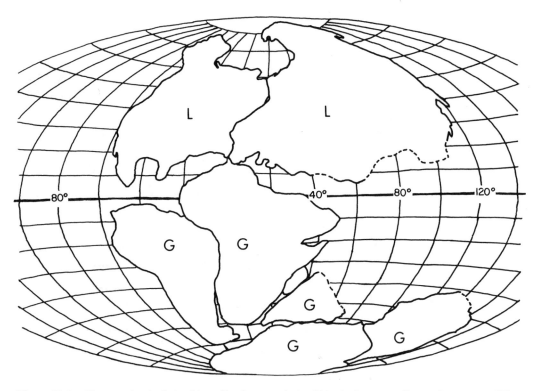

Figure 16-1. The postulated relationships of land masses during Triassic time, according to the concept of Plate Tectonics and Continental Drift, and the evidence of the distributions of land-living tetrapods. Abbreviations : G, Gondwana; L, Laurasia.

Permian times. Lands were emergent and widely distributed through the early portion of the Mesozoic era, with the result that the land-living tetrapods ranged around the world through almost all degrees of latitude. Labyrinthodont amphibians can be found in Triassic sediments from Spitsbergen, through the middle latitudes of the northern and southern hemispheres, to the tips of the southern continents and beyond, in Antarctica. Closely related reptiles enjoyed worldwide ranges, having been discovered in all the continents, extending north and south almost as far as the uttermost limits of the land masses. Evidently there were extensive continental connections during the Triassic period, enabling the land-living tetrapods to spread from one region to another. At the same time climates were sufficiently amenable, even though locally varied, so that amphibians and reptiles became distributed far and wide.

In at least two regions of the world the lower Triassic continental sediments and their fossils form a record continuous with upper Permian sediments and faunas, showing that life on the land went on with scarcely any interruption, even though other evidence indicates a major break in earth history at this point. These two regions are South Africa, where in the Karroo series the Triassic Middle Beaufort beds, containing a great host of mammal-like reptiles, succeeds the Permian Lower Beaufort beds with an orderly sequence of sediments and faunas, and northern Russia, where the lower Triassic Zone V of the Dvina series follows the upper Permian Zone

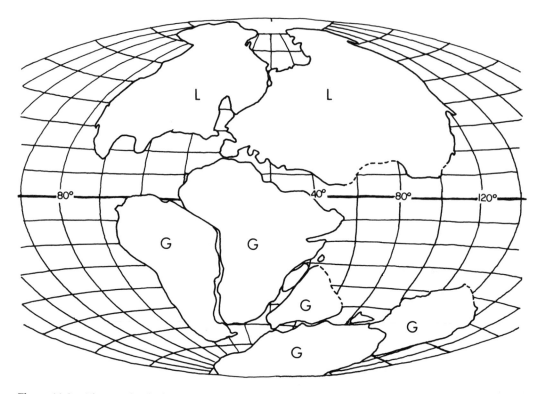

Figure 16-2.   The postulated relationships of land masses during Jurassic time, according to the concept of Plate Tectonics and Continental Drift, and the evidence of the distributions of land-living tetrapods. Abbreviations: G, Gondwana; L, Laurasia.

IV. In these areas we are given a glimpse of the truly continuous history of the earth and its life that would be general rather than exceptional if the geologic record were complete.

The Middle and Upper Beaufort beds of South Africa are commonly designated as two formations or zones named from characteristic reptiles. They are the Katberg formation or *Lystrosaurus* zone below and the Burgersdorp formation or *Cynognathus* zone above. (It had long been thought that there is a third zone—the *Procolophon* zone between these two; recent evidence seems to indicate, however, that the *Procolophon* zone is essentially a local facies of the *Lystrosaurus* zone.)

Of particular significance has been the discovery in recent years of *Lystrosaurus*, *Procolophon*, *Thrinaxodon*, and other characteristic African *Lystrosaurus* zone reptiles in the Fre-mouw Formation of Antarctica, thus affording strong evidence for the close connection of Africa and Antarctica within Gondwana. Quite recently *Cynognathus* has been found in Antarctica to strengthen the Lower Triasic similarities between the south polar continent and South Africa. In central India the Panchet beds have yielded *Lystrosaurus* identical with that reptile in South Africa, while the overlying Yerrapalli beds contain fossils that are closely correlative with those of the *Cynognathus* zone. Similarly, mammal-like reptiles that are identical with some of the genera in the *Cynognathus* zone have been found in the Puesto Viejo formation of Argentina, an indication of a very close link in early Triassic time between South Africa and South America. We see in these examples the paleonological evidence for Gonwana, within which there were wide distributions of land-living tetrapods.

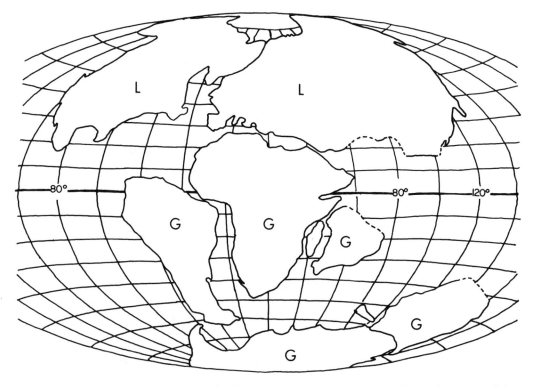

Figure 16-3.   The postulated relationships of land masses during Cretaceous time, according to the concept of Plate Tectonics and Continental Drift, and the evidence of the distributions of land-living tetrapods. Abbreviations: G, Gondwana; L, Laurasia.

In most parts of the world there are, of course, breaks between the Permian and the Triassic sediments and their faunas that cause contrasts in the life of the two periods. The classic threefold sequence of the Triassic, which gives this name to the period, is found in central Europe, where lower, middle, and upper Triassic sediments are represented respectively by the Bunter, the Muschelkalk, and the Keuper series. In the Bunter and the Keuper are found characteristic land-living tetrapods that are described in a preceding chapter, whereas the Muschelkalk, being a marine facies in this area, has yielded some of the early marine reptiles. In England only the lower and upper Triassic are present.

In North America the Moenkopi formation of the southwestern United States has in recent years produced a fauna composed largely of labyrinthodont amphibians, the age of which is generally regarded as early to middle Triassic. In Australia, which in the Mesozoic was evidently connected to Antarctica, some Triassic vertebrates have been found in the lower Triassic Narrabeen beds and Blina shales, the middle Triassic Hawkesbury beds, and the upper Triassic Wianamatta beds.

Middle Triassic tetrapods are less well known than those of lower and upper Triassic age, possibly owing to the spread of marine environments during the middle phase in this period of earth history. The Muschelkalk of central Europe has been mentioned. In Greenland and in Nevada there are middle Triassic marine sediments containing the fossils of primitive ichthyosaurs, and in the alpine region of Europe there are sediments of this age containing fish faunas.

However, recent work in the southern hemisphere has revealed tetrapod faunas that very probably represent the middle Triassic land-living reptiles that were excluded from the type Triassic area of central Europe by the spread of the Muschelkalk seas. In East Africa the Manda fauna appears to be more advanced than the Upper Beaufort fauna of South Africa and, therefore, perhaps of middle Triassic age. Of particular interest is the sequence of Triassic faunas in South America. The abundant Santa Maria fauna of Brazil and the slightly later Ischigualasto fauna of Argentina contain numerous mammal-like reptiles that are not only more advanced than those of the South African Beaufort beds but also show affinities to some of the similar reptiles in the Manda formation. Furthermore, some of the thecodont reptiles from the Santa Maria formation are very close indeed to a rare but beautifully preserved thecodont skeleton recently described from the type middle Triassic of Europe. Thus, we are finally afforded a key to the nature of middle Triassic land-living tetrapods. It should be mentioned that the Santa Maria and Ischigualasto faunas contain primitive dinosaurs, an indication that dinosaurs first appeared earlier than previously had been thought, or that perhaps these sediments especially the Ischigualasto range up into the basal portion of the upper Triassic.

It is perhaps in the upper Triassic sediments that the early Mesozoic tetrapods reach their widest distribution and their most varied development. Throughout the world, sediments of late Triassic-early Jurassic age reveal closely related faunas that contain numerous stereospondyls, varied thecodonts in the culminating phases of their evolutionary development, and early dinosaurs. Such is the Keuper fauna itself, found in central Germany and in England. Such are the upper Triassic faunas of North America—the Newark faunas of the eastern seaboard, the Chinle faunas of the southwestern states, the Dockum fauna of Texas, and the Popo Agie fauna of Wyoming. From these sediments come some of the most complete fossils of upper Triassic tetrapods,

to give a rather well-documented record of what land life was like at that time. These faunas contain aquatic facies too, as revealed not only by the amphibians but also by numerous fresh-water fishes. Related faunas are found in the Maleri beds of India, in the Lufeng series of Yunnan, China, and in the Stormberg series of the South African Karroo sequence.

As we pointed out in Chapter 11, the close of the Triassic period was marked by significant extinctions that brought an end to various groups of characteristic Triassic vertebrates, especially those that had persisted from the Permian into the Triassic period. Consequently there were distinct faunal breaks between vertebrate animal assemblages of the Triassic period and those of the succeeding Jurassic period. However, in the southwestern states the sedimentary transition between the two periods was a gradual one, so much so that it has been a debated question for years as to where Triassic sediments end and Jurassic beds begin. Unfortunately the fossil record of the lower Jurassic sediments in this area is so scanty that the nature of the faunal transition cannot be accurately determined.

In Europe there is an horizon and a fauna that in a sense is transitional between typical Triassic conditions and those characteristic of the Jurassic period. This is the Rhaetic of Germany and England, a level marked by a stratum containing numerous small bones and fragments of land-living reptiles. Some students regard the Rhaetic as representing the end phase of Triassic sedimentary history, others as the beginning of Jurassic events, and still others consider it as truly transitional between the Triassic and the Jurassic.

JURASSIC ENVIRONMENTS AND FAUNAS

The opening of the Jurassic period is marked by a remarkable paucity of land-living tetrapods in the geologic record. It can be attributed in part to the fact that with the advent of the Jurassic period there seems to have been a spread of marine waters, so that lands that had been extensive during late Triassic times became restricted, whereas shallow seas advanced across many continental regions. Moreover, lowermost Jurassic continental sediments, where preserved, are frequently almost barren of fossils. For instance, in the western part of North America there is an extensive formation of Jurassic age, specifically the Wingate sandstone, composed of ancient dune sands in which fossils are rarely found. It is therefore obvious that in addition to the spread of lower Jurassic seas there were in some regions prevalent desert environments, in which reptiles were probably comparatively rare, and in which conditions were not favorable for the burial and fossilization of those reptiles that did live in such habitats.

So it is that much of the good evidence for lower and middle Jurassic vertebrates comes from Europe, mainly from marine sediments along the Channel coast of England. Here is the well-known sequence of Lias, Dogger, and Malm, representing the lower, middle, and upper Jurassic respectively. Along the Dorset coast of England, across the channel in the vicinity of Caen, France, and in southern Germany at Holzmaden are lower Jurassic beds containing fossils of ichthyosaurs and plesiosaurs, together with many marine invertebrates. Lower Jurassic vertebrates are known from some localities in North America and Australia. The Kota formation of central India recently has yielded an abundance of large sauropod dinosaurs, above which are found characteristic Liassic fishes. An important middle Jurassic horizon is the Stonesfield slate of southern England, in which are found jaws and teeth of primitive mammals.

In late Jurassic times there were evidently broad lowlands in various parts of the world. The Morrison formation, exposed over great areas in western North America, consists of stream channel and lacustrine sediments that were deposited over a large portion of this continent. In Europe it would seem that there were probably numerous low islands and perhaps long peninsulas from mainland areas, so that this region was like the modern East Indian archipelago. Here, various land-living and marine tetrapods lived, died, and were buried in close association with each other. For instance, the Malm of southern England, containing in ascending order (among other horizons) the Oxford clays, the Kimmeridge beds, and the Purbeck beds, has yielded marine reptiles, theropod, and sauropod dinosaurs, stegosaurians, crocodilians, and other land-dwelling or semiaquatic tetrapods. The Purbeck is notable especially because it is an horizon from which a considerable fauna of primitive mammals has been collected. The Tendaguru beds of Tanzania, with a remarkable association of giant dinosaurs, are related to the Purbeck of England and to the Morrison of North America.

The upper Jurassic of Europe, the Morrison, and the Tendaguru contain dinosaurian faunas that show strong similarities. In these widely separated regions are large carnivorous theropods, numerous gigantic sauropods, small camptosaurs, and stegosaurs. In addition these faunas contain various crocodilians and turtles. It is evident that in spite of the general spread of Jurassic seas across parts of Pangaea there were intercontinental connections in late Jurassic times, by means of which the large dinosaurs migrated from one continent to another. From the nature of these faunas and from other evidence it would seem that environmental conditions were closely similar in North America, Europe, and Africa; that in these regions lands were low and covered with tropical jungles, that swamps were extensive and temperatures uniformly warm. In brief, this was a period of worldwide climatic and environmental uniformity, as shown by the land-living reptiles, and, therefore, can be contrasted sharply with the late Triassic, when lands were emergent and environmental conditions were varied.

Interesting light is thrown upon late Jurassic landscapes and environments by the famous lithographic limestone deposits at Solnhofen, Germany. Here are sediments that were obviously deposited in the still, shallow waters of a coral lagoon, in an environment rather similar to the coral atolls and lagoons for our modern South Sea islands. In these sediments, various invertebrates found their final resting places, as occasionally did some of the flying reptiles that fell into the quiet waters. And providentially some Jurassic birds also dropped into the Solnhofen lagoon, to be preserved as the several known skeletons of *Archaeopteryx*.

Such surroundings as typified Jurassic landscapes were most favorable for the evolution and deployment of the giant dinosaurs. Vegetation was lush and abundant, furnishing an ample food supply for the plant-eating dinosaurs. As a result the sauropods grew to great size, as to a lesser degree did the stegosaurians. The carnivorous dinosaurs that preyed upon these large herbivores also became giants. It was an age of giants.

THE EARLY CRETACEOUS WORLD

It would appear that the extensive marine incursions and the low, tropical lands of late Jurassic times extended into the lower portion of the Cretaceous period. Consequently land-living faunas are not numerous, although perhaps more so than those of late Jurassic age. Of particular interest are the reptiles, especially the dinosaurs, found in the lower Cretaceous Wealden beds of southern England and Belgium. These are the beds containing *Iguanodon*, an ornithopod of historic interest, and one known from numerous skeletons. Dinosaurs are known from other regions; from the Cloverly formation of western North America, from the Nequen formation of southern Argentina, from the Uitenhage beds of South Africa, from the dinosaur beds of Shantung, China, and from the Rolling Downs beds and the Broome Sandstone of Australia. Even so, our knowledge of lower Cretaceous tetrapod faunas is imperfect.

THE DINOSAUR FAUNAS OF LATE CRETACEOUS TIMES

In contrast to the incomplete record and our consequent inadequate knowledge of tetrapod faunas during some of Jurassic and early Cre-

taceous times, the record of the late Cretaceous is comparatively abundant and extensive. This was a time of earth change, when lands were beginning to rise, and continents not only were being extended in area but also were experiencing mutual changing relationships, owing to very active drift. It was a time of modernization in the plant world, when the flowering plants and the deciduous trees became established and widely distributed over the lands. It was probably a time of comparatively varied environmental conditions, at least as contrasted with much of Jurassic history. All in all, conditions were favorable for the spread of land-living tetrapods and for their preservation as fossils. In addition, marine and lacustrine deposits were accumulated on an extensive scale, and some of these contain not only the record of late Cretaceous marine reptiles but also evidence as to the great radiation of modern types of bony fishes—one of the important aspects of the history of life during Cretaceous times.

In upper Cretaceous sediments of several continents are preserved the richest known dinosaurian faunas, exhibiting these great reptiles in the culminating phases of their evolutionary history. Also, these sediments contain abundant records of other reptiles and of other vertebrates. The upper Cretaceous dinosaurian faunas of North America, especially those preserved in the western states and provinces of the United States and Canada, are very probably the most extensive of any in the world, and they give us a comprehensive picture of the course of tetrapod evolution during the final phases of Mesozoic history. In Alberta are the so-called Belly River and the Edmonton series, and in Wyoming and adjacent states the Lance formation; and these horizons, together with correlative formations in other western areas, represent in a general way a sequence showing a succession of genera and faunas through late Cretaceous times. In various instances it is possible to follow the evolution of phyletic lines of dinosaurs and other reptiles from the Oldman formation of the Belly River through the Edmonton into the Lance, this last representing the final stage of Cretaceous history. This sequence of dinosaur-bearing formations along the Canadian border is paralleled by related sediments, containing dinosaurs and other reptiles, in Colorado, New Mexico, and across the Rio Grande River in Mexico. To supplement this record of land-living tetrapods, there are marine formations, notably the Niobrara limestone and the Pierre shale, giving evidence of the plesiosaurs, ichthyosaurs, and mosasaurs of that time, as well as the developing marine teleost fishes. These sediments make western North America a happy hunting ground for the student of Cretaceous vertebrates, the scene of numerous expeditions and collecting trips, many of them famous in the history of paleontology, that have taken place during the last century.

North America was probably a continental region during late Cretaceous times, with a long, narrow, north-to-south inland sea separating its eastern and western sections during much of the period.

Europe, on the other hand, was more broken up, as it had been in the Jurassic period, and it is likely that there were many islands dotting the tropical seas in that region. Certainly marine deposits predominate in the upper Cretaceous of Europe, and here is found the chalk of England from which the name for the Cretaceous period is derived. Marine reptiles are known from various localities, perhaps the most famous being the mosasaurs from the Maestrichtian deposits of Belgium. In Transylvania are sediments containing dinosaurs, and fossils of these reptiles have also been found during recent years in France and in Portugal.

As a result of expeditions during the last seven decades, a notable array of dinosaurs has been found in Cretaceous beds in Mongolia, in the formations known as the Ondai Sair, Oshih, Iren Dabasu, Djadochta or Bain Dzak, and Nemeget. It may be that the Mongolian formations make a sequence in approximately the order given above, but this is a point that cannot be clarified at the present time.

Certainly, the Nemeget beds in southern Mongolia, in which have been found spectacular dinosaurs of Lance type, are of very late Cretaceous age. These deposits were laid down in separate inland basins of the Asiatic Cretaceous continent; consequently there is no direct way of linking them, to determine in what order they may have been accumulated. Moreover the fossils have not thrown any truly significant light on this problem. We may say that they are all of Cretaceous age—possibly all of late Cretaceous age.

Other upper Cretaceous sediments of importance are the red beds of Patagonia, from which a considerable dinosaurian fauna has been described, the Baurú formation of Brazil, with varied dinosaurs and other reptiles, the Baharije beds of north Africa, and the Opal beds of Australia. In recent years dinosaurs have been discovered in French Morocco.

In the limestones of Mount Lebanon, Syria, are found great quantities of teleost fishes constituting one of the best-known Cretaceous fish faunas.

The correlative relationships of Mesozoic vertebrate-bearing sediments, briefly described and discussed in the foregoing paragraphs are illustrated by the accompanying chart.

THE END OF THE MESOZOIC ERA

With the close of the Mesozoic era the years of the dinosaurs came to an end. These great reptiles, dominant on the land for more than one hundred million years, failed to survive the transition from Mesozoic to Cenozoic times, as did likewise certain marine reptiles, the plesiosaurs, and the mosasaurs. In addition the pterosaurs or flying reptiles became extinct. It is probable that these several groups of ruling reptiles did not die out at the same time; and evidence seems to indicate that the ichthyosaurs vanished before the end of the Cretaceous period. Nevertheless the extinction of the dominant Cretaceous reptiles was, on the whole, sudden in a geological sense, and dramatic. A great host of reptiles, many of them giants, vanished into limbo, leaving as

CORRELATION OF MESOZOIC VERTEBRATE-BEARING SEDIMENTS

		Europe	North America		South America	Africa	Asia	Australia Antarctica (A)
Cretaceous	Upper	Transylvania Maestricht	Lance Hell Creek Edmonton Oldman Judith River Two Medicine Pierre Niobrara / Animas Kirtland Fruitland Mesa Verde / Aguja Difunta	Monmouth Matawan Magothy Raritan	Patagonia / Baurú	Baharije / Madagascar	Nemegt Bain Shire Lamenta Trichinopoly Barun Goyot Mt. Lebanon Bain Dzak Iren Dabasu Oshih	Opal Beds / Griman Creek / Broome
Cretaceous	Lower	Wealden	Dakota Cloverly Trinity	Arundel	Nequen Bahia Botucatú	Uitenhage	Shantung	Rolling Downs
Jurassic	Upper (Malm)	Purbeck Kimmeridge Oxford Solenhofen	Morrison			Tendaguru		
Jurassic	Middle (Dogger)	Stonesfield						
Jurassic	Lower (Lias)	Holzmaden	Navajo Kayenta Moenave Wingate			Madagascar Clarens Elliot	Kota Lufeng	Talbragar Durham Downs

Triassic	European stage	Western / Eastern North America	South America	South Africa	East Africa	India / China	Australia / Antarctica
Upper	Rhaetic; Keuper	Hosselkus–Chinle; Dockum–Popo Agie; Newark	Colorados; Ischigualasto	Stormberg; Elliot Red Beds; Molteno		Maleri; Lufeng	Wianamatta
Middle	Muschelkalk		Santa Maria; Chañares	Upper Beaufort; Burgersdorp; *Cynognathus* zone	Upper Ruhuhu; Manda	Shansi	Hawkesbury; Walloon; Marburg; Kirkpatrick (A)
Lower	Bunter; Dvina-V	Moenkopi; Red Peak	Puesto Viejo	Middle Beaufort; Katberg; *Lystrosaurus* zone		Sinkiang; Yerrapalli; Panchet	Narrabeen; Blina; Arcadia; Fremouw (A)

survivors only the turtles, the crocodilians, the lizards and snakes, the very restricted sphenodonts and for a short time during early Cenozoic times the champosaurs.

What brought about the extinction of the great mesozoic reptiles? This is a most difficult question for which no ready answer is apparent. We would think that if the crocodiles could survive the transition into Cenozoic times, certainly some of the smaller dinosaurs might have continued. The environmental changes from late Cretaceous into early Paleocene times were not sudden nor were they drastic, and it would seem that some of the dominant reptiles might have adapted themselves to new conditions as these changes took place. Indeed, many changes, such as the establishment and deployment of the angiosperms or flowering plants, and the beginnings of continental uplifts, had already occurred during the latter part of the Cretaceous period, and to such developments the dinosaurs had adjusted themselves with no seeming difficulty.

Yet the fact remains that all the dinosaurs, the ichthyosaurs, the plesiosaurs, and the flying reptiles had vanished from the earth by the beginning of the Cenozoic era. All that can be said is that conditions changed, and for some reason the ruling reptiles were unable to adapt themselves to the changing world. These changes probably involved the establishment of more varied temperatures than had existed during much of the Mesozoic era and the sharpening of zoned climates from the equator to the poles, with consequent effects upon the complexion and distribution of plant life. However that may be, the result was to bring an end to the ruling reptiles of the Mesozoic era.

There are today two sets of theories to explain the extinctions that took place at the end of the Cretaceous time. One such set comprises the various terrestrial theories that seek an explanation in events that occurred on the earth without outside influences. The other set is composed of the extraterrestrial theories that look to forces beyong the earth as the primary causes for the Cretaceous extinctions. The terrestrial theories envision environmental changes and other earthbound events as being primarily the cause of the extinctions, which, many paleontologists believe, were for the most part rather gradual. In other words, the dinosaurs and some of their contemporaries were "going down hill" during later Cretaceous time, so that the terminal extinctions were the final step in a long-continuing phenomenon. According to the extraterrestrial theories, on the other hand, the terminal Cretaceous extinctions were sudden, the result of a catastrophic collision of a huge meteor or perhaps a comet with the earth. This theory is at the moment very popular because, among other things, it offers a relatively simple explanation to account for the disappearance of the dinosaurs and certain other reptiles at the close of Cretaceous time. Yet it does not explain the selectivity of the Cretaceous extinctions; why so many animals contemporaneous with the dinosaurs survived from Cretaceous into early Cenozoic time.

As soon as the dinosaurs and the other ruling Mesozoic reptiles vanished, the mammals came into their own. Although the first mammals appeared during the Triassic period, and although modern placental mammals inhabited the Cretaceous world, these vertebrates remained as small, minor members of the middle and late Mesozoic faunas. It would seem that the presence of the varied reptiles of the Mesozoic "held down" the first mammals, and it was not until the dominant reptiles had vanished, to vacate numerous ecological niches, that the mammals enjoyed their first great burst of evolutionary adaptation. By the opening of the Paleocene epoch the world was abundantly inhabited by the mammals, and from that time until today the mammals have reigned supreme.

Sphenodon

Surviving Reptiles

REPTILES SINCE THE CRETACEOUS

In late Cretaceous times varied reptiles belonging to ten orders inhabited almost every conceivable environment on the continents, and ranged far and wide over the surface of the oceans. At the beginning of the Tertiary period the reptiles, now overshadowed by the mammals and birds, were included within five orders, of which one, the Choristodera or champsosaurs, was to disappear not long after the advent of this new age, leaving four orders of reptiles that have survived from early Cenozoic times to the present.

These four orders of modern reptiles are the Chelonia or turtles, the Crocodilia or crocodilians, the Sphenodonta, and the Squamata or lizards and snakes. Of these orders, the sphenodontans are now represented by the single genus *Sphenodon*, confined to a very limited range in New Zealand. Since the evidence of the fossils would seem to indicate that the sphenodontans have probably been a restricted group of reptiles ever since the end of Jurassic times, the truly abundant reptiles that survived the close of the Mesozoic era have therefore been the tropical and subtropical crocodilians, and the widely distributed turtles, lizards, and snakes. Consequently these three orders of reptiles, even though not dominating the earth as did the reptiles of Mesozoic days, have nevertheless been very successful tetrapods, persisting through a long span of geologic time as competitors of the "higher" vertebrates, the birds and mammals.

The origins of the surviving reptiles go far back in geologic history. The turtles appeared during the Triassic period, as did the sphenodontans, when these latter reptiles enjoyed a European distribution. The crocodilians arose at the close of the Triassic period, to occupy the ecologic niche vacated by the phytosaurs and large labyrinthodonts. The ancestors of the Squamata had appeared in Triassic times, and from these ancestors the lizards became defined and well established during the Jurassic period. By Cretaceous times the snakes had developed

the highly specialized adaptations so characteristic of them. Therefore it is necessary to look back to early and middle Mesozoic times, or earlier, to begin the survey of surviving reptiles that will form the subject of this chapter.

TURTLES

From the Permian beds of South Africa there has been described a rare and all too fragmentary fossil that gives a possible inkling of the origin of the turtles. This is the genus *Eunotosaurus*, a very small reptile only a few inches in length. Little is known of the skull, but the palate and lower jaw are sufficiently well preserved to indicate that *Eunotosaurus* was well supplied with small teeth around the margins of the jaws and on the palate. The dorsal vertebrae and the ribs, which are nicely fossilized, indicate a remarkable degree of specialization in the trunk region of this reptile, for the elongated vertebrae between the shoulder region and the pelvis in *Eunotosaurus* were limited in number to nine, whereas there were but eight ribs, each so broadly expanded that it was in contact with its neighbor before and behind. There is no certainty on the basis of this fragmentary evidence that *Eunotosaurus* was ancestral to the turtles, but such adaptations in the vertebrae and the ribs might provide an indication of the specializations of turtle ancestors.

The first true turtles made their appearance by the late part of the Triassic period, by which time they were far advanced along the lines of adaptive radiation typical of modern turtles. In *Proganochelys*, a characteristic Triassic genus, the bones of the skull had been reduced in number, teeth were absent from the margins of the jaws, and the body was protected by a heavy shell. These are the basic turtle adaptations, and evolution among the turtles since Triassic times has been mainly a matter of refining the characters that were established in *Proganochelys*. For instance, the later turtles were completely toothless, whereas the Triassic

Cryptodire

Pleurodire

Proganochelid

Figure 17-1.   Three phases in the evolution of the turtles. The first turtles were the proganochelids of Triassic age, here represented by *Proganochelys*. In these turtles the head could not be withdrawn into the shell. The pleurodires, or side-neck turtles, and the cryptodires, or vertical-neck turtles, followed separate paths of evolutionary radiation during late Mesozoic and Cenozoic times. Prepared by Lois M. Darling.

forms still retained teeth on the palate. Also, the more advanced turtles developed the power to retract the head, legs, and the tail within the shell, which may not have been possible in *Proganochelys*.

The turtles are probably direct descendants of the captorhinomorphs, and through their evolutionary history they have tended to retain the solid skull roof of the primitive reptiles, though with a reduced complement of bones. In many advanced turtles there have been secondary emarginations and reduction of the skull roof. In other respects the turtles

show various adaptations that have served them well in their long struggle for survival. The jaws early became devoid of teeth, as mentioned, and covered with a horny beak. This beak makes a strong shearing mechanism, as we know, equally effective for eating meat or plants. The limbs are heavy, and in the land-living forms the feet are short, with the bones of the toes reduced in number from the primitive reptilian condition. In the marine turtles the feet have become modified as large, flipperlike paddles.

But the truly characteristic specialization in

the turtles is the development of the shell. In these reptiles the ribs, by a process of differential growth, have enveloped the girdles and upper limbs segments to support the protective bony carapace. On the ventral surface the bony plastron has been developed. Both carapace and plastron are covered with horny sheaths, and are connected to each other along the sides. In this fashion the turtles have evolved into completely armored reptiles. Turtles are frequently cited as examples of slowness and general stupidity in the animal world, yet their adaptations for heavy protection at the expense of mobility have stood the test of time, for these are among the most ancient of the existing tetrapods.

Turtles beyond the Triassic proganochelyids belong to two suborders of turtles evolved during late Triassic to late Jurassic times, to continue until the present day. One of these, the pleurodires or side-neck turtles, is characterized, among other things, by the fact that the neck is bent laterally when the head is pulled into the shell. The pleurodires were rather widely distributed in Cretaceous and early Tertiary times, but during the later phases of their evolutionary history they have been restricted primarily to the southern hemisphere. For example, *Podocnemis* is a pleurodire, now living in South America and Madagascar, that during Cretaceous times ranged through many continental regions. At the present time pleurodires inhabit South America, Africa, southern Asia, and Australia.

By far the most numerous and succesful of the turtles have been the cryptodires or vertical-neck turtles, which today are found throughout the world. In these turtles the head is withdrawn into the shell by a vertical, S-shaped flexure of the neck, and the cervical vertebrae are highly specialized to accomplish this. These turtles have been prominent in tetrapod faunas since Cretaceous times, and have evolved along many lines of adaptive radiation. Thus there are cryptodires that live in rivers and marshes, whereas others are suited for life on dry land. Some live in forests, others on plains and in deserts. Some turtles, spend their lives in the sea, coming on land only for the purposes of breeding. Some are very small; others, like the modern Galapagos tortoises, *Testudo*, have attained very great size. Some are completely carnivorous, some completely herbivorous, and others omnivorous. By reason of their structural adaptations and the wide range of their adjustments to the changing environments of the world, the turtles have endured for many millions of years; and, if man does not disturb them completely, it looks as if they will be on the earth for a long time to come.

CROCODILIANS

In looking at a crocodile or an alligator we are in a way looking back into the Age of Dinosaurs. Frequently the largest and most aggressive of modern reptiles and the close relatives of the dinosaurs, they give us a faint idea of what the dominant reptiles of the Mesozoic era may have been like in the flesh. Furthermore, we are fortunate in being able to trace the history of the crocodilians in detail back through the days when dinosaurs ruled the lands, even to the time when dinosaurs were in the early stages of their evolutionary history. From their beginnings, the crocodilians have primarily been denizens of rivers and lakes, and even of the oceans—environments in which their bones have been abundantly preserved and fossilized through the ages. Consequently, like the turtles, of which many have been aquatic, the fossil record of the crocodilians is a relatively full one.

The earliest crocodilians belong to the late Triassic to early Jurassic order Protosuchia. Members of this group have skeletal structure indicating they were terrestrial predators rather than aquatic ones as are most crocodilians. Protosuchians are known from North and South America, East Asia, Europe, and South Africa.

One of the best-known early crocodilians is the genus *Protosuchus*, found in Arizona in rocks of early Jurassic age. *Protosuchus* was a moderately small reptile, some three or four feet in length, and obviously of thecodont ancestry. It was quadrupedal in pose, but the hind limbs were considerably larger than the forelimbs, indicating the descent of this animal from bipedal ancestors. The legs were strong, and evidently *Protosuchus* was rather well adapted for running around on the land. It may have been something of a swimming animal too, although there is no direct evidence to prove this, beyond certain obvious relationships with the later crocodilians. Its archosaurian skull had certain crocodilian characters, notably a flattening of the skull roof, a reduction in size of the upper temporal opening, and the suppression of the preorbital fenestra. The teeth were sharply pointed, indicating that this reptile was carnivorous. In the front foot the proximal bones of the wrist were elongated, as in all the crocodilians, whereas in the hind foot there was a large calcaneum and astragalus, again a crocodilian trait. Of particular interest is the fact that the proximal end of the elongated pubis did not form any portion of the acetabulum, this being a character peculiar to the crocodilians. Along the midline of the back was a double row of bony, protective scutes.

Protosuchus makes a suitable ancestor for the later crocodilians. With the beginning of Jurassic times the crocodiles quickly occupied the ecological niche formerly held by the phytosaurs and have continued in this position through successive ages and faunas down to the present day. The adaptations of the crocodilians have involved phylogenetic growth of the body, so that these reptiles range from medium-sized animals to veritable giants, and the development of specializations for swimming and for an aggressive, carnivorous mode of feeding. Thus the crocodilians as a group have been characterized through their history by strong and frequently elongated jaws equipped with sharp teeth, by rather short but strong legs terminating in broad, webbed feet, by a long, deep, powerfully muscled tail making a strong propeller for swimming, and by

heavy armor over the back and the sides, in the form of bony scutes, covered by horny external plates.

The first crocodilians to evolve from protosuchian ancestors were the mesosuchians, the most numerous and generally dominant crocodilians of Mesozoic times, living abundantly from early Jurassic times to the end of the Cretaceous period, and straggling into the Tertiary period. The mesosuchians differed from the protosuchians and from modern crocodilians in that the upper temporal opening was very large, perhaps as an adaptation for the accommodation of particularly strong temporal muscles. The external nostrils, as in all crocodilians, were at the tip of the snout, whereas the internal nares were located at the posterior borders of the palatine bones. Between these two openings the nasal passage was separated from the mouth cavity by a median junction of the premaxillary, maxillary, and palatine bones, forming a bony tube to enclose the nasal passage.

Since the early and middle portion of the Jurassic period was a time of great marine inundations, when lands were restricted, it is not surprising to find that many of the early crocodilians were well adapted for a life either along the shores of the seas or entirely in the open oceans. Such are the genera *Teleosaurus* and *Steneosaurus*. One family of mesosuchian ancestry, the Metriorhynchidae or geosaurs, became highly adapted for marine life. They lost the heavy bony armor, the limbs were modified into paddles, and the vertebrae of the tail were abruptly downturned to form a reversed heterocercal tail fin. In spite of their high specializations this group of marine crocodilians died out during early Cretaceous times.

Another line of mesosuchians is represented by a number of genera, among which are *Sebecus* and *Baurusuchus*. Although both these forms are found in South America, *Sebecus* in the early Tertiary of Patagonia and *Baurusuchus* in the Cretaceous of Brazil, it is probable that the sebecosuchians were of world-wide dis-

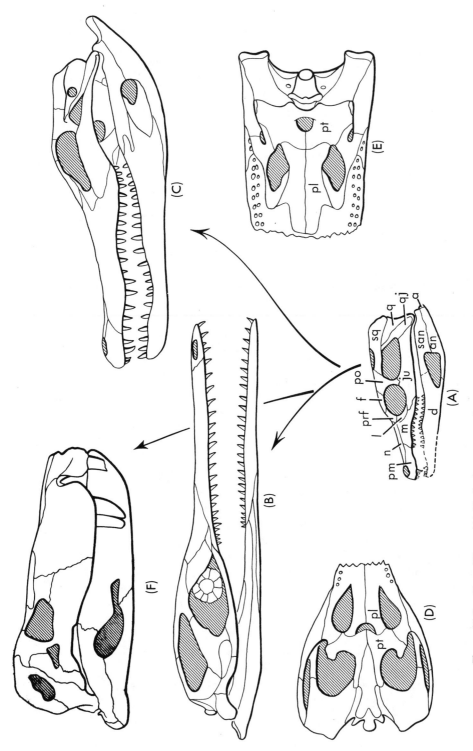

Figure 17-2. Crocodilians. **(A)** Skull of *Protosuchus*, one of the ancestral crocodilians, one-third natural size. **(B)** Skull of *Geosaurus*, a marine crocodilian of Jurassic age, one-fourth natural size. **(C)** Skull of *Alligator*, a Cenozoic crocodilian, one-fifth natural size. **(D)** Portion of the palate in *Steneosaurus*, a Jurassic mesosuchian, showing the internal nares bordered by the palatine and pterygoid bones. **(E)** Portion of the palate in *Alligator*, showing the internal nares completely enclosed by the pterygoid bones, as is characteristic of the eusuchian crocodile. **(F)** Skull of *Baurusuchus*, a Cretaceous sebecosuchian, one-eighth natural size. For abbreviations, see pages 447–449.

tribution. These crocodilians differed from other members of the order largely by reason of the fact that the skull was very much compressed from side to side and rather deep, and the teeth were like-wise laterally compressed. In *Baurusuchus* the teeth were much reduced in number, and some of the anterior teeth were greatly enlarged, like canines, giving to the skull an appearance superficially similar to that of a mammal-like reptile.

The more familiar line of advanced crocodilians is the line of the eusuchians, the crocodiles of the modern world. Eusuchians made their appearance in Cretaceous times, and evolved rapidly to displace their ancestors, the mesosuchians. In the eusuchian crocodiles the internal nares have moved far back in the palate so that they are completely enclosed by the pterygoid bones. This position causes the nasal passage to traverse a long tube from the external nostrils to the back of the throat, completely separated from the mouth cavity. In modern crocodiles there is a special flap on the back of the tongue, which together with a fold of skin from the back of the palate can separate the respiratory passage from the mouth, a useful adaptation for an aquatic animal.

Since the beginning of the Cretaceous period the eusuchians have evolved along three lines, characterized by the crocodiles, the gavials, and the alligators. The crocodiles are narrow-snouted crocodilians with a wide distribution at the present time in the tropical and subtropical zones of the earth. The alligators are wide-snouted crocodilians, today characteristic of North and South America, with one species in China. The gavials are very narrow-snouted crocodilians, now living in India.

During Cretaceous and Cenozoic times the eusuchians were much more widely distributed than they are today, an indication of the greater extent of tropical and subtropical conditions in past ages as compared with the present. Perhaps the culmination of eusuchian evolution was attained by the late Cretaceous genus, *Deinosuchus*, the largest of all croco-

diles, known from fossils collected along the Rio Grande River in Texas. The huge skull was six feet in length and broad in proportion, making this reptile the possessor of some of the largest and most powerful jaws of all late Cretaceous reptiles. *Deinosuchus* must have had a total length of forty or fifty feet, and it is quite likely that this giant crocodile preyed upon the dinosaurs with which it was contemporaneous.

The largest modern crocodilians reach lengths of twenty feet or more, but such dimensions are rare. Even a ten-foot or a twelve-foot alligator or crocodile is a formidable animal, well able to hold its own in this modern world against all adversaries except man. Unfortunately, the great crocodilians, the impressive remnants of the Age of Dinosaurs, are being constantly reduced in numbers and range by modern firearms and human encroachment into their habitats.

SPHENODONTANS

On a few islands near the coast of New Zealand lives *Sphenodon*, also known as the tuatara, the sole surviving sphenodontan. Before the coming of white men *Sphenodon* inhabited much of New Zealand, but with the advance of modern civilization it disappeared at an alarming rate and was on the verge of extinction. Happily the tuatara is now rigidly protected. This reptile, about two feet in lenth, looks very much like a large lizard; but, unlike any lizard, it has large, fully-formed, upper and lower temporal fenestrae. The teeth are fused to the jaws rather than being in sockets. In *Sphenodon*, as in all sphenodontans, the front of the skull forms a sort of overhanging "beak," set with teeth.

The sphenodontans made their appearance near the end of the Triassic period. They probably were derived from eosuchians, primitive lepidosauromorph diapsids. According to the fossil record, the Triassic and Jurassic was the time of greatest deployment and variety among sphenodontans.

After the late Jurassic, the sphenodontans were very much restricted, and it appears that they continued as a limited evolutionary line from that time until the present day. The genus *Homoeosaurus*, of late Jurassic age, is close to the modern *Sphenodon*, indicating that there has been little change among the sphenodontans over a lapse of more than one hundred million years.

LIZARDS AND THE SNAKES

The lizards and the snakes are by far the most numerous and diverse of modern reptiles. Whereas there is a single species of sphenodontan, about twenty-five species of living crocodilians, and some four hundred species of turtles, there are probably about thirty-eight hundred species of lizards and three thousand species of snakes. Moreover, it seems likely that this great preponderance of the Squamata among the reptiles has pertained since the end of the Age of Dinosaurs.

In lizards, only the upper temporal opening, above the postorbital-squamosal bar, is fully developed. Below this bar the cheek region is open, since there is no lower bar formed by the quadratojugal between the jugal and quadrate bones, as in other reptiles. It appears, therefore, that the lizards (like the distantly related euryapsids) are modified diapsids—that the open temporal region below the postorbital-squamosal bar is, in fact, the lower diapsid opening, no longer enclosed below by a bony bar. Since the lower bar is lacking in these rep-

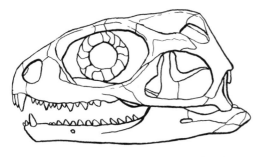

Figure 17-3. Skull of the modern *Sphenodon*.

tiles, the quadrate bone is free at its lower end, and this freedom gives great mobility to the joint between the skull and the lower jaw. In addition to the quadratojugal, certain other skull bones generally are absent in the lizards, notably the lacrimal, the postparietal, and probably the tabular. In contrast to these specializations in the bones of the skull, the lizards commonly retain the pineal opening, which is a primitive character among tetrapods. The teeth in lizards are not set in sockets, but are fused to the edges or to the inner sides of the jaws.

Lizards are quadrupedal reptiles in which the fourth toe of the hind foot is commonly elongated. These reptiles, when running, usually give a strong lateral push with the long, outer side of the foot, and this is the chief propulsive force. Some lizards, such as the collared lizard *Crotaphytus* of western North America, rise up on the hind legs when running and, thus, they parallel many of the extinct bipedal archosaurians in this mode of locomotion.

The beginnings of the Squamata go back to Permo-Triassic times, particularly to forms that foreshadow the basic specializations that were to characterize the lizards of later ages. The earliest lizards, such as *Paliguana* and *Saurosternon* from the upper Permian or lower Triassic of South Africa, combine the features that would be expected in primitive lizards, notably small size, a slender skeleton, and an open cheek region with a movable quadrate bone. Other characters of the skull and skeleton, such as the adaptation of the teeth for an insectivorous diet, the development of the ear, the form of the shoulder girdle, and the general development of the limbs and feet, are similar to those typical of the lizards.

It would seem that these forms, known collectively as paliguanids, probably were ancestral to two lizards of late Triassic age, *Kuehneosaurus* from Britain and *Icarosaurus* from North America. These two genera, still retaining primitive characters in the skull such as a prominent lacrimal bone and numerous

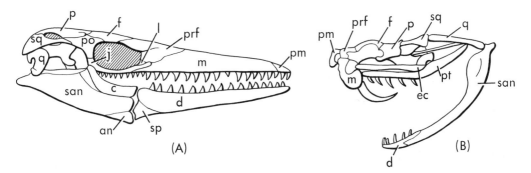

Figure 17-4. The skull in the Squamata. **(A)** *Tylosaurus*, a mosasaur or giant marine lizard. This skull is about three feet in length. Notice the free distal end of the quadrate bone, and the well-defined joint in the middle of the lower jaw. **(B)** *Crotalus*, a rattlesnake. Notice the reduction of bone in the skull, and the long, movable quadrate. A forward movement of the lower end of the quadrate in the snake increases the gape of the jaws. In the rattlesnake, and many other poisonous snakes, a special muscle pulls the pterygoid bone forward and rotates the maxilla, thus erecting the large poison-carrying fang for striking. For abbreviations, see pages 447–449.

teeth on the palate (which are not found in all later lizards), were nevertheless highly specialized by the great elongation of the ribs in the thoracic region of the body. As mentioned in Chapter 14, the only logical explanation for this remarkable feature is that these ribs supported in life a membrane that enabled the animals to glide from tree to tree. Thus, it appears that very special and truly unexpected adaptations marked the initial stages of squamate evolution.

From this beginning in late Permian and Triassic times, the lizards became established during the Jurassic period to evolve along the multitudinous lines of adaptive radiation that mark their evolution through middle and late Mesozoic and Cenozoic history. Today lizards share with their cousins, the snakes, the distinction of being the most successful of modern reptiles if range of adaptations, the multiplication of species, geographic distribution, and the abundance of individuals are criteria of success.

Lizards in the modern world vary in size from tiny animals only a few inches in length to the large monitors, *Varanus*, some of which attain lengths of six or eight feet. These reptiles are too numerous for detailed consideration in a book such as this. A major paleontological interest, perhaps, is in the fact that

during Cretaceous times certain varanid lizards became adapted to life in the sea and grew to gigantic size, reaching lengths of thirty feet or more. These were the mosasaurs, already described briefly in Chapter 15.

The snakes, the last of all major groups of reptiles to evolve, are in essence highly modified lizards. Unfortunately the fossil history of the snakes is very fragmentary, so that it is necessary to infer much of their evolution from the comparative anatomy of modern forms.

In the snakes the limbs are lost, and locomotion is accomplished by various movements of the body—by sinuous side-to-side movements, by a straightforward motion in which the body muscles acting in rhythmic waves pull the snake along, and even by a sort of corkscrew looping motion, the so-called sidewinding of some western American rattlesnakes. The body and tail are greatly elongated by an enormous multiplication of vertebrae, and of course with such an elongated body shape there have been numerous adaptations in the form and arrangement of the internal organs. No less remarkable than the adaptations of the body are those of the skull, which has become a highly kinetic structure, with a long, movable quadrate bone. This gives a sort of double joint to the back of the jaw. In addition, the premaxillae in the skull and the anterior

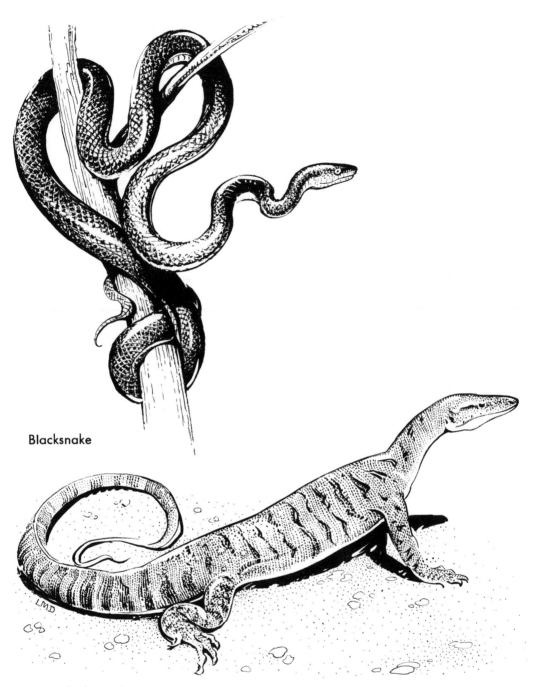

Blacksnake

Tree–climbing Goanna

Figure 17-5. Modern squamates: a snake and a lizard. Prepared by Lois M. Darling.

ends of the dentaries in the lower jaw are not coalesced, but loosely connected by ligamental bonds, so that they can be spread apart to a considerable degree. When the snake kills its prey, it engulfs the victim whole, spreading the jaws to swallow an animal often considerably larger in diameter than the snake itself.

The snakes are remarkable by reason of the development of highly toxic poisons in many genera, such as the Old World vipers (*Vipera*) and cobras (*Naja*) and the New World rattlesnakes (*Crotalus*), a late phase of snake evolution possibly not appearing before Miocene times. It is because of this characteristic that snakes are held in such great dread by most people all over the world.

At about the time that snakes arose from their lizard ancestors the dinosaurs were in the final stages of their evolutionary history. Then the dinosaurs and the other dominant reptiles of the Mesozoic became extinct, and reptiles were forced to compete with highly efficient and very active tetrapods, the birds and mammals. This they did, and in so doing they succeeded by diverse means. The turtles continued through the Age of Mammals because of their highly developed protective shells, making them extraordinarily invulnerable to attack. The crocodilians persisted by living in tropical environments, where as aggressive beasts of prey they were relatively immune to attack from all except the largest and boldest of the mammals. The sphenodontans continued by retiring to an isolated part of the world, where there were no mammals to bother them. The lizards succeeded largely by following a life spent in the protection of dense foliage and rocks, where they could escape from their enemies. And the snakes succeeded and even perfected their adaptations during the age of mammalian supremacy, in part by becoming secretive animals, living in foliage and among rocks, burrowing under the ground, and retreating to the water, and in part by developing adaptations that made many of them the most dreaded of vertebrates, the carriers of violent death by poison.

Mesozoic Mammals

Beginning of the Mammals

ORIGIN OF THE MAMMALS

The first mammals, descendants of some of the mammal-like reptiles, appeared in Triassic times. These earliest mammals of the late Triassic and Jurassic periods, known for the most part from a few skeletons and from a considerable number of skulls, jaws and teeth, were very small and comparatively insignificant, and all through the remainder of the Mesozoic era the mammals continued as small and minor members of the Jurassic and Cretaceous faunas, completely overshadowed by the numerous and ubiquitous reptiles that populated the lands and waters of the earth. Yet in spite of the littleness of the Mesozoic mammals, theirs was an especially important contribution to the evolutionary history of life, for it was during middle and late Mesozoic times that the mammals were going through the initial stages of their development to establish the basic mammalian stocks from which arose the tremendous variety of mammals that lived during the Cenozoic era.

As we have seen in Chapter 9 of this book, certain mammal-like reptiles evolved very far along lines leading to the mammals. Several groups of theriodonts had advanced toward the mammalian stage, and with some of these highly evolved reptiles it is truly a matter of definition, based on a few characters, whether they should be retained within the reptile class or regarded as ancestral mammals. Certainly it was but a short step from these animals to the first undoubted mammals.

It is not easy to determine the precise line of mammalian ancestors among the theriodont reptiles. Some theriodonts were far advanced toward the mammals in certain characters, but comparatively primitive in others; and among all the theriodonts the mixtures of advanced and conservative characters are so various that it is difficult to point to any one particular group and define it as progressing most positively in the direction of the mammals. It seems reasonably certain, however, that the ancestry of the mammals should be sought among the more advanced theriodonts, particularly among the cynodonts.

In past years there has been a great deal of argument as to whether the mammals had a monophyletic or a polyphyletic origin—that is, whether they were ultimately descended from a single group of mammal-like reptiles, or whether they had their origins from several such reptilian ancestors. In recent years many students of this problem have tended to favor the monophyletic origin of the mammals, with the cynodont genus, *Probainognathus*, selected as representative of what the ultimate mammalian ancestor may have been like. This concept is based upon the evidence of numerous fossils collected in recent years, and as Crompton and Jenkins have shown, logically replaces the polyphyletic theory for mammalian origins which was based upon limited fossil materials.

Probainognathus, a small cynodont reptile from the Triassic sediments of Argentina, shows characters in the skull and jaws far advanced toward the mammalian condition. Thus it had teeth differentiated into incisors, a canine and postcanines, a double occipital condyle and a well developed secondary palate, all features typical of the mammals, but most significantly the articulation between the skull and the lower jaw was on the very threshold between the reptilian and mammalian condition. The two bones forming the articulation between skull and mandible in the reptiles, the quadrate and articular respectively, were still present but were very small, and loosely joined to the bones that constitute the mammalian joint, namely the squamosal and dentary respectively. Furthermore it seems evident that these two latter bones participated in the jaw articulation, as did the quadrate and articular bones. Therefore in *Probainognathus*, there was a double articulation between skull and jaw, and of particular interest, the quadrate bone, so small and so loosely joined to the squamosal, was intimately articulated with the stapes bone of the middle ear. It quite obviously was well on the way toward being the

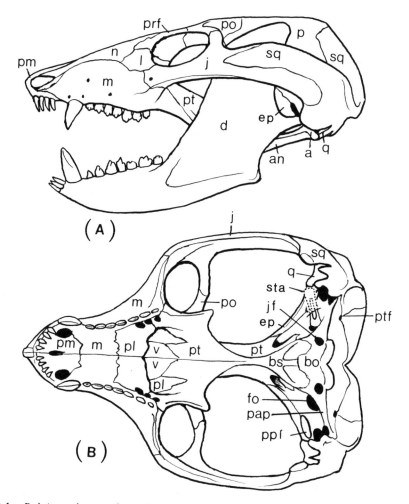

Figure 18-1. *Probainognathus*, an advanced cynodont reptile from the Middle Triassic of South America. **(A)** Lateral view of skull and jaw. **(B)** Palatal view of skull. Natural size. For abbreviations, see pages 447–449.

incus bone of the three bone complex that characterizes the mammalian middle ear.

As for the postcranial skeleton, other cynodonts closely related to *Probainognathus* show various features prophetic of the mammalian skeleton. In the genera *Thrinaxodon* and *Cynognathus*, for example, the vertebral column was distinctly differentiated into cervical, thoracic and lumbar vertebrae, thus delineating the three regions of the backbone in front of the pelvis so characteristic of the mammals. Although the cervical ribs were still defined in such cynodonts they were very short and

might well have been antecedent to the mammalian condition, in which the cervical ribs have become fused to become integral parts of the vertebrae. The lumbar ribs, too, were very short; indeed in *Thrinaxodon* they were in the form of small, flat plates, instead of being elongated ribs. Such a distinct lumbar region in these mammal-like reptiles suggests that there was a diaphragm, a diagnostic mammalian feature that would seem possibly to have become established before the mammalian condition was reached.

In these cynodont reptiles there was an out-turned ridge along the front of the scapula or shoulder blade, a structure quite clearly ancestral to the scapular spine of the mammals. And in the pelvis the ilium was very much enlarged, pointing to the expanded ilium of the mammals. Although the limb bones of these cynodonts retained certain primitive characters, the ulna of the lower front limb probably had an expanded cartilaginous olecranon process, "virtually necessitated by . . . evidence for well-developed triceps musculature." (Jenkins, 1971). The upper bone of the hind limb, the femur, was characterized by its upper end being set at a considerable angle to the shaft of the bone, so that the articulation of the head of the femur within the socket of the pelvis brought the hind limb into a position parallel to and beneath the vertebral column, rather than sprawling to the side as in primitive reptiles. Finally, the feet in these cynodonts were trending toward the mammalian condition, with well-integrated wrist and ankle bones, there being a projecting tuber on the calcaneum or heel bone for the strong attachment of muscles, thereby giving power to the thrust of the hindlimb. Finally some of the advanced cynodonts show a mammalian arrangement of the foot bones, with two phalanges in the first digit and three in the other four digits, an advance beyond the reptilian phalangeal formula.

A cynodont similar to *Probainognathus*, if not this particular genus, might very well have been directly ancestral to the Triassic mammals, *Megazostrodon*, *Erythrotherium* and *Eozostrodon*. *Megazostrodon*, most definitely a mammal, as shown by the characters of its skull and jaws and especially the jaw articulation, by its differentiated dentition, by its mammalian vertebral column with well distinguished neck, thoracic and lumbar regions, by the structure of the limb girdles, and by the mammalian

limbs and feet, was a very small animal, about ten centimeters in length. It is typical in this respect of other Triassic mammals—an indication that these first descendants from cynodont ancestors probably lived furtive and crepuscular lives and thus were able to survive in a world dominated by active and aggresive reptiles. This was to be the pattern for survival among the mammals through more than one hundred million years of Mesozoic history, when dinosaurs ruled the continents.

ESTABLISHMENT OF MAMMALIAN CHARACTERS

The diagnostic characters of the mammals are various. Briefly, modern mammals are active tetrapods in which the body temperature is generally comparatively uniform and the basic metabolism high, for which reason they are often referred to as "warm-blooded" animals. There is typically a protecting and insulating covering of hair. The young are usually born alive (the monotremes lay eggs) and are nourished during the early stages of their life on milk supplied by the mother.

As for the hard parts, which are of particular concern to the student of fossils, there are various diagnostic characters. Thus the mammals have a double occipital condyle forming the articulation for the skull upon the first vertebra of the neck, there is a secondary hard palate of bone separating the nasal passage from the mouth, and the external nasal opening is a single orifice in the front of the skull. Furthermore, the joint between the skull and the lower jaw is formed by the squamosal and dentary bones, respectively. The quadrate and articular bones, the articulating elements between the skull and jaw in the reptile, have in the mammal retreated into the middle ear to become transformed into two of the three ear

Figure 18-2. Evolution of the bones in the mammalian middle ear from certain bones involved with the reptilian jaw articulation, and from the reptilian stapes. (**A**) and (**B**) *Lycaenops*, a mammal-like reptile from the Upper Permian of South Africa. (**C**) and (**D**) *Diarthrognathus*, an advanced mammal-like reptile from the Upper Triassic of South Africa. (**E**), (**F**) and (**G**) *Didelphis*, the modern North American opossum.

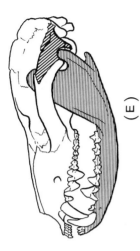

(A)

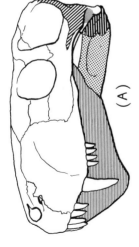

(B)

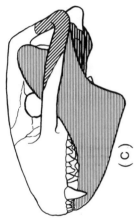

(C)

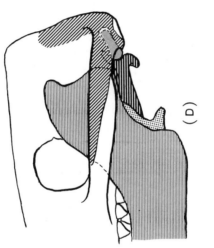

(D)

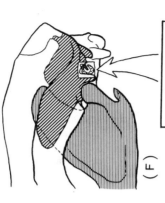

(E)

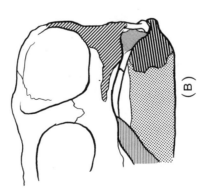

(F)

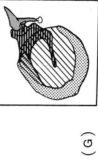

(G)

— Angular

— Squamosal

— Quadrate

— Dentary

— Articular

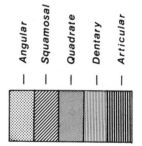

ossicles, the incus and malleus, respectively, that together with the stapes (inherited from the reptilian stapes) make a chain to transmit vibrations from the eardrum to the inner ear. This is one of the most remarkable transformations of anatomical structures from one function to another in the history of vertebrate evolution. Of particular importance is the comparatively large braincase in all but the very primitive mammals, a reflection of the growth of the brain and the greatly increased intelligence in these tetrapods. Continuing, the mammals have a differentiated dentition of incisors, canines, and cheek teeth, of which the latter generally have a crown consisting of several cusps, held in the jaw by two or more roots. The mammals have the ribs of the neck permanently fused to the vertebrae, to form integral parts of these bones, whereas in the back part of the body the lumbar vertebrae are free of ribs. In the mammals there is a strong spine running down the middle of the shoulder blade or scapula. The bones of the pelvis, the ilium, ischium, and pubis are fused to form a single bony structure. Finally, the toe bones are reduced in number so that there are but three phalanges in each digit, except for the first one, in which there are two phalanges. Other characters define the mammalian skeleton, but these are some of the more important and obvious ones.

A comparison of these typically mammal characters with those of advanced mammal-like reptiles can be outlined somewhat as follows:

Mammal-Like Reptiles	Mammal
Double occipital condyles (cynodonts, ictidosaurs, tritylodonts)	Double occiptal condyles
Secondary palate (except in primitive forms)	Secondary palate
Separate external nares (except in tritylodonts and ictidosaurs)	Single external narial orifice
Enlarged dentary	Dentary alone forming the lower jaw
Quadrate-articular joint (plus squamosal dentary joint in ictidosaurs)	Squamosal-dentary joint
Reduced quadrate-articular	Incus-malleus
Restricted braincase	Enlarged braincase
Differentiated dentition	Differentiated dentition
Separate cervical ribs	Fused cervical ribs
Lumbar ribs	Free lumbars
Front border of scapula turned out (cynodonts)	Scapular spine
Pelvic elements separate	Pelvic elements fused
Phalangeal formula 2-3-3-3-3 (in some therocephalians)	Phalangeal formula 2-3-3-3-3

We may only infer some of the other changes that took place in the transition from reptile to mammal. The generally constant body temperature and the high level of activity in the mammals are correlated with the four-chambered heart in these tetrapods, in which the arterial blood is kept entirely separated from the venous blood. Perhaps this stage had been attained in some of the theriodonts; certainly we may suppose that it was present in the primitive Triassic mammals. It seems probable that the insulating coat of hair evolved along with the elevated body temperature in the mammals, to protect them from the cold and from the heat as well. The mammals are characterized by a diaphragm, which separates the thoracic portion of the body cavity from the abdominal region, and assists in drawing air into the lungs and forcing it out. No such structure exists in modern reptiles, and it is reasonable to suppose that the diaphragm developed as a new device that made possible a large degree of oxygen intake for active animals, and that this took place at a late stage during the transition from reptile to mammal. It is, however, well within the realm of probability that some of the advanced theriodonts possessed hair and a diaphragm. As to the development of new methods of reproduction, it seems likely that these appeared after mammals were firmly established on the earth. The modern monotremes lay eggs, al-

Figure 18-3.   Some contrasting characters of reptiles and mammals.

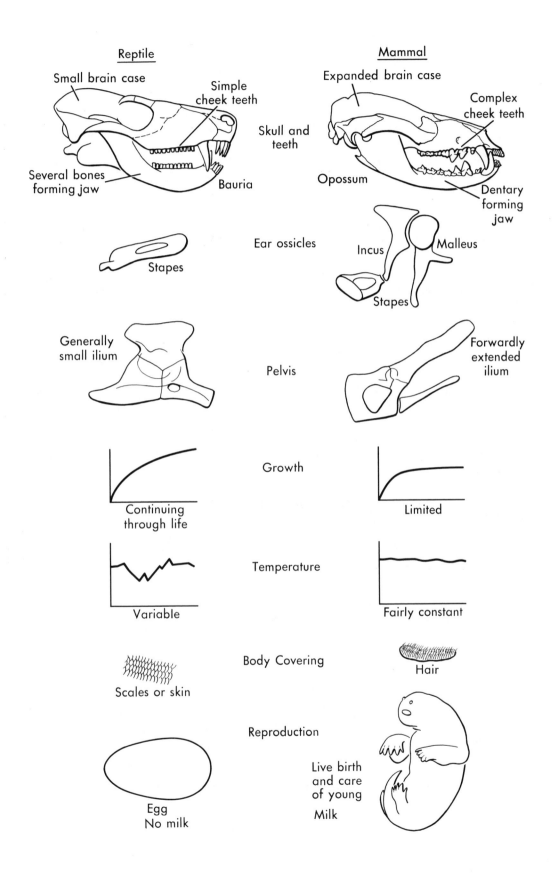

Reptile   Mammal

Small brain case   Expanded brain case
                    Simple   Complex
                    cheek teeth   cheek teeth

                              Skull and
                              teeth

Several bones   Opossum
forming jaw   Bauria   Dentary
                                                                        forming
                                                                        jaw

                              Ear ossicles   Incus   Malleus
        Stapes
                                                         Stapes

Generally   Forwardly
small ilium   Pelvis   extended
                                                               ilium

                              Growth
Continuing   Limited
through life

                              Temperature
Variable   Fairly constant

                              Body Covering   Hair
Scales or skin

                              Reproduction
                                        Live birth
                                        and care
                                        of young
Egg   Milk
No milk

though they suckle their young with milk that is secreted from modified sweat glands on the under surface of the body. Perhaps this was the stage of reproductive development in many of the primitive mammals of middle Mesozoic times.

The marsupials or pouched mammals, in which the young are born alive as tiny, larval animals, to be nurtured for some weeks in the mother's pouch until they have reached a stage of independence, are first found in upper Cretaceous sediments. Likewise the placental mammals, in which the young are born in a comparatively advanced stage of development, first appear in strata of late Cretaceous age. It is therefore evident that these highly developed mammals became established during Cretaceous times.

TRIASSIC AND JURASSIC MAMMALS

Five orders of mammals are known from sediments of late Triassic and Jurassic age, and are represented by fossils discovered in Eurasia, North America, and Africa. These are the Docodonta, the Triconodonta, the Symmetrodonta, the Eupantotheria, and the Multituberculata. These orders of very ancient mammals, some of which seem to be quite unrelated to the others, are an indication that the transition from the reptilian to the mammalian stage of evolutionary development most probably took place along a broad front of adaptive radiation. The interrelationships

of these orders of primitive mammals have been and still are a subject of much dispute—a fact that must be kept in mind during the discussion that follows.

Among the earliest of the mammals are triconodonts known as morganucodonts, represented by the genus *Morganucodon* (which probably is the same as an earlier-named genus, *Eozostrodon*) from the upper Triassic of Europe and by the genus *Megazostrodon* from the Upper Triassic of South Africa. Originally, these forms were known from scattered teeth and a few jaw fragments but, within recent years, fossil bones in large numbers have been recovered from Triassic fissure fillings within Carboniferous limestones in South Wales. Skulls with associated skeletons are known from South Africa. In addition, some closely related materials have been found in southwestern China.

The morganucodonts were tiny mammals with slender lower jaws. In these early mammals the lower jaw was of mammalian form, and functionally was composed of a single bone, the dentary, at the back of which there was a large and high coronoid process for the attachment of strong temporal muscles, and a well-formed condyle for articulation with the squamosal bone of the skull. But, significantly, on the inner side of the jaw was a groove within which was preserved the remnant of the articular bone—a sort of paleontological reminder of the old reptilian jaw

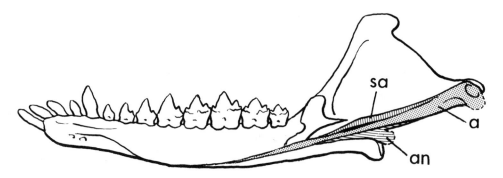

Figure 18-4.   The lower jaw of *Morganucodon* a Jurassic mammal. Note the reptilian bones still retained. Abbreviations: a, articular; an, angular; and sa, surangular. Lingual view, about five times natural size.

joint, still preserved in this ancient mammal. Since the quadrate bone of the skull was also preserved, obviously these animals possessed both jaw joints, as did various advanced mammal-like reptiles, some of which are described in Chapter 9. Here we see examples of the gradual transition from reptile to mammal. *Diarthrognathus*, from the upper Triassic of South Africa, is on the reptilian side of the line because, although it had the double jaw joint, the quadrate-articular articulation was still dominant. *Morganucodon*, from the upper Triassic of Europe and Asia is on the mammalian side of the line because, although it too had the elements of both articulations, the squamosal-dentary joint was the dominant one.

The morganucodonts had a mammalian type of dentition, with small incisor teeth, a single large, sharp canine, and postcanine premolars and molars, which were multicuspidate teeth with double roots. The cusps of the teeth were arranged more or less linearally along the median axis of the tooth.

The triconodonts, which together with the docodonts may be placed in the Subclass Eotheria, ranged in size from tiny beasts, no larger than mice, to animals like *Triconodon* of late Jurassic age, that were as large as cats. The rather long jaws had a considerable array of differentiated teeth— as many as four incisors, a canine, and nine postcanine teeth. In the development of the dentition, the triconodonts illustrate very nicely the specializations of tooth formula and form that were attained by even the earliest and most primitive of the mammals over the mammal-like reptiles. In the mammal-like reptiles, for example, the postcanine teeth (commonly designed as cheek teeth) either were all alike or formed a series that gradually increased in complexity from front to back. In the earliest mammals, on the other hand, the postcanine teeth are clearly divisible into anterior premolars and posterior molars, the latter being more complex in structure than the former. The differentiation of the teeth into incisors, canines,

premolars, and molars constitutes one of the very important characters of mammals that is first seen in the primitive Triassic and Jurassic forms and is carried through from them to the most specialized members of the class. No mammals have had more than a single canine in the tooth series, but the other teeth have varied in number. A triconodont with nine postcanine teeth had four premolars and five molars. The dental formula in the lower jaw therefore would be indicated as 4-1-4-5, meaning four incisors, one canine, four premolars, and five molars.

The skull and skeleton of the triconodonts showed advances beyond the cynodont condition, as has been mentioned above in the remarks concerning morganucodonts. It is now evident that the braincase was specialized toward the mammalian stage, that the vertebrae were more mammalian than those of the most advanced cynodonts, that the pelvis was distinctly mammalian in form with a forwardly elongated ilium—not unlike what is seen in the pelvis of a modern monotreme such as *Tachyglossus*, that the articular heads of the humerus and of the femur were strongly rounded as in the mammals, that the ankle was advanced, and that the phalanges, showing the mammalian formula, were adapted for grasping and climbing. In contrast, the shoulder girdle had advanced but little beyond the cynodont condition.

All in all it is reasonable to think that these very early mammals were warm-blooded and insectivorous, and that they probably were adept climbers.

The docodonts, possibly descended from morganucodonts, and typified by *Borealestes* and *Docodon* were small, mouse-sized mammals of middle and late Jurassic age, known primarily from jaw fragments and teeth. No postcranial bones have as yet been described. These persistently primitive mammals, which left no descendants, are notable particularly because of the independent evolution of the molar teeth into rather complex structures. It would seem that this development was an ad-

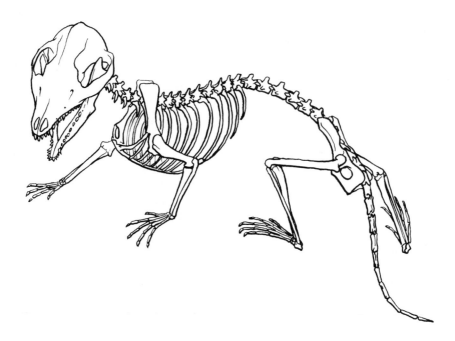

Figure 18-5. *Megazostrodon*, a Lower Jurassic mammal from South Africa. About natural size.

aptation to an omnivorous- frugiverous diet— something of an advance in feeding habits over the probably primitive insectivorous diet of the triconodonts. The jaws of the docodonts were slender, and although the articulation between skull and jaws was of the mammalian squamosal-dentary type, vestiges of the quadrate-articular joint were still retained. These early mammals may be regarded as one of the mamalian "experiments" that developed during Mesozoic time but failed to survive.

The symmetrodonts are so named because in them the molar teeth were composed each of three principal cusps arranged in a rather symmetrical triangle when looked at in crown view. In the upper molar the point of the triangle was on the internal side next to the tongue; in the lower molar the point was external next to the lips. Thus, the triangles sheared past each other when the jaws were brought together. These mammals, typified by the upper Jurassic to lower Cretaceous genus *Spalacotherium*, and known only from teeth and jaw fragments, were of very small size and probably were of insectivorous habits. The earliest symmetrodonts, such as *Kuehneotherium*, of late Triassic or early Jurassic age would seem to show a full complement

Figure 18-6. (A) Cynognathus, a Triassic therapsid reptile, showing the differentiated dentition and the cusped cheek teeth. (B) Priacodon, a Jurassic triconodont, with these characters advanced beyond the therapsid condition. (C–K) Crown views of cheek teeth of early mammals, in each case an upper molar above and a lower molar below. (C) Multituberculate. (D) Triconodont. (E) Morganucodont. (F) Docodont. (G) Symmetrodont. (H) Metatherian (marsupial). (J) Eupantothere. (K) Eutherian (placental). Not to scale. C through F represent various Mesozoic "experiments" in mammalian molar patterns. C (the multituberculates) being successful through a considerable span of geologic time. G (the symmetrodonts) represents an approximation to the molar pattern that later was to be successful through Cenozoic time. J (the eupantotheres) is the beginning of this successful pattern, which led to the molar patterns in H (the marsupials) and K (the placental mammals).

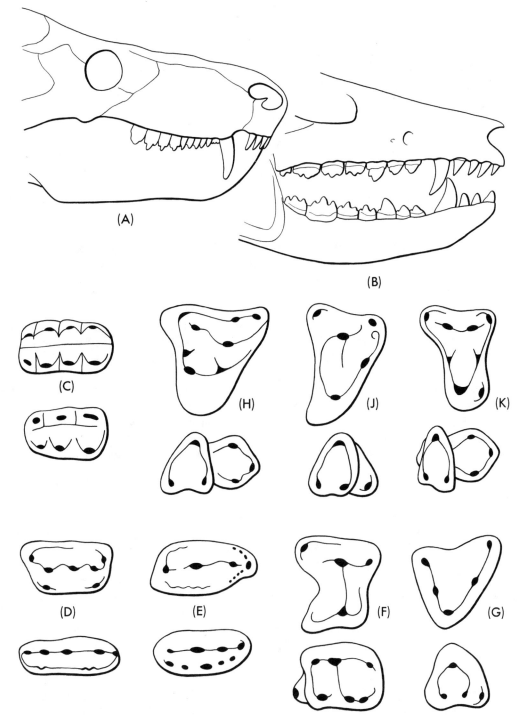

of reptilian bones in the lower jaw, whereas the last of the symmetrodonts, as represented by *Symmetrodontoides* of late Cretaceous age, show the lower jaw consisting of a single bone, the dentary. Thus there would seem to have been a transformation from the reptilian to the mammalian type of jaw articulation during the Mesozoic history of this group of archaic mammals. They represent, like the triconodonts, an "experiment in evolution" that failed to survive beyond the limits of Cretaceous time. They paralleled the more successful early mammals, the eupantotheres, and are commonly included with this latter order in the infraclass Pantotheria.

Of particular importance are the eupantotheres, often referred to as the trituberculates. In these middle and upper Jurassic mammals the jaws were long and slender, and there was a large array of cheek teeth. In *Amphitherium*, for instance, there were four incisors, a canine, four premolars, and seven molars. The upper molars in the eupantotheres were of triangular shape in crown view, with a prominent cusp on the inner apex of the triangle and with various cusps and cuspules on the outer side of the tooth. A row of these molars formed in essence a series of triangles or trigons, with the apices pointing in toward the midline of the mouth. Naturally there were spaces between the teeth along the inner edges of the dental series, and into these spaces there fitted the triangular lower molars, in which the apices of the triangles were directed out toward the cheek. These triangles of the lower molars are known as trigonids. In addition, there was a low heel or talonid, on the back of each lower molar, which received the pointed inner cusp of the upper molar. It can be seen that this arrangement of upper and lower molar teeth formed a rather complicated mechanism for cutting by the shear of the trigons and trigonids past each other, and for grinding by the action of the internal cusp of the upper molar in the talonid like a pestle in a mortar. Such an arrangement and mechanism of the molars, more complex and sophisticated

than the arrangement in the symmetrodonts, is exactly what we find in the primitive representatives of the later mammals, and for this reason it is generally considered that the eupantotheres were the direct ancestors of the marsupial and placental mammals of Cretaceous and Cenozoic times.

It is significant that the only known skeleton of an eupantothere (these early mammals are known almost entirely from teeth and jaw fragments) shows the presence of epipubic bones, as in marsupials. This skeleton is characterized by well-developed claws and a strong tail, which suggest that it was of arboreal habits.

The one other order of Mesozoic mammals was the multituberculates, included within the subclass Allotheria, a quite separate group of highly specialized animals. These would seem to have been the first of the herbivorous mammals, showing as early as Jurassic times various adaptations for eating plants. The skull was heavy, with strong zygomatic arches for the attachment of powerful muscles to move the massive lower jaws. In the front of the skull and in the mandible was a pair of large, elongated incisors, behind which there was a considerable gap. The molars above and below were characterized by longitudinally arranged rows of cusps, paralleling similar adaptations in the cheek teeth of some of the tritylodont reptiles of the Triassic period, previously described. In the primitive multituberculates there were two parallel rows of longitudinal cusps in both upper and lower molars; in the later forms there were three rows of cusps in the upper teeth. The last lower premolar tooth was generally a greatly enlarged shearing blade, with strong vertical ribs on its inner and outer surfaces. In general these adaptations of the skull and dentition in the multituberculates were broadly similar to those seen in the later rodents, and it is reasonable to think that these early mammals lived a type of life that was imitated many millions of years later by the rodents. Indeed, since the multituberculates persisted into early Cenozoic time, as we

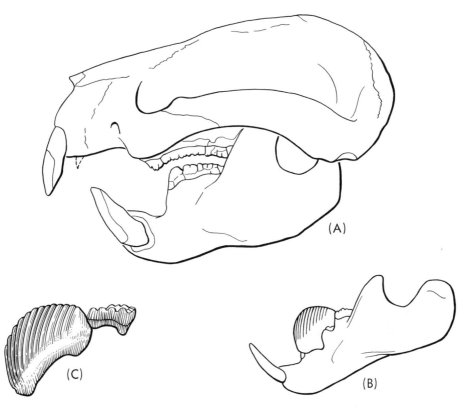

Figure 18-7.   Paleocene multituberculates. **(A)** Skull and jaw of *Taeniolabis*, two-thirds natural size. **(B)** Mandible of *Ptilodus*, somewhat enlarged. **(C)** Last lower premolar and first lower molar of *Ptilodus*. Four times natural size. The premolar was probably used for cutting food, the molar for crushing.

shall see, it is quite possible that their extinction was brought about by competition from the early rodents. The Jurassic multituberculates, typified by the genus *Plagiaulax*, were rather small.

CRETACEOUS MAMMALS

The multituberculates carried on the adaptations that had been established by their Jurassic forebears, living and expanding through the complete span of the Cretaceous period and into the beginning of Cenozoic times. The Cretaceous and early Cenozoic evolution of the multituberculates was marked by certain refinements of the adaptations already outlined for these animals, especially by an increase in size. The culmination of evolution in these mammals was reached after the close of the Cretaceous period in the Paleocene genus *Taeniolabis*, an animal as large as a beaver, with a skull six inches in length and with large, chisel-like teeth, and in certain other forms that survived into early Eocene times.

A few triconodonts, symmetrodonts, and eupantotheres survived into early Cretaceous times. Eupantotheres were mentioned above as being probably ancestral to the marsupial and placental mammals that first appeared in the Cretaceous period. Since these mammals will be described in detail in subsequent chapters they need not be considered here, except to be placed among the Cretaceous mammals. Marsupials related to the modern American opossum are found in upper Cretaceous sediments of North America. Placental insectivores, related to modern shrews and hedgehogs,

Platypus

Echidna

Figure 18-8.   The two monotremes, or egg-laying mammals, of Australia. Prepared by Lois M. Darling.

have been discovered in upper Cretaceous beds of Mongolia and North America, while in this latter continent are found early condylarths, these being archaic hoofed mammals, and the earliest supposed primate, *Purgatorius*.

THE MONOTREMES

It is pertinent at this place to mention briefly the monotremes, constituting the subclass Prototheria, known in the Australian fossil record from a single lower jaw with cross-crested molar teeth, of Cretaceous age and named *Steropodon*; from isolated lower molar teeth of Miocene age, named *Obdurodon;* and from Pleistocene materials. The teeth of *Steropodon* show resemblances to the teeth of *Obdurodon*, and may be compared with the vestigial teeth of the living Australian monotreme, *Ornithorhynchus*.

This last-mentioned genus, known by the popular names of platypus or duckbill, is one of three living monotremes, the other two being the spiny echidnas, or anteaters, *Tachyglossus* of Australia and *Zaglossus* of New Guinea. Superficially these monotremes are highly specialized, the platypus for a life in streams and in underground burrows along the banks, and the anteaters for a hedgehoglike existence in deep forests. In the duckbills the front of the skull and lower jaw are flattened into a ducklike beak for burrowing in the mud of streams in search of worms and grubs. The teeth are shed and replaced by hard pads in the adults. The feet are modified as webbed paddles. The anteaters are protected by sharp spines that cover the body. In them the jaws are toothless and elongated into a long, tubular snout, with which they probe ant hills.

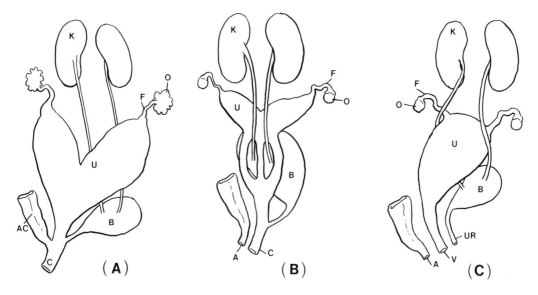

Figure 18-9.   Urogenital systems in **(A)** a modern monotreme, **(B)** a modern marsupial, and **(C)** a modern placental mammal. In the monotreme the passages from the uterus, the bladder and the intestine open into the cloaca, as in reptiles. In the marsupial the passages from the uterus and bladder open into a cloaca, but the anal exit is separate. In the placental mammal there are separate openings for the uterus, bladder and intestine.
Abbreviations: A = anus, AC = anal canal, B = bladder, C = cloaca, F = Fallopian tubes, K = kidney, O = ovaries, U = uterus, UR = urethra, V = vagina.

In spite of these specializations the recent monotremes are basically very primitive mammals. They reproduce by laying eggs, which are hatched in burrows. The young are suckled on milk that is secreted, as mentioned above, by modified sweat glands that are homologous to the mammae or breasts in the higher mammals. The skeleton and soft anatomy show the persistence of various reptilian characters. For instance, the shoulder girdle is very primitive with a persistent interclavicle, large coracoids, and no true scapular spine. The cervical ribs are unfused. Various reptilian characters persist in the skull. The rectum and urinogenital system open into a common cloaca as in reptiles, not separately as in mammals. There are no external ears or pinnae as in most other mammals.

The primitive shoulder girdle of the monotremes is closely comparable to the shoulder girdle of *Morganucodon*, the triconodont found in the Triassic fissure fillings of South Wales. The monotremes may have had their origins in docodont ancestors, in turn derived from morganucodant-like progenitors. On the other hand, some authorities feel that the monotremes may be included within the Theria, the subclass which embraces a majority of the mammals, rather than being isolated in a separate subclass, the Prototheria. However that may be, there is good reason to think that the monotremes represent an ancient line of descent from the mammal-like reptiles, continuing in an isolated corner of the world, where they have been able to survive as basically primitive mammals, with an overlay of certain specializations. In many respects the monotremes give us an excellent view in the flesh of mammals intermediate in their stage of evolution between the mammal-like reptiles and the higher mammals.

BASIC MAMMALIAN RADIATIONS

This short survey of the primitive mammals of Mesozoic times (plus the monotremes) outlines the first evolutionary developments in the mammals prior to their great expansion, which took place at the beginning of Cenozoic times. Although Mesozoic mammals are not numerous or of great variety, they are of the utmost importance because of their bearing on the later history of the mammals. They lived for a long time—a longer time span than the time span since the close of the Cretaceous period (the Age of Mammals as we usually think of it). And in this long expanse of late Triassic, Jurassic, and Cretaceous time, the primitive mammals established the primary lines of evolutionary radiation that determined subsequent mammalian history.

We may think of the history of the mammals as consisting of three phases of adaptive radiation.

The first embraced the late Triassic and Jurassic periods, during which five mammalian orders appeared and developed; these orders are the Docodonta, Triconodonta, Symmetrodonta, Eupantotheria, and Multituberculata.

The second phase of mammalian radiation occurred in the Cretaceous period, during which all of the Mesozoic orders except the docodonts continued their evolutionary development, but during which the triconodonts, symmetrodonts, and eupantotheres became extinct—their extinction taking place at the close of early Cretaceous history. It seems evident, however, that the eupantotheres gave rise to the two great mammalian groups that were to inherit the earth—the marsupials and the placentals. The development of these mammals from eupantothere ancestors very probably occurred during early Cretaceous times. While these significant evolutionary events were taking place, the multituberculates carried on in full vigor, to survive into early Cenozoic times.

The third and final phase of mammalian radiation took place in the Cenozoic era, during which marsupials and placentals reached unprecedented heights of evolutionary development, and the multituberculates became extinct. The monotremes, known only during the final stages of Cenozoic history, very probably persisted through all three periods of mammalian radiation. Needless to say, the Cenozoic radiation of the mammals is particularly interesting to students of evolution, since it is during this phase of mammalian history that these tetrapods evolved along the numerous lines that have led to the varied and successful mammals of Tertiary, Quaternary, and recent times. It is to the third phase of mammalian radiation that we shall devote the remainder of this book.

Perhaps the radiation of the mammals as described above can be indicated by the following simple diagram.

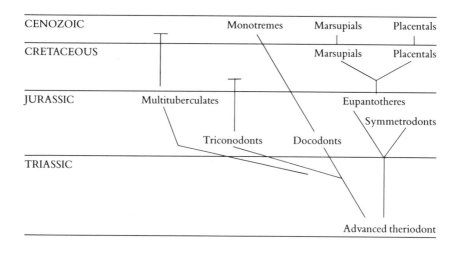

Kangaroos

Marsupials

CENOZOIC CONTINENTS AND MARSUPIAL HISTORY

In order to understand the evolution and the modern distribution of the marsupials it is necessary to give particular attention to the history of the continental land masses during the Cretaceous and Cenozoic eras. Otherwise it may seem strange that the fossil record of these mammals is so fragmentary and restricted, and their modern distribution so peculiarly limited to the western world and to Australia.

As is pointed out in Chapter 16, the rifting of the great supercontinent of Pangaea was a continuing event during the Mesozoic Era, so that by the end of the Cretaceous period South America either had become completely separated from Africa to become an island continent, or it was near to such a separation, with perhaps a tenuous connection between the northeastern bulge of South America and the Mauretanian area of Africa. In either case, there would seem to have been a connection of sorts between the tip of South America and the peninsular portion of Antarctica, which latter continental block was still firmly connected with Australia.

The oldest marsupials presently known in the geologic record are from late Cretaceous sediments in North America, the age of which predate the end of Cretaceous history by about fifteen million years. A few marsupials similar to the late Cretaceous marsupials of North America have been found in Peru and Bolivia, in Cretaceous beds even later in age than the North American occurrences. Mesozoic marsupials are not known in other parts of the world. Therefore it can be argued that the marsupials had their origins in North America and reached South America at the very end of Cretaceous history, possibly as accidental waifs. (Such an introduction of marsupials into South America from the north is perhaps reinforced by the presence in Peru of a lower jaw of very late Cretaceous age indicating the introduction from North America of an archaic notoungulate—a predecessor of the hoofed mammals that were so typical of Cenozoic history in South America.) Sometime during the Eocene epoch marsupials reached Europe, probably from North America, where they persisted through Miocene time and then became extinct. Also during the Eocene epoch some marsupials reached Antarctica, almost certainly from South America, probably on a long migration route to Australia. Marsupials became extinct in North America, the probable center of their origin, during Miocene time, but were reintroduced from South America during the great interchange of faunas that took place between the two continents of the western hemisphere in late Pliocene and early Pleistocene times.

Once isolated, South America became a refuge for the marsupials, since there was no opportunity for advanced placentals to enter the continent from the north. Therefore the marsupials were able to continue their evolutionary development as active competitors with the descendants of the ancient placentals that had been stranded on this continent. They lived on with a fair degree of success until the close of the Tertiary period, when South America once again became connected by an isthmian link with North America. Then there was an influx of highly specialized placentals from the north, and most of the marsupials were exterminated as a result of competition from these invaders. (Likewise, most of the original placentals of South America also disappeared under the wave of new mammals coming in from the north). Australia evidently retained a connection with Antarctica until Eocene time, as did South America, so that Antarctica formed a bridge whereby marsupials were able to invade Australia. Then Australia broke away to become an island continent drifting to the northeast with its marsupial passengers. Here the marsupials prospered, with virtually no placental competition. Here they developed through a wide range of adaptive radiation, continuing to the present time as the dominant mammals in their isolated environment.

MARSUPIAL CHARACTERS

It was mentioned in Chapter 18 that the marsupials are mammals in which the young are generally nurtured by the mother in a special pouch or pocket. Hence the name Marsupialia for the order, derived from the Latin word, *marsupium*, meaning a pouch. Marsupials are born as larval animals after a short period of gestation, and immediately they find their way into the mother's pouch. Here they become attached to the teats, which are located within the pouch. They remain attached to the teats for some time, receiving the milk that is literally pumped into them by the mother. As they develop and take form they finally achieve a degree of independence. Then the young marsupials live a life partly in and partly out of the mother's pouch.

Of course this very specialized method of reproduction and natal care is known to us from the modern marsupials. However there are various osteological characters that distinguish the marsupials, making it possible to identify them and trace their history as it is inadequately preserved in the fossil record.

The braincase is relatively small in the marsupials, a character that readily distinguishes them from the placentals and at the same time reflects their lower mentality. Consequently the sagittal crest, for the attachment of the temporal muscles that close the jaws, is frequently high and strong. The opening for the eye is confluent with the temporal opening, as it generally is among primitive mammals. The bony palate is perforated by well-defined openings, seldom solid as it is in the placental mammals. The posterior angle of the lower jaw, beneath its articulation with the skull, is usually turned in or inflected, as it rarely is in the placental mammals.

The teeth in the marsupials vary in number. There may be as many as five upper and four lower incisor teeth on each side as compared with the maximum of three in the placental mammals. In primitive marsupials there are commonly three premolars and four molar teeth on each side of both upper and lower jaws, whereas in the placental mammals there are four premolars and three molars. The molar teeth in the marsupials show a triangular pattern of cusps, evidently inherited from the eupantothere tooth pattern described above; thus there is an upper trigon with two outer cusps and an inner one that shears past a lower trigonid with one outer and two inner cusps. In addition, the lower molar has a basined talonid, to receive the point of the inner upper cusp. In these mammals there is less replacement of milk teeth by permanent teeth than there is in the placental mammals; indeed, only the posterior premolars are replaced.

The postcranial skeleton of the marsupials is greatly modified from the skeleton in the mammal-like reptiles, and in general is comparable to the skeleton in the placentals. There are seven cervical vertebrae, followed by about thirteen ribbed dorsals and commonly seven lumbar vertebrae, these latter free of ribs. There is generally a large clavicle, but the coracoid is reduced. There is a strong spine on the shoulder blade. In the pelvis there are commonly epipubic or "marsupial" bones that help to support the pouch. The toes are always clawed. Many marsupials show a strange specialization of the hind foot, whereby the second and third toes are very slender, closely appressed and enclosed in a single sheath of skin at their base. These two toes in effect "balance" the fifth toe, and the fourth toe, the largest of the series, serves as the axis of the foot. The slender second and third toes, known as syndactylous digits, make a sort of two-pronged comb with which, as can be seen in modern marsupials, the animals groom their fur.

THE AMERICAN OPOSSUM

The American opossum, *Didelphis*, is in some respects a Cretaceous survivor in the modern world. Although this interesting mar-

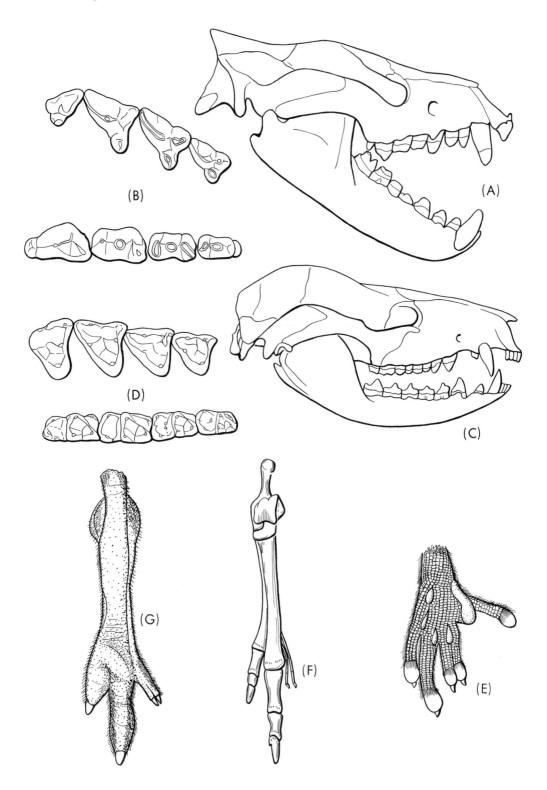

Figure 19-2. Skeleton of the American opossum, *Didelphis*, a generalized marsupial. Note the marsupial bones, for support of the pouch, projecting forwards from the pelvis. About one-third natural size.

supial shows certain significant changes from related didelphids of Cretaceous age, it nonetheless retains many of the characters of skull, dentition, and post-cranial skeleton that typify the earliest and most primitive of the pouched mammals. Therefore we are fortunately able to obtain a very fair idea of a primitive mammal by studying the opossum. This familiar inhabitant of our southern states is a small animal, about the size of a house cat. It is covered with coarse, light gray hair and it has a naked, ratlike tail. The opposum tail is prehensile, which means that it can be used for grasping limbs of trees, like a "fifth hand." The feet are primitive, with all the toes present in an unreduced state, and well provided with claws. The clawed feet and the prehensile tail make the opossums great climbers, and they spend much of their time in trees.

The skull is generalized, showing the various marsupial characters already listed, such as the small braincase, the perforated palate, and the inflected angle of the jaw. The teeth, too, are primitive, with a dental formula of five upper and four lower incisors, a canine, three premolars, and four molars. The pattern of the molar teeth is of the primitive type already described, a series of opposing triangles with basined talonids in the lower molars.

This animal may be considered as representing in a general way what the ancestral marsupials were like. From an opossumlike base the marsupials have evolved, especially in the Americas and in Australia, along various lines of adaptive radiation.

Figure 19-1. Marsupials. **(A)** Skull and jaw of *Borhyaena*, a large Miocene carnivorous marsupial from South America, about two-fifths natural size. **(B)** Upper and lower shearing molar teeth of *Thylacinus*, the recent Tasmanian "wolf," about natural size. **(C)** Skull and jaw of *Didelphis*, the modern North American opossum. **(D)** Upper and lower molar teeth of *Didelphis*. C and D about two-thirds natural size. **(E)** Hind foot of the opossum, showing the divergent first toe, an adaptation for climbing. **(F)** Bones of the hind foot, and **(G)** the hind foot of kangaroo, showing the syndactylous second and third toes, for grooming the fur. Greatly reduced.

Borhyaena and Litoptern

Thylacosmilus

Figure 19-3. Middle and upper Tertiary carnivorous marsupials of South America. *Borhyaena* was as large as a wolf, *Thylacosmilus* equaled a jaguar in size, and paralleled in a remarkable way the large, saber-tooth cats. Prepared by Lois M. Darling.

EVOLUTION OF THE AMERICAN MARSUPIALS

The marsupials that, like the Virginia opossum, have remained comparatively primitive throughout the course of their history are the didelphids of North and South America. It is probably because the didelphids have continued as small and generalized marsupials that they have been able to survive, in spite of having to share their environments with numerous highly specialized placental mammals. The common opossum, *Didelphis*, has already been described. Among its predecessors were *Alphodon* and *Eodelphis*, from the Cretaceous of North America, and *Peratherium*, from the early Tertiary of North America and Europe. Another persistent primitive opossum is the "mouse opossum," *Marmosa*, of Central America. *Chironectes*, the modern yapok of Central America, is an opossum that has become adapted to life in streams and rivers.

From didelphid ancestors certain South American marsupials specialized as aggressive carnivores during Tertiary times. These were the borhyaenids, of which the Miocene genus *Borhyaena* was typical. This was a rather large marsupial, perhaps as large as a wolf or a large dog. The skull was very doglike, with the canines enlarged as piercing and stabbing teeth, and some of the molars modified into shearing blades. The body was long, and the back was strong and supple. The tail was long. The limbs were strong, and the feet had sharp claws. All in all, *Borhyaena* was like a large, long-tailed dog in its general adaptations, and it obviously preyed on other mammals, both marsupial and placental, with which it was contemporaneous.

In Pliocene times some of the borhyaenids became remarkably specialized. One genus, *Thylacosmilus*, an animal as large as a tiger, had a short skull and a tremendously elongated, bladelike upper canine tooth, whereas in the lower jaw there was a deep flange of bone to protect this tooth when the mouth was closed. Here we see an uncanny resemblance to the saber-tooth cats of the Pleistocene of North America. In spite of such advanced specializations *Thylacosmilus* and the other carnivorous borhyaenids were unable to withstand the influx of the carnivorous placentals from North America when the two continents became reunited during the Pliocene epoch. Therefore these marsupials, showing close patterns of convergence with the placental carnivores of other continents, became extinct.

In another line of marsupial evolution in South America are the caenolestoids, small and in most respects generalized marsupials that range in age from the Paleocene to the present day. The modern representative of this phylogenetic line is *Caenolestes*, the "opossum rat," of Ecuador and Peru, a marsupial without a pouch. There has been an enlargement of the middle lower incisor teeth in these marsupials, with a correlative reduction or complete suppression of the other incisors. In one family of caenolestoids, the Polydolopidae, the

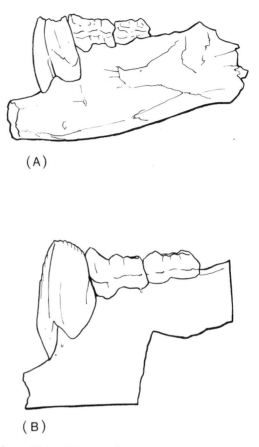

(A)

(B)

Figure 19-4. **(A)** *Antarctolops,* lower jaw with molars, of a marsupial from the Upper Eocene of Antarctica. **(B)** *Polydolops,* lower jaw with molars of a marsupial from the Upper Paleocene-Middle Eocene of South America. Both three and one-half times natural size.

first lower molar is remarkably enlarged, to form a high cutting tooth. This character, so typical of *Polydolops* from the Paleocene and Eocene of Argentina, is repeated in the genus *Antarctodolops* from the Eocene of the Antarctic peninsula. The occurrences of these two genera in southern South America and Antarctica respectively affords putative, if not definitive, evidence for the influx of marsupials into Australia by way of an Antarctic bridge from South America, as briefly mentioned above.

Except for the caenolestoids, all the American marsupials can be designated as *polyprotodonts*, which means that they had several incisor teeth on either side in the upper and lower jaws. The caenolestoids, however, seem to show some relationship in this respect to the *diprotodonts*, the marsupials of Australia that have a single incisor on each side of the lower jaw.

ADAPTIVE RADIATION OF THE AUSTRALIAN MARSUPIALS

Australia is (or was until the coming of the white man) the great home of the marsupials, harboring an array of these mammals not equaled elsewhere in the world. The beginning of marsupial evolution on this continent has been determined from fossils found during recent years in late Oligocene sediments. As a supplement to a growing fossil record, ranging through the middle and late Cenozoic, inferences based upon studies of recent forms are most valuable for helping us to understand the course of evolution among the pouched mammals of Australia. The range of adaptations in the modern Australian marsupials is great, and indicates variety of evolutionary lines during Tertiary times.

There are three groups of marsupials in Australia, the dasyurids, the parameloids, and the diprotodonts. A fourth group of Australian marsupials may be indicated by some recently discovered fossils of Miocene age, described as a new genus, *Yalkaparidon*; however more information is needed before unequivocal decisions can be made as to the relationships of these interesting remains.

The dasyurids are polyprotodont marsupials that show adaptations for various modes of life. Some of the dasyurids are carnivorous, like the thylacine or Tasmanian wolf, *Thylacinus*, in which there seems to be a remarkable parallelism with the extinct borhyaenids of South America. The Tasmanian devil, *Sarcophilus*, is a smaller but aggressive carnivore; and still other meat-eaters among the dasyurids are represented by the numerous "native cats," of which *Dasyurus* itself is an example. Other dasyurids are *Myrmecobius*, an ant-eating marsupial, and *Notoryctes*, a molelike animal, living in burrows under the ground.

The parameloids are also polyprotodont marsupials, mostly of small size, in which the hind feet have the second and third toes modified as syndactylous digits, used for combing the fur. These are the bandicoots, long-snouted animals with long hind legs for hopping, rabbit fashion. Some bandicoots feed upon insects or plants, and some of them are omnivorous, eating anything they can get.

The diprotodont marsupials are, of course, the ones in which there is but a single lower incisor on each side, and in which the hind feet have syndactylous digits. They form a large, varied group that includes some of the most characteristic of the Australian marsupials. Here are the numerous phalangers or Australian opossums, many of them showing squirrel-like adaptations. One of the well-known diprotodonts is the koala or native "bear," *Phascolarctos*, an animal strangely adapted for feeding on the leaves of certain eucalyptus trees and nothing else. Another is the wombat, *Phascolomys*. Finally, there are the wallabies and kangaroos, so well known as to need no description. These are the herbivores of Australia, adapted for browsing and grazing upon a variety of plants. Instead of running, as do large plant-eating mammals in other parts of the world, the kangaroos sail across the landscape in graceful, prodigious leaps, propelled by their powerful hind legs.

A Pleistocene relative of the modern diprotodont marsupials was the giant animal *Diprotodon*. This marsupial was as large as a large ox, and was a rather clumsy, four-footed beast. In life it may have looked something like a huge wombat.

There were other giant diprotodont marsupials living in Australia during Pleistocene times, among which were gigantic kangaroos. *Thylacoleo*, about the size of a modern lion, has been interpreted by many paleontologists as having been something like a cat-like carnivore, a conclusion founded on the large, shearing teeth in the skull and lower jaw.

Tasmanian "Wolf"

Marsupial "Mouse"

Wombat

Bandicoot

Koala

Phalanger

Marsupial "Mole"

Figure 19-5. Adaptive radiation among the recent marsupials of the Australian region. All drawn to approximately the same scale. Prepared by Lois M. Darling.

Other students have suggested that this interesting marsupial may have been a fruit-eater. The structure of the feet indicate that it probably was an adept climber in trees.

The modern marsupials of Australia constitute a remarkable example of adaptive radiation and evolutionary convergence. On this island continent, protected from invading placentals by isolation from other landmasses, the marsupials have evolved in various directions for many differing modes of life, so that they occupy the ecologic niches that in other continents have been usurped by the Cenozoic placentals. Thus the carnivorous thylacines (which in recent times have been limited to the island of Tasmania, and today may be extinct) can be compared with wolves of other regions, the small dasyures with small cats, weasels, and martens. These marsupial carnivores prey upon other marsupials, the thylacines on the larger plant-eaters, the dasyures on the numerous small marsupials that inhabit the grasslands, the thickets, and the trees. The wallabies and kangaroos can be compared with deer, antelopes, and gazelles in other continents, even though they have no physical resemblance to the hoofed animals. Nevertheless these are the larger plant-eaters, most of which live on the plains or in the open forest, and escape by their ability to cover ground at a very rapid pace. The wombats can be compared with large rodents, like woodchucks, and there are numerous small marsupials comparable to the small rodents of other regions. Phalangers can be compared with squirrels, bandicoots with rabbits, and so on.

THE EVOLUTIONARY POSITION OF THE MARSUPIALS

It has been common practice for zoologists to regard the marsupials as constituting a sort of lower grade in the hierarchy of mammals—an intermediate step in the evolution of mammals between the ancestral mammals of the Jurassic period and the Cenozoic placentals. However the evidence indicates that although the placentals may have gone through a stage in their early history somewhat comparable to that of the marsupials, the two groups probably arose independently from a common eupantotherian ancestry, to evolve side by side. Certainly the fossil record of the placentals is as ancient as that of the marsupials.

It is probably valid to think of the marsupials and placentals as arising at about the same time, during the Cretaceous period. They developed two quite dissimilar methods of reproduction, as well as various anatomical differences. During the early stages of their evolutionary histories they were probably well matched, so that marsupial adaptations were about as efficient in evolutionary terms as placental adaptations. But as time went on, and especially with the opening of the Cenozoic era, the placentals became dominant. There were probably various factors that led to the dominance of the placentals over the marsupials, but of these it is likely that the superior intelligence of the placental mammals was of particular importance.

Dinosaurs and Insectivores

Introduction to the Placentals

20

BASIC PLACENTAL CHARACTERS

The Cenozoic era is commonly referred to as the Age of Mammals. We might with reason call it the Age of Placental Mammals, because these have been the overwhelmingly dominant mammals in almost all parts of the earth ever since the transition from Cretaceous into Cenozoic times. During mammalian history the monotremes have been represented by 2 fossil and 3 recent genera, belonging to a single order; the order of multituberculates, by 17 genera of early Tertiary age; the marsupials by 127 genera, fossil and recent, also contained within a single order, and the placentals by 2648 genera belonging to 28 separate orders. On a percentage basis, the Cenozoic placentals total 95 per cent; the non-placentals 5 per cent. These figures, based on the classification of mammals made some years ago by George Gaylord Simpson, illustrate placental dominance without the need of further discussion. A classification including the most recent discoveries would give somewhat different figures but essentially the same proportions between categories.

The placental mammals or eutherians are, of course, those mammals in which the young go through a considerable period of prenatal growth, to be born in a comparatively advanced stage of development. In these mammals one of the membranes inherited from the old reptilian egg, the allantois, is in contact with the uterus, within which is the embryo; and through this area of contact, the placenta, nourishment and oxygen are carried from the mother to the developing embryo. As a result of this embryonic growth, the placental mammal when born is to a greater or lesser degree a miniature replica of its parents. Even the most helpless of newborn placentals, like baby rodents or carnivores or human beings, are incomparably more advanced than the larva-like newborn of the marsupials. And in many placentals, such as the hoofed mammals or the whales, the young are born as active little animals, fully able to follow their mothers within a few hours or less after their birth.

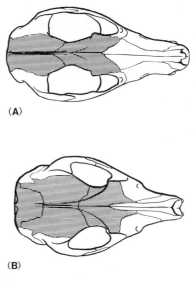

(A)

(B)

Figure 20-1. A comparison of the braincase in **(A)** *Didelphis*, a marsupial, and **(B)** *Gymnurechinus*, an insectivore placental, to show the increase of cranial capacity in a placental mammal skull as compared with a marsupial skull. Skulls drawn to unit lengths.

In osteological characters the placentals show various distinguishing traits. Perhaps the most important character to be seen in the skeleton of these mammals is the expanded braincase, reflecting the superior intelligence of most placentals as contrasted with the marsupials. Additional features that characterize the skull in the placentals are the solid bony palate, as contrasted with the generally pierced palate of the marsupials, and the usual lack of an inturned or inflected angle in the lower jaws. In the postcranial skeleton the placental like the marsupial is characterized by seven cervical or neck vertebrae, behind which is a thoracic series of vertebrae bearing ribs, and after that a ribless lumbar series. The girdles and limbs in the placentals are basically similar to those of the marsupials, already described, although of course there are many placental specializations that result from adaptations to varied modes of locomotion. The placentals lack epipubic or marsupial bones on the pelvis.

Of all the hard parts, however, the teeth are of particular importance in a study of the placental mammals. It is very probable that if all the placental mammals were extinct, and represented only by fossil teeth, their basic classification would be essentially the same as the classification now drawn up on the knowledge of the complete anatomy in these mammals. This is indeed fortunate. Even though many fossil mammals are frequently preserved only as fragments of jaws containing teeth, it is possible to study them and reach reasonably valid conclusions as to their evolutionary history and their relationships. For this reason it may be useful at this place to introduce a discussion of the teeth in the placental mammals.

THE TEETH OF PLACENTAL MAMMALS

The basic tooth formula for the placental mammals is three incisors, a canine, four premolars, and three molars, on each side in both upper and lower jaws. This formula, expressed by the notation I3/3-C1/1-P4/4-M3/3, appears in the first placentals of Cretaceous age, and it persists in many modern mammals. Of course there has been extreme specialization of the teeth in many placentals, and this has frequently led to divergences from the primitive dental formula.

It should be mentioned that in the Jurassic eupantothere, *Peramus,* which by reason of its molar morphology may be antecedent to the metatherian and eutherian mammals, there are eight cheek teeth on each side in the upper and lower jaws, these having been interpreted as four premolars and four molars or five premolars and three molars. It seems probable that one of the premolars was lost, while there were certain changes in dental morphology, leading to the three premolars and four molars, basic for marsupials and the four premolars and three molars, basic for placentals.

In most of the placental mammals the incisor teeth tend to be comparatively simple, single-rooted pegs or blades, adapted to nipping food. In some mammals the incisors are enlarged, in others they are reduced or suppressed. In a few they become complex, with comblike crowns. Yet in spite of their various specializations, the incisors always retain the single root that holds them in the jaws.

The canines in primitive mammals are enlarged teeth of spikelike form, obviously used for stabbing and piercing functions. These teeth, like the incisors, remain usually single rooted throughout their many divergent adaptations, but they may show specializations in the crowns, especially with regard to shape and size.

The premolars of the placentals are generally of complex structure, and they commonly increase in complexity from front to back. Thus the first premolar may be a narrow-crowned tooth with two roots, whereas the last member of the series may be broad, with several cusps making up the crown and with three or more roots. It is not uncommon for the posterior premolars to show close resemblances to the molar teeth in many specialized mammals.

It is in the molar teeth, however, that we find a key to eutherian relationships. Indeed, much of our knowledge of the evolution of mammals through geologic time is based upon the close study of the molar teeth.

In discussing the Jurassic mammals it was shown that the eupantotheres can be regarded as the probable direct ancestors of the marsupials and placentals by virtue of the construction of their molar teeth. Among the early marsupials and placentals the upper dentition consisted of trigons that sheared past the trigonids of the lower dentition. In addition to this shearing action of the upper and lower molars there was a crushing function brought about by the biting of the inner cusp of the trigon into a basin-like talonid, located behind the trigonid of the lower molar. Molars of this type are frequently said to be of tritubercular, tuberculosectorial, or tribosphenic form. Perhaps the last of the foregoing terms is the preferable one; it will be used in this book. The tribosphenic molar constituted a base from which evolved the multitudinously varied molar teeth of the higher mammals.

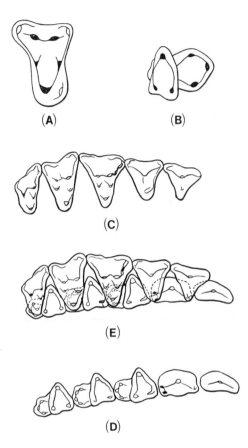

Figure 20-2. **(A)** Left upper, and **(B)** left lower tribosphenic molars, to illustrate the spatial relationships of the tooth cusps. **(C)** Right upper tooth row, containing the last two premolars and three molars. **(D)** Occlusal diagram of upper and lower tooth rows. **(E)** Right lower tooth row, containing the last two premolars and three molars.

The primitive nature of the tribosphenic molar pattern was established many years ago by the great American paleontologist, Edward Drinker Cope, and the nomenclature of the cusps was proposed by his disciple, Henry Fairfield Osborn. Cope and Osborn considered the tribosphenic molars as opposed, reversed triangles. So Osborn named the three cusps of the upper molar the protocone, the paracone, and the metacone, the first being on the inner side of the tooth, the other two on the outer side. In addition there are often two intermediate cusps in the upper molar, between the main cusps; these were designated the protoconule and the metaconule. In the lower molar the outer cusp was designated the protoconid, and the two inner cusps were called the paraconid and the metaconid. In the lower molar there are commonly three cusps around the talonid or posterior basin. The outer one of these cusps was called the hypoconid, the inner one the entoconid, and the posterior one, on the back of the basin, the hypoconulid.

These basic cusps, which constitute the upper and lower tribosphenic molars, can be regarded as having had a common derivation, thus making them homologous in all marsupials and placentals. Their spatial relationships can be indicated as on the facing page.

In many of the more advanced mammals there is a fourth main cusp, the hypocone, occupying the back inner corner of the upper molar. This cusp appears as an addition to the tooth during the evolutionary history of various orders of mammals; consequently it is doubtful that the hypocone is homologous in those mammals having the cusp. Finally, in the molars of many mammals there are various crests or ridges, which have been designated as lophs in the upper molars and lophids in the lower molars, whereas certain small accessory cusps around the edges of the teeth have been termed styles and stylids in the upper and lower molars, respectively.

The names of the cusps in mammalian molar teeth as proposed by Osborn were based upon his theory that the protocone and protoconid were the ancestral primary cusps, derived from reptilian teeth. Later authors have disputed this theory and have proposed other names for the tooth cusps. However the Osbornian names are so firmly established in the literature that it seems only practical to retain them as positional designations, without any implications as to the manner or the order in which they developed.

In the placental mammals the motions of the jaws involve four types of action between the upper and the lower molar teeth, three of which are present in the tribosphenic molars

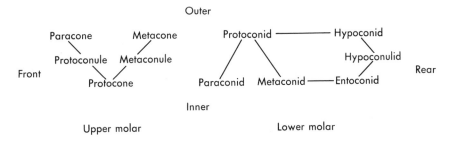

Upper molar   Lower molar

of the primitive mammals. In the first place there is the alternation of cusps, the biting past each other of these elements in the upper and lower molars, to hold and tear the food. For instance, along the outer edge of the dentition the protoconid of the lower molar alternates with the metacone and paracone of successive upper molars, while the hypoconid of the lower molar alternates with the paracone and metacone of a single upper molar. Along the inner edge of the dentition the protocone alternates with the metaconid and paraconid of successive lower teeth. Second, there is the shearing of the edges of teeth, or of lophs or crests, past each other to cut the food. In the tribosphenic molars the front and back edges of the trigons of the upper molar shear past the back and front edges of the trigonids of

successive lower molars. In the third place there is the opposition of certain parts of the teeth, which serves to pulverize or crush the food. Such is the action of the protocone biting into the basin of the talonid. Finally, there is the grinding of opposite tooth surfaces against each other like mills, to grind the food. This action is seen in the expanded molar crowns of many of the specialized mammals.

An important part of the discussion in the rest of this book will have to do with the adaptive radiation of teeth from the tribosphenic type in response to the need for preparing the food by the actions of alternation, shear, opposition and grinding. It may seem that a great deal of emphasis is being given here to the teeth in placental mammals. The fact is, however, that much of the success of mammals

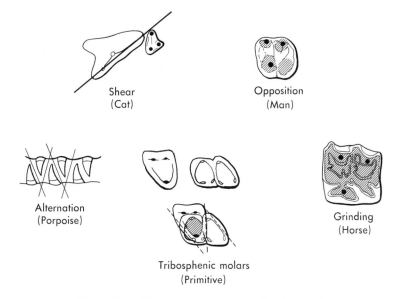

Figure 20-3.   Types of action in mammalian cheek teeth.

during their Cenozoic history can be attributed to the adaptations of the teeth.

A CLASSIFICATION OF THE PLACENTAL MAMMALS

About thirty-three orders of placental mammals or eutherians evolved during the Cenozoic era, and of them eighteen orders exist at the present time, according to the classification used in this book. The orders of placental mammals can be listed as follows.

Insectivora: the insectivores. Shrews, moles, and hedgehogs.
Macroscelidea: the elephant shrews.
Scandentia: the tree shrews.
Chiroptera: the bats.
Dermoptera: the colugos or "flying lemurs."
Taeniodonta: the taeniodonts, an extinct group.
Tillodontia: the tillodonts, an extinct group.
Primates: the primates. Lemurs, tarsiers, monkeys, apes, and men.
Edentata: the edentates. Anteaters, tree sloths, ground sloths, armadillos, and glyptodonts.
Pholidota: the pangolins.
Rodentia: the rodents. Squirrels, beavers, mice and rats, porcupines, cavies, and chinchillas.
Lagomorpha: the rabbits and hares.
Cetacea: the porpoises and whales.
Acreodi: large carnivores, ancestral to cetaceans. Extinct.
Creodonta: the primitive carnivores. Extinct.
Carnivora: the carnivores or beasts of prey. Dogs, wolves and foxes, bears, pandas, raccoons, weasels, mink, otters, wolverines, badgers, skunks, civets, hyaenas, cats, sea lions, seals, and walruses.
Condylarthra: the condylarths. Primitive hoofed mammals. Extinct.
Pantodonta: large, bulky hoofed mammals. Extinct.
Dinocerata: gigantic hoofed mammals with saber-like canines, and horns. Extinct.
Tubulidentata: the aardvarks.
Notoungulata: the notoungulates. Primitive hoofed mammals of South America. Extinct.
Litopterna: the litopterns. Hoofed mammals of South America. Extinct.
Astrapotheria: the astrapotheres. Large South American mammals. Extinct.
Trigonostylopia: the trigonostylopids. Early South American ungulates. Extinct.
Pyrotheria: the pyrotheres. Very large mammals of South America. Extinct.
Xenungulata: the xenungulates. Large hoofed mammals of South America. Extinct.

Perissodactyla: the perissodactyls or odd-toed hoofed mammals. Horses, titanotheres, chalicotheres, tapirs, and rhinoceroses.
Artiodactyla: the artiodactyls or even-toed hoofed mammals. Dichobunids, entelodonts, pigs, peccaries, anthracotheres, hippopotamuses, oreodonts, camels, tragulids, deer, giraffes, pronghorns, antelopes, goats, sheep, muskoxen, and cattle.
Hyracoidea: the conies or dassies of Africa and Asia Minor.
Proboscidea: the proboscideans. Moeritheres, dinotheres, mastodonts, mammoths, and elephants.
Desmostylia: relatives of the sea cows and the proboscideans. Extinct.
Sirenia: the sea cows, or sirenians.
Embrithopoda: the arsinoitheres. Large hoofed mammals of Egypt. Extinct.

The earliest and most primitive eutherians were insectivores known as the Proteutheria, and it seems obvious that they were probably the ancestors of all other placental mammals. From an insectivore stem the eutherians have radiated in such a variety of directions along so many evolutionary lines that it is not easy to group the orders into larger categories.

Ever since the days of Cuvier, attempts have been made to group the orders of mammals, with varying degrees of success, as might be imagined. There is an almost insuperable problem here, posed by the isolated position of a large proportion of mammalian orders; thus, although some orders may be brought together in natural groups, the association of others (no matter how it is done) is forced and unnatural. Consequently, no attempt is made here to bind together the orders of mammals into large formal taxonomic units—but instead they are treated independently. However, it may be helpful at this point to discuss briefly the general relationships between the orders that show such relationships, and to mention the possibilities with regard to the isolated orders.

On this basis, about eight categories among the orders of mammals may be recognized. They are as follows:

1. First there are the orders that show definite relationships to the primitive placentals, exemplified by the insectivores. Included here, of course, are the in-

sectivores themselves, the macroscelideans, the scandentians, and the bats, which are aerial mammals whose specializations have been developed over an insectivore base. Moreover, the Dermoptera, represented today by the strange, gliding colugos of the Orient, are of insectivore derivation; so much so, in fact, that some authors would include them as a suborder of Insectivora. Again, the Primates, no matter how advanced some of them (particularly Man) may seem to be, are clearly of insectivore origin. Indeed, it is very difficult, when dealing with the fossil materials, to draw a definitive line between the very primitive primates and the insectivores. Finally, it appears that the Taeniodontia, a restricted order of early Tertiary mammals, have an insectivore ancestry.

2. The insectivore ancestry of the earliest meateating mammals, the creodonts also is obvious to such a degree that some of the early Tertiary fossils, which at one time were included among the insectivores, are now placed within the Creodonta. The Carnivora, the carnivores that persist in the modern world, may be linked closely with their primitive creodont forerunners.

3. Although the whales and porpoises are distinct from all other mammals because of their extreme specializations for life in the sea, it seems probable that they may have an origin from creodontlike ancestors, the Acreodi.

4. The Condylarthra, primitive hoofed mammals, appear to have origins in common with those of certain early creodonts. The Condylarthra, in turn, seem to occupy a position in the phylogeny of mammals that apparently is basic to a considerable array of hoofed mammals (to use the word "hoofed" in a very broad sense). Thus, the Perissodactyla, the odd-toed hoofed mammals, probably have a condylarth ancestry. The same appears to be true for at least three orders of hoofed mammals that lived in South America during much of Cenozoic time. These are the Notoungulata, the Litopterna, and the Astrapotheria. The evidence indicates that these mammals evolved independently, after South America became an island continent in early Tertiary time, from condylarth ancestors that had migrated from the north before South America was thus isolated. Two other South American groups of extinct hoofed mammals, (here classified as orders) the Xenungulata and the Pyrotheria, also may have condylarth origins. Again, the Pantodonta and Dinocerata, two orders of large, primitive hoofed mammals, especially characteristic of North America and Asia, may be derived readily from a condylarth base. The Artiodactyla, the even-toed hoofed mammals, likewise may be of condylarth ancestry. The Tillodontia, a very restricted order of early Tertiary mammals, can logically be related in a general way to the condylarths. Finally, the tubulidentates or aardvarks, which seem like no other mammals in our modern world, have various osteological characters that may indicate their ultimate condylarth origin.

5. The Lagomorpha, the hares and rabbits, superficially rodentlike in their appearance and habits, constitute a very distinct and isolated mammalian order that may be of condylarth derivation.

6. The rodents themselves, the most numerous of modern mammals, not only in numbers of individuals but also in numbers of species, stand in baffling isolation. All attempts to relate the ancestry of the rodents to other groups of mammals have been in vain. The earliest rodents are clearly rodents, although it is probable that their ancestors must have been related to the basic eutherian-insectivore stock.

7. Another isolated order of mammals is represented by the Edentata—the anteaters and armadillos and their relatives. The Old World pangolins, possibly derived from the early Cenozoic palaeanodonts of North America, stand as an isolated mammalian group.

8. Finally, there is a group of mammals, whose members, although quite varied as to external appearances and adaptations, are nevertheless linked by their morphological structures. They are the Proboscidea, the elephants and mastodonts; the Sirenia or sea-cows; the Desmostylia, an extinct order of semimarine mammals; the Hyracoidea or Old World conies; and possibly the Embrithopoda, represented by a single genus from the Oligocene of Egypt.

These are the possible groupings of the orders of placental mammals. So far as we can determine, all of the mammalian orders extend back into early Tertiary times, and probably quite a number of them had their origins during the Cretaceous period. For this reason the interrelationships of some of them are not always obvious. Of course, as has been outlined, we can see how many orders approach each other as they are traced back through time to the early phases of Tertiary history. But other orders are of more ancient ancestry, and for their derivations and basic relationships we must hopefully await the discovery of new fossils in beds of Cretaceous and very early Tertiary age.

In the remaining chapters of this book the various orders of placental mammals will be discussed, but not always according to the sequence by which they have been listed in the outline classification above. It is not easy to be completely logical when considering a group of animals as extensive and varied as the

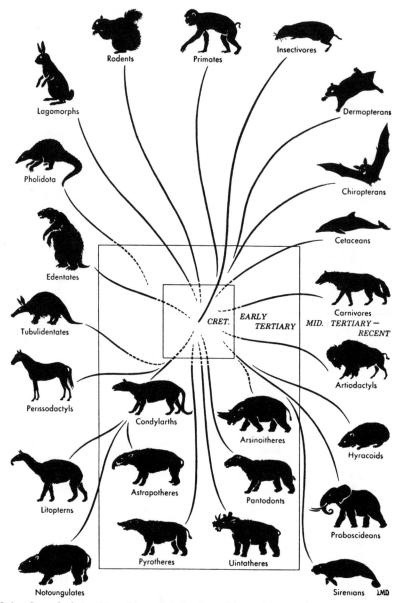

Figure 20-4.   General relationships of the principal orders of placental mammals. Prepared by Lois M. Darling.

mammals, particularly when the time dimension of their evolutionary history is taken into consideration. Therefore it may be expedient in discussing some orders to take up the evolutionary history of certain orders of mammals according to their geographic distribution or their position in time, rather than according to the strict concept of relationships. Let us hope that this will not be unduly confusing to the reader.

Mole   Glyptodonts, Ground Sloth

The Diverging Placentals

21

BASIC PATTERNS OF
EUTHERIAN EVOLUTION

The earliest placental mammals seem to have appeared during Cretaceous history as primitive insectivores, the Proteutheria, and, as has been pointed out in the last chapter, there was a wide range of adaptive radiation diverging from this base, through the extent of Cenozoic time. Two other placental orders are represented in Upper Cretaceous sediments, namely the Condylarthra, as exemplified by the genus *Protungulatum*, and the Primates known from the single genus *Purgatorius*. Although the Cretaceous eutherians were rather similar to each other, as far as we can determine from all too scanty remains, the divergence of early placentals proceeded at a rapid rate during the initial phases of Cenozoic history. With the disappearance of the dinosaurs (the dominant land-living tetrapods of the Cretaceous period) there was, in effect, an "evolutionary explosion" of mammals to fill most ecological niches abandoned by the dinosaurs. The placentals became by far the most numerous and varied of these mammals. The effects of this explosion, if it may be called one, were various. Some of the placental groups that made their appearances in basal Cenozoic times were rather short-lived and restricted to the early phases of Tertiary history; other groups continued from their ancient beginnings to modern days. But whether the various lines of eutherian radiation were short-lived or long-lived, the total effect in the early Cenozoic was that of primitive mammals becoming adapted to many aspects of a varied environment.

Some of these ancient eutherians displayed quite plainly their insectivore heritage; others, even in the early stages of their evolutionary history, seem to have departed so far in their morphological adaptations from the basic eutherian stock that they retained virtually no vestiges of their ultimate ancestry. It has been shown that these circumstances make it possible to group some of the eutherian orders, while others stand in isolated positions.

In this chapter we consider briefly the insectivores as they are known from fossil materials and from recent examples. These mammals, here grouped in a single order, exemplify the primitive eutherian conditions even into recent times; thus, they afford a broad view of what the most ancient and generalized placental mammals were like—a view that is not limited to the dry bones but, thanks to the survival through long geologic epochs of persistently primitive types, includes much knowledge of the soft anatomy and the habits of these interesting and crucially important animals. Also, we consider some mammals that retained or still retain, in spite of their specializations, a palimpsest of insectivore features, as well as certain mammals that show little if any trace of their distance ancestry.

Two groups of primitive mammals, formerly included within the Insectivora, are now generally recognized as separate orders. These are the Macroscelidea or African elephant shrews, and the Scandentia or Oriental tree shrews. Both orders are scantily represented in the fossil record.

The bats, to be described here, are essentially insectivores that developed remarkable specializations for flying. The colugos are, in many respects, primitive mammals, adapted for gliding. In the primates, which we will discuss in the following chapter, a high degree of mental development and manual dexterity has evolved, although these mammals (quite obviously derived from the insectivores) generally are comparatively primitive in the development of the postcranial skeleton. The edentates, which we include in this chapter for want of a better place to discuss them, are quite distinct descendants of the insectivores in which specializations have been in the direction of climbing and digging, of restrictions in the diet with a consequent simplification or suppression of the teeth, and, among some, of the development of heavy armor for protection. The Pholidota, as we shall learn, have evolved independently in a manner similar to some of the edentates. And the extinct taeniodonts and tillodonts, which were short-lived orders—the former seemingly of insectivore origins and the later perhaps derived from the condylarths—are included in this chapter as a matter of convenience.

THE INSECTIVORES

One of the very primitive Cretaceous insectivores belonging to the suborder Proteutheria was the genus *Zalambdalestes* from Mongolia. This was a small mammal, with a low skull less than two inches in length. There was a pair of enlarged, piercing incisors in the upper and lower jaws, as is common among the insectivores. The molars were tribosphenic, with the paracone and the metacone of the upper teeth widely separated on the outer border of each tooth, and with the protocone forming the inner point of the triangle. These molars appear to be intermediate in structure between the eupantothere molar and the molar of the more advanced placentals.

Zalambdalestes can be regarded as approximating in structure the ancestral insectivores and through them the basic placentals. None of the insectivores have evolved into large or even medium-sized mammals, and most of them have been and are small. The common shrews are the smallest of all modern mammals, and probably the secretive habits and the minute size of these insectivores have been instrumental in their long survival with little change from Cretaceous times to the present. All the insectivores have had a small and primitively constructed brain, as we might expect. The teeth have retained much of the ancestral tribosphenic molar pattern throughout the adaptive radiation of these mammals. The tympanic bone, which in higher mammals is frequently inflated to form a sort of capsule around the middle ear, is in the insectivores a simple ring. The skeleton has continued the generalized form typical of the primitive mammal, with an overlay among some groups of specializations for very definite modes of life. For instance, in the moles the limbs and feet are highly adapted for digging. Certain insectivores have been characterized through their evolutionary development by triangular upper molars, in which there is a high central cusp, perhaps a fusion of the paracone and metacone, a low inner cusp, perhaps the protocone, and various cuspules along the outer edge. Although these teeth appear to be primitive, they probably are not, but instead represent specializations from the typical tribosphenic mo-

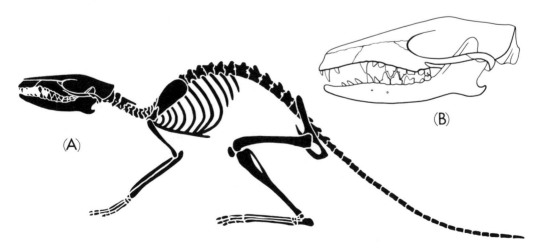

Figure 21-1.   **(A)** The skeleton of *Zalambdalestes*, a Cretaceous mammal from Mongolia, illustrating the structure of a primitive placental. About two-thirds natural size. **(B)** The skull of *Zalambdalestes*. Somewhat larger than natural size.

lars of the ancestral insectivores. Among these insectivores are the giants of the order that are exemplified particularly by *Centetes*, the modern tenrec of Madagascar, an animal almost two feet in length. Among most of the insectivores evolution in the upper molars has been marked by a wide separation of the paracone and metacone on a W-shaped outer ridge or ectoloph. In addition, some of these latter insectivores have a hypocone on the back inner corner of the upper molar. The Old World hedgehogs, typified by *Erinaceus*, illustrate this type of molar structure.

During the past two decades a number of students who have devoted many years of research to the study of primitive eutherian mammals, have essayed to classify the insectivores. These scholars have found the task a frustrating one, despite their authoritative knowledge of insectivores. The difficulty is, of course, that the primitive placentals show numerous close resemblances, so that it becomes an exacting exercise to sort them into suborders and lesser taxonomic groups. Yet the fossil evidence often shows how the various early insectivores evolved in directions toward later, quite distinct lines of mammalian evolution.

One plan is to sort the insectivores, fossil and recent, into three general groups that can be considered suborders. They are as follows:

1. The Proteutheria, or menotyphlans (the latter name is one, however, whose meaning is now confused because of the various ways in which it has been used). Here are included the leptictids and pantolestids of late Cretaceous and early Cenozoic age, among which is *Zalambdalestes*, mentioned above. Other Cretaceous genera of particular evolutionary importance are *Gypsonictops*, *Cimolestes* and *Kennalestes*.
2. The Soricomorpha, including various shrews—fossil and recent, the tenrecs—known as fossils in East Africa and Madagascar and today found in those regions, and the chrysochlorids or golden moles, also of Africa.
3. The Erinaceomorpha, the morphologically primitive hedgehogs that were widely distributed during Cenozoic history and persist today in Europe, Asia and Africa.

Much fault can be found with this particular subdivision of the insectivores, but faults will be discerned in any classification of the insectivores that is presented, because of the differences of opinion that exist among the people who are most familiar with these mammals. The important fact to remember is that some of the insectivores, particularly certain fossil forms, show us very nicely what the earliest eutherian mammals were like, and others demonstrate the specializations that have taken place, as additions or modifications to the basic eutherian structure, within varied lines of adaptive radiation through the long span of Cenozoic history.

MACROSCELIDEA

The macroscelideans or elephant shrews were long considered as being contained within the Order Insectivora, but in recent years many authorities have regarded them as constituting a small but separate order of mammals. The relationships of the macroscelideans with other mammalian orders are unclear, in part because of a scanty fossil record of these particular mammals. Modern elephant shrews, distinguished from the insectivores by various features of the skull, are specialized by reason of their long hind limbs for rapid bounding. The nose in the elephant shrews is greatly elongated and flexible—hence the name, elephant shrew. Their habitat is Africa.

SCANDENTIA

The tupaids or Oriental tree shrews have been variously considered as insectivores or alternatively as very primitive primates. As in the case of the macroscelideans, these mammals are poorly represented in the fossil record. Modern tree shrews look something like squirrels with very primate-like external ears. They also show many "lemuroid-like" characters, such as a complete post-orbital bar in the skull and large hemispheres in the brain. They are now generally regarded as constituting a separate mammalian order.

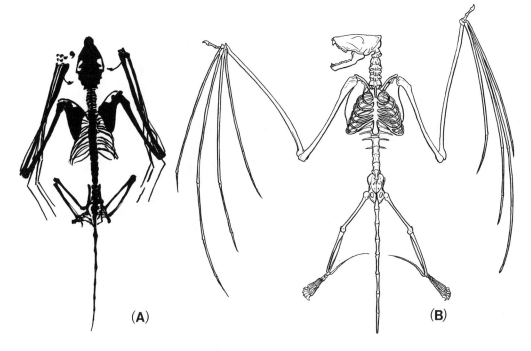

Figure 21-2. **(A)** *Icaronycteris*, an Eocene bat, compared with **(B)** *Nyctalus*, a modern Oriental bat. This shows that the earliest known bat had become as fully adapted for flight as are modern bats. Not to scale.

THE BATS

Because of their habits, bats like birds were not commonly interred in sediments and fossilized. Therefore the history of these mammals is inadequately known, even though bats are numerous being among recent mammals second only to the rodents in numbers, both of species and of individuals, and of worldwide distribution in modern times. These, the only mammals to have mastered true flight, probably originated at a relatively early date, and they must have experienced an initial stage of very rapid evolution, because the first known bats of Eocene age, as particularly exemplified by the type of *Icaronycteris*, a beautifully preserved skeleton from Wyoming, were highly developed and not greatly different from their modern relatives. There are no known intermediate stages between bats and insectivores.

In the bats, as in other flying vertebrates, the front limbs are modified to form wings. The limb bones are elongated, as are the fingers except the thumb, and these support the membrane that forms the wing. The thumb is a free digit with a claw on the end. The hind limbs are weak, so that bats are rather helpless on the ground. However the feet have clawed digits, and with these the bats hang themselves upside down when they sleep. The skull may show various specializations, but the molar teeth are often primitive, rather like those of some of the insectivores, with a W-shaped outer ridge or ectoloph.

Modern bats are commonly nocturnal. In these mammals the sense of hearing is remarkably acute, as indicated by the enormous external ears or pinnae, so characteristic of many bats. Recent experiments show that as they fly bats emit a constant stream of ultrasonic squeaks. The ears catch the echoes or reflections of the sound from nearby ob-

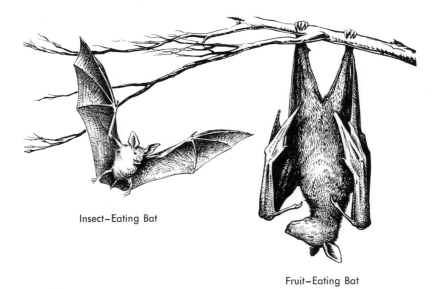

Insect–Eating Bat

Fruit–Eating Bat

Figure 21-3.   Modern bats, representative of the Microchiroptera (left) and the Megachiroptera (right). Prepared by Lois M. Darling.

jects, and thus the bats are guided by the sensitivity of their hearing. They are able to fly at night and pursue insects without any reliance upon the sense of vision. This sonar or echolocation device of the modern bats probably evolved at a fairly early stage in the history of these mammals, and was very likely an important factor in their successful evolution.

The bats may be divided into two large categories or suborders. The Megachiroptera, perhaps the more primitive of the bats, are the large fruit-eating bats of the Old World and the Pacific region, among which are the "flying foxes," largest of all bats. Most of the bats, however, are contained within the suborder Microchiroptera, a group of worldwide distribution. Most of these bats are insectivorous, taking their food on the wing, as they have probably done through the history of the order. Certain modern microchiropteran bats, the so-called vampires of tropical America, have become highly specialized for sucking blood from large mammals.

DERMOPTERA

Another descendant from the insectivores is the colugo (*Galeopithecus*) or "flying lemur" of the East Indies. This animal is not a lemur, and it does not fly. It is an herbivorous, tree-living mammal, about the size of a large squirrel, with broad folds of skin extending between the legs and onto the tail, with which it is able to glide for long distances from one tree to another. A few Paleocene and Eocene fossils from North America may belong to early relatives of the colugo.

THE EDENTATES

The word "edentate," which means "without teeth," has been variously applied to mammals in which the teeth are greatly simplified, reduced, or suppressed, as an adaptation to very specialized diets. In the modern sense, the term Edentata is restricted to an order of mammals that evolved largely in South America and never got beyond the limits of the New World.

Truly ancestral edentates, as yet unknown in the fossil record, possibly appeared in South America during late Cretaceous or early Paleocene times. Fossils from the Upper Paleocene and Eocene of that continent indicate that primitive armadillos had at those early stages of Cenozoic history become established as the forerunners of what was to be a large and successful group of Neotropical mammals. The sloths appeared in the Oligocene epoch, the anteaters in the Miocene epoch. Then the edentates evolved on the southern continent in complete isolation from the rest of the world. They quickly became prominent members of the South American faunas, and have remained so to the present day.

Certain distinctive characters, developed in the South American edentates, led to their early specialization for restricted modes of life, and masked the primitive features that had been inherited from their insectivore ancestors. In these mammals were developed accessory xenarthrous articulations between the lumbar vertebrae of the back, making this region of the backbone very strong. In addition the neck was increased among some of the edentates to include as many as nine cervical vertebrae, instead of the usual seven that are almost universal among the mammals. The feet had large claws. In the skull the brain was comparatively small and primitive, and it remains so in the modern edentates. The zygomatic arch or cheek bone was usually incomplete. Finally, the teeth were much reduced and simplified, or completely suppressed.

During Tertiary times the higher edentates evolved along two broad lines of adaptive radiation. One of these lines included the ground sloths, now extinct, the tree sloths, and the anteaters. This group is generally designated as a suborder—the Pilosa. The other evolutionary line, also of subordinal rank, was that of the Cingulata, the armored armadillos and the glyptodonts.

The most distinctive evolutionary development in the armadillos and glyptodonts was the growth of extensive armor to protect these animals against attack. This armor was formed of heavy bony plates covered with horny scutes. In the modern armadillos there are solid shields of bone over the shoulders and the hips, and between these areas are transverse movable bands of scutes, which gives flexibility to the back. There are also scutes on top of the skull and on the tail. The teeth in the armadillos consist of simple pegs without enamel, along the sides of the jaws. The armadillos, which have been very successful mammals from early Tertiary times to the present day, are scavengers, eating insects, carrion, and almost any other food they can find on the ground.

The late Cenozoic glyptodonts like *Glyptodon* were giant cousins and probable descendants of the armadillos, in which the body armor formed a solid bony carapace, like the shell of a turtle. In the largest of these ungainly mammals the shell was a massive, heavy dome, often five feet or so in length. Not only was the body heavily armored, but also the top of the head was covered with a thick shield of coalesced plates, and the tail was encased with concentric rings of bony armor. In some of the glyptodonts the end of the tail was provided with an enlarged, spiked knob of bone, remarkably like the heavy maces that were carried by medieval knights; with this weapon these glyptodonts were able to flail their enemies with deadly force. The legs and feet were very heavy, to support the great mass of the animal. The skull and lower jaw were extraordinarily deep, and the teeth, limited to the sides of the jaws, were tall, each tooth consisting of three columns, joined along a fore to aft line.

Huge ground sloths evolved in South America during late Cenozoic times side by side with the glyptodonts. An evolutionary trend in these edentates was to gigantic size; they never developed any bony armor, although there was a sort of cobblestone pavement of bones in the skin. In these animals

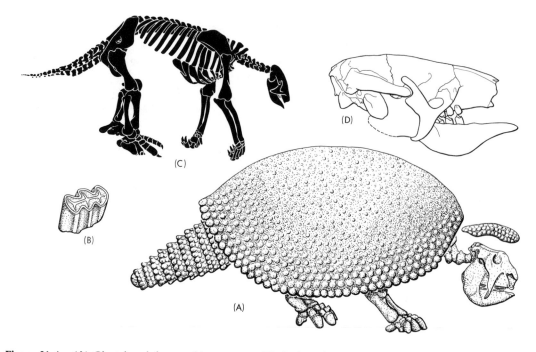

Figure 21-4. **(A)** *Glyptodon*, skeleton and bony armor. **(B)** A glyptodont tooth, **(C)** Skeleton and **(D)** skull of *Nothrotherium*, a ground sloth. A and C, about one-thirty-second natural size; D, about one-seventh natural size.

the skeleton became very heavy throughout, and some of the bones of the legs grew to be enormously broad. The feet were very large, with claws that were probably used for digging. It is evident that the ground sloths, like some of their living relatives, walked on the sides of the hind feet and on the knuckles of the front feet. The skull was somewhat elongated, and the teeth, limited to the sides of the jaws, were peglike. The largest of the ground sloths, such as *Megatherium*, were as large as small elephants, and in life they must have weighed several tons. These were probably plant-eaters that lived largely on the leaves of trees and bushes.

In modern tropical America are the tree sloths, the close relatives of the extinct ground sloths. The tree sloths are small to medium-sized, leaf-eating edentates, that live curiously inverted lives hanging upside down from the limbs of trees by their long, hooklike claws.

Time means nothing to the sloths, which are proverbially slow and clumsy.

Also in the tropics of America are the ant-eaters, which evolved from the same ancestors as the tree sloths. *Myrmecophaga* is a ground-living edentate, in which the skull is enormously elongated by the extension of the snout into a long tube, which is used as a probe to investigate ant hills and termite nests. There is a very long protusible tongue that is utilized for lapping up the insects on which these animals feed. The teeth are completely suppressed. The claws are greatly enlarged and very sharp, and are used not only for digging into the homes of ants and termites, but also as formidable weapons. Like the extinct ground sloths, the anteaters walk on the sides of the hind feet and on the knuckles of the front feet.

Thus we see the edentates evolving in South America during Tertiary times as the armadillos and the glyptodonts, on the one hand,

and as the ground sloths, the tree sloths, and the anteaters on the other. In late Pliocene times the isthmian link between North and South America was reestablished for the first time since the beginning of the Tertiary period, and there was an influx of immigrants from the north into the South American continent. These immigrants brought about the extinction of many South American mammals of ancient lineage, as we have seen, but not of the edentates. These mammals not only withstood the impact of the northern immigrants, but in turn they spread to the north, to invade Central and North America. Consequently certain edentates became characteristic members of the late Pliocene and Pleistocene faunas of North America, and they were so successful that their bones are among the commonest of Pleistocene mammalian fossils.

The glyptodonts and the giant ground sloths ranged widely during Pleistocene times in both North and South America, and the evidence shows quite clearly that these mammals survived until comparatively late dates. There is now ample proof that early man in the New World was contemporaneous with the ground sloths. In addition, the discovery of partial mummies of these great edentates, with patches of skin and hair intact, is an indication of the fact that the ground sloths probably persisted to within the last few thousand years.

The anteaters and the tree sloths never got north of the tropical region, but the armadillos entered the southern portion of North America, where they live at the present time. In fact, there has been a decided northward and eastward spread of these edentates within the last century. Armadillos were formerly limited in the United States to the area immediately north of the Rio Grande, but they have now extended their range as far north as Oklahoma and eastward along the Gulf coast. These are adaptable mammals, and it looks as if they will continue to prosper and spread, in spite of the perils of modern civilization.

PHOLIDOTA

The pangolins or scaly anteaters live in the tropical regions of Asia and Africa. These mammals, belonging to the genus *Manis*, of Pleistocene and Recent Age, have a remarkable body covering of overlapping, horny scales, which makes them look something like reptiles or perhaps like large, animated pine cones. The skull is elongated and rather similar in some respects to the skull in some of the American edentates. There are no teeth, and these animals subsist on ants and termites. There are large, powerful claws on the feet for digging, and the long tail is prehensile, as an aid to climbing in trees.

The fossil record of the pangolins, though fragmentary, affords some clues as to the origin of these strange mammals. A few pangolin bones of Oligocene and Miocene age have long been known in Europe, and within recent years these have been supplemented by the discovery of a partial pangolin skeleton in the Oligocene beds of Wyoming. These materials, named *Patriomanis*, show close affinities with the recent pangolin, *Manis*, of Asia and Africa.

It now seems probable that several early Cenozoic North American genera, commonly known as palaeanodonts, may be ancient members of the Pholidota, rather than of the Edentata, to which group they have long been assigned. These ancestral Pholidota are exemplified by the Eocene genera *Palaeanodon* and particularly *Metacheiromys*, a small animal about forty centimeters or a foot and a half in length. The legs of *Metacheiromys* were rather short, and the feet had sharp claws; the tail was long and heavy. The skull was low and somewhat elongated. Of particular significance is the fact that the dentition was modified; the incisors and the cheek teeth were almost completely suppressed, but the canines were retained as relatively large, sharp blades. In some other palaeanodonts the cheek teeth remained but were reduced to single pegs with almost all the enamel missing. The palaeanodonts

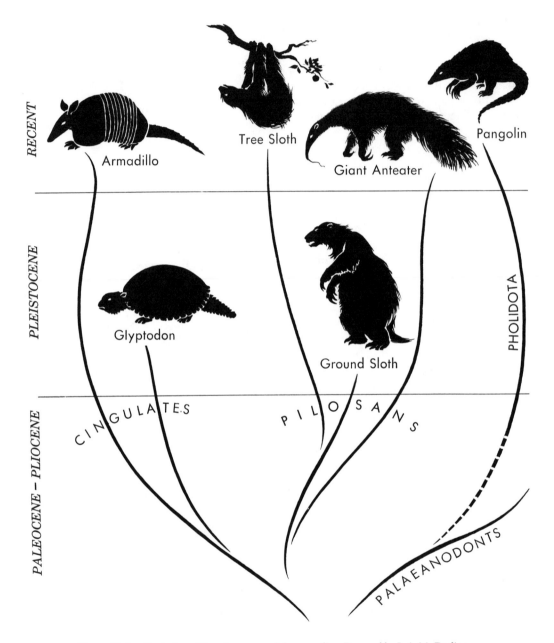

Figure 21-5. Evolution of the edentates, and the pangolins. Prepared by Lois M. Darling.

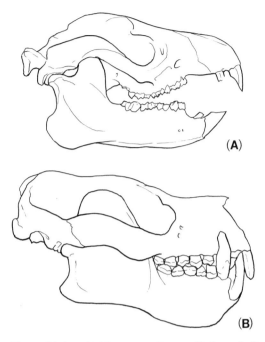

Figure 21-6. **(A)** *Trogosus*, an Eocene tillodont, skull and lower jaw. About one-third natural size. **(B)** *Psittacotherium*, a Paleocene taeniodont, skull and lower jaw. About one-third natural size.

survived in North America until about the end of the Oligocene epoch.

THE TAENIODONTS AND THE TILLODONTS

Here we may consider two "orphan orders" of mammals, primitive placentals that evolved during Paleocene and Eocene times and then became extinct. It seems probable that these mammals were never very abundant; they were early "experiments in evolution" destined to short phylogenetic histories.

The taeniodonts evolved as rather large mammals, especially for early Tertiary times. In the most advanced of these animals, like the Eocene genus *Stylinodon*, the skull and jaws, a foot or more in length, were deep and powerful and the teeth were modified as long, rootless pegs, with a limited band of enamel on each tooth. The limbs were heavy, and on the feet were strong claws, perhaps useful for digging up subterranean tubers. For these reasons it has been suggested that the taeniodonts were related to the edentates. However, as we go back in taeniodont history, we see that the teeth of the earlier forms, like *Conoryctes* of the Paleocene, approached the ancestral tribophenic type. Therefore it would seem likely that these mammals represent an early and independent specialization quite separate from the edentates.

The tillodonts were also large animals for early Tertiary times, and some of the last of them, such as *Tillotherium*, were the size of large bears. They were like bears, too, in the strong skeleton with heavy, clawed feet. The skull, however, had a curious resemblance to the skull in rodents, with enlarged, chisel-like incisor teeth in both upper and lower jaws. The taeniodonts and tillodonts were successful for a time, but with the rise of large, progressive mammals during the Eocene epoch these strange mammals disappeared from the earth.

Stone Age Man

Evolution of the Primates

22

THE ORIGIN OF THE PRIMATES

It is common in discussions of vertebrate evolution to consider the primates—the lemurs, tarsiers, monkeys, apes, and men—last of all. This practice is an outgrowth of the idea that man is the crowning achievement of evolutionary history, that man is at the top of the heap, the ruler of the world, and the arbiter of his destiny. According to this view the grand climax of the evolutionary story is achieved with a description of primate evolution. In a sense this is true, for there is no doubt that the higher primates surpass all other animals in their mental development, and man is a phenomenon unique in the history of life on the earth. On the other hand, the evidence indicates that the primates are probably descended from insectivore ancestors, and for this reason it is quite logical to include them at this point in our discussion.

Although an insectivore ancestry for the primates is probable, the precise derivation of this order of mammals is uncertain. In Upper Cretaceous sediments of Montana there has been found a tooth, named *Purgatorius*, that would seem to be definitely that of a primate, thus carrying the history of the order back into Mesozoic time (*Purgatorius* is known from fairly numerous specimens of Paleocene age, and these corroborate its primate affinities.)

Perhaps an impression of the most ancient primates may be obtained by looking at the modern tree shrews of the Orient, characterized particularly by the genus *Tupaia*. As we have seen, the exact relationships of this mammal have been a subject for controversy; some authorities would place it among the primates, others would include it among the insectivores, while still other scholars would set it off in a category by itself. Morphologically the tree shrews are so close to the line of demarcation between insectivores and primates that, even if they are included in the former mammalian order, they, nevertheless, give us some insight into the adaptations that were leading from primitive eutherian mammals toward the first primates.

Tupaia is a small mammal about the size of a squirrel, with a long snout and a long tail. Most species of *Tupaia* are well adapted for climbing among the lofty branches of tropical trees in search of the insects and fruits on which they feed. The fact that the tupaiids are partially frugivorous (in fact, many of them are dominantly fruit-eating rather than insect-eating animals) points to the manner in which the earliest primates may have changed from a carnivorous or insectivorous diet to a diet composed largely of fruits. This is an important key to the origin of the primates; their separation from an insectivore stem very possibly was brought about, largely, by a change in behavior patterns, whereby their search for food was concentrated on the ever-present and abundant fruits of tropical forests rather than on insects. This behavior would in time affect through selection the morphology and function of the feeding mechanism, and eventually the morphology and function of the whole animal. It is interesting to note that in the tree shrews, such as *Tupaia*, a bony postorbital bar separates the eye from the temporal region, as is characteristic of the primates. In addition, the middle ear region is similar to that of the lemurs and, as in the primates, the thumb and the great toe are set somewhat apart from the other toes. Perhaps, of particular significance is the fact that the brain is relatively large, and that its olfactory region is small. This may be convergence in evolution between the tree shrews and the primates, but it points to the evolutionary trends that, above all, determined the successful direction of primate radiation— the development of a large brain and the dominance of vision among the senses.

From an ancestry illustrated in effect by the tree shrews, the primates evolved along various lines of adaptive radiation during the course of Cenozoic times. This evolutionary development involved an initial radiation of primitive primates in Paleocene and Eocene times. These first primates are the Plesiadapiformes, to be described below. After the appearance of the Plesiadapiformes there was

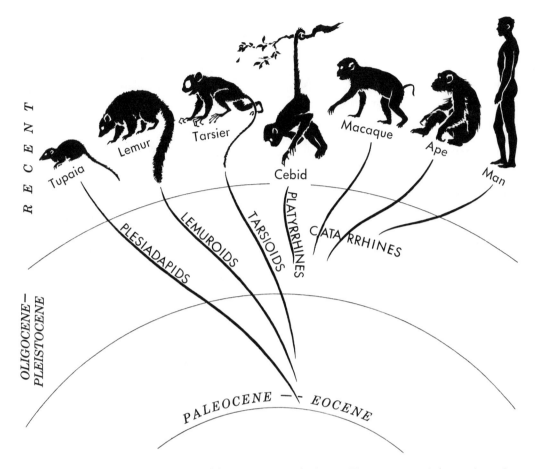

Figure 22-1. Evolution and relationships of the major groups of primates. *Tupaia* represents in a general way the type of primitive placental (either insectivore or primate) from which the basic stock of the primates was derived. Prepared by Lois M. Darling.

a second and definitive radiation of primates, beginning in Eocene time and leading to the primates that have survived into modern times, the lemurs and lorises, the tarsiers, and the monkeys, apes and men.

Because the several large groups of primates resulting from this Eocene to modern evolutionary radiation have been so successful as to be represented in Recent faunas, we are able to follow the successive steps in primate evolution not only from the study of fossil materials but also from firsthand acquaintance with living primates in various stages of advancement. This is indeed fortunate, since throughout their history most of the primates have

been forest loving animals, dwelling in environments in which they frequently have not been abundantly buried and preserved as fossils.

PRIMATE CHARACTERS

In many respects the primates are not highly specialized mammals, so it is not surprising that these mammals should retain many generalized eutherian characters. Nevertheless they show definite specializations along certain lines, and of these the basic underlying trend running through all primate history has been the adaptations for life in the trees primarily in tropical environments. Consequently there has been an emphasis in the primates on those characters important to arboreal life.

Quick reactions are important to active, tree-living mammals. The primates are notable by reason of their activity and restlessness, and especially because of their curious concern with what goes on around them. They are very much aware of the world they live in; they explore their environment with alert eyes. Binocular vision is of particular importance to the primates and is probably more highly developed in them than in any other mammals, giving them a keen visual appreciation of their treetop world. The sense of smell, however, is of no great value to tree-living animals, and so in most of the primates the nose and the olfactory portion of the brain are reduced.

Although there has been a reduction of the olfactory region in the brain of most primates, there has been a converse increase in the rest of that organ, especially the cerebrum. The large brain and the increase in intelligence are the most significant aspects of primate evolution, and they have been of prime importance to the successful progress of these mammals. Primates see things clearly and in detail, they can interpret what they see, and because of their high degree of intelligence and nervous control they are able to use their hands for skillful explorations and manipulations of the things around them. The brain, the eyes, and the hands have made the primates what they are.

Of course these adaptations are reflected in the bony structure. The braincase is large in the primates, and in the more advanced forms it grows to such an extent that it makes up the bulk of the skull. The eyes are also large, and in most of the primates they are directed forward, instead of laterally. They are separated from the temporal region by a bony bar, and in the more advanced primates there is a complete bony wall enclosing the orbit. Since the sense of smell is so greatly curtailed in the primates, the nose is generally small; we might say that it is crowded and pinched between the large eyes.

The jaws are frequently short; the teeth are usually primitive or generalized. The third incisor is lost, as is likewise the first premolar in all but the most primitive primates. In the higher primates the two anterior premolars are lost; therefore the dental formula is generally 2-1-3-3, or 2-1-2-3. The cheek teeth are commonly low crowned, with blunt cusps, an adaptation that permits the primates to eat a great variety of food. Most of the primates are indeed strongly omnivorous—an outcome of the probable early trend, mentioned above, among the most primitive primates to turn away from an atavistic insect diet in favor of fruits and other foods. The ability of a majority of the primates to eat almost anything edible has been, together with the development of the brain, the eyes, and the hands, a significant factor in their success.

The limbs and feet are mobile. Moreover, the joints of the limbs allow much rotation of the bones. The hands and feet, which generally have nails rather than claws, retain five digits. The thumb and first toe in most primates are set apart from the other digits and aid in grasping limbs and manipulating objects.

Although the hands are immensely useful for manipulating objects and as aids to feeding, the majority of the primates, nonetheless, rely on hands as well as feet for locomotion. In fact, some of the primates largely depend on the arms and hands for locomotion—swinging most skillfully through trees. Some of the higher primates are partially bipedal—man is completely so. In many of the primates the tail is long, and used as a balancing organ during their progress through the trees; in the South American monkeys it is even prehensile, and serves as an aid to climbing. In the apes and man, the external tail, of course, is lost.

In addition to these distinctive features, the primates are characterized by certain other traits, of which particular importance may be accorded to the relatively slow postnatal development that necessitates a long period of intimate association between the mother and her baby. Family life is therefore well developed in many of the primates and, of course, reaches its highest stage of development in modern man. In this context it should be said that most primates are highly vocal—they are among the noisiest of mammals.

These are some of the general characters typical of all the primates. Adaptive radiation, however, has been so diverse in these mammals that there are many other characters restricted to the various divergent primate groups. Some of them are further perfections of the characters mentioned above; others are special features peculiar to certain groups.

A CLASSIFICATION OF THE PRIMATES

Several systems for the classification of the primates are presently recognized. Without attempting to argue the merits of these systems, it is here proposed to recognize five basic divisions within the primates of subordinal rank. These are the Plesiadapiformes, the ancestral primates of Paleocene and Eocene age and reasonably well exemplified by *Plesiadapis*, to be described; the Strepsirhini, very primitive primates including the adapids of Eocene to Miocene age and their descendants, the modern lemurs and lorises; the Haplorhini, the modern tarsiers and their ancestors, the omomyids of Eocene and Oligocene age; the Platyrrhini or South American monkeys, ranging in age from the Oligocene epoch to modern times; and the Catarrhini, the Old World monkeys, apes and men, variously extending from the Oligocene to Recent time.

PLESIADAPIFORMES

These, the most primitive primates, are exemplified by the Paleocene genus, *Plesiadapis*, known from fairly complete skeletal materials. Although this genus is not on the direct line to later primates it does give some indication as to what the earliest primates were like. *Plesiadapis* was a small mammal, rather squirrel-like in appearance, with a somewhat elongated skull in which the orbit was confluent with the temporal region rather than being separated from the back of the skull by a bony bar or a wall as is usual among the primates, with a primitive, generalized skeleton, and with clawed feet. The long bones of the lower limbs, namely the radius and ulna, and the tibia and fibula, were completely separate, thus enabling the feet to be rotated, while the digits were elongated to such a degree as probably to enable this mammal to grasp the branches of trees.

Although the Plesiadapiformes are well represented by a considerable array of genera in the Paleocene and Eocene deposits of North America and Europe, these primates failed to survive the transition from Eocene to Oligocene times. As their numbers were diminished during later Eocene history they were replaced by the adapids and the omomyids, these the ancestors of the lemuroids and the tarsiers, next to be considered.

STREPSIRHINI—THE ADAPIDS, LEMURS AND LORISES

The adapids and lemuroids, the omomyids and tarsiers, collectively may be designated as prosimians—these being the primates that preceded the monkeys, apes and men in time, and evolved along lines apart from the higher primates of late Cenozoic and Recent times. At this place those prosimians belonging to the suborder Strepsirhini will be considered.

The adapids arose in early Eocene time, probably as descendants from plesiadapiform ancestors, experienced a certain amount of adaptive radiation in North America and Eurasia during Eocene time, and then continued in diminished numbers through the Oligocene and Miocene epochs. In many respects these early primates continued the functional morphological adaptations that had characterized their plesiadapiform predecessors, especially in the postcranial skeleton. But advances may be seen in the skull and jaws.

Such adapids are nicely represented by *Smilodectes* and *Notharctus* of the North American Eocene. These were small mammals in which the skull was advanced toward the later Primate condition by virtue of a bony bar enclosing the orbit. In *Notharctus* the face was

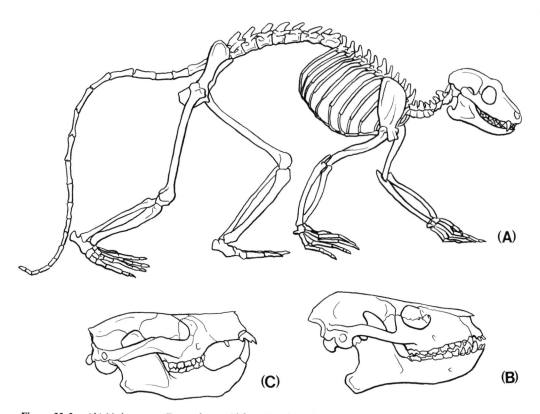

Figure 22-2. **(A)** *Notharctus*, an Eocene lemuroid from North America. About two-fifths natural size. **(B)** The skull and jaw of *Notharctus*, compared with **(C)** the skull and jaw of *Plesiadapis*, a Paleocene plesiadapid primate from North America. (B) One-half natural size. (C) About four-fifths natural size.

rather long, but in *Smilodectes* it had become very much shortened, as is so characteristic of many of the higher primates. In this latter genus the orbits were directed forward to a considerable degree, thus affording the animal perfected stereoscopic vision. In both forms the teeth made a continuous series, without any appreciable gaps between the incisors and canine and the cheek teeth. The small incisors were of normal shape; there were small, tusklike canines, and the molars were low crowned and rather quadrangular, especially by reason of the development of a hypocone in the upper molars. The back was supple, and the tail was very long. The legs were long and slender, and capable of a great range of movement in all directions; and in the hands and feet the first digit was even at this early

date set apart from the other toes. The toes were supplied with nails, rather than the claws which had typified the plesiadapiformes. It is evident that *Smilodectes* and *Notharctus* were able climbers, grasping limbs and branches with their hands and feet, stretching and reaching with their legs in almost any direction necessary for clambering among the tree tops, and using their long tails as balancing organs. In Eocene times these adapids lived in the tropical and subtropical forests that covered much of North America and Eurasia, and established a mode of life that proved to be very successful.

The lemurs and lorises would seem to have had their origins from adapid ancestors, and some of the persistently primitive lemurs, such as *Lemur* itself, are morphologically very similar to their Eocene progenitors. The modern lemurs, living in Madagascar, are small animals with very long tails, elongated flexible limbs, and grasping hands and feet. The muzzle is elongated and pointed, giving the face a foxlike appearance. The eyes are large. The molar teeth are primitive, but in the front of the jaws the upper incisors may be reduced or absent, whereas the lower incisors are long and horizontally directed. These teeth form a sort of comb, with which the lemurs groom their fur. Modern lemurs are noctural animals, roaming through the trees at night in search of insects and fruits.

As might be expected, there have been various departures from the central lemur stock among the lemurs of Madagascar during their long period of isolation on that island. Some of these lemurs are very small; others, like the indris, are rather large and monkeylike, walking on their hind legs. In Pleistocene and sub-Recent deposits on Madagascar are found the remains of a giant lemur, *Megaladapis*, as large as a chimpanzee. This lemur lacked a tail, and it had a large, heavy head.

Among the strangest of the lemurs is the aye-aye of Madagascar, *Daubentonia*, in which the incisor teeth are rodent-like, the digits elongated, and the third finger very long and slender. This peculiar finger is used to dig insects from the bark of trees.

The lorises, now living in India and Africa, are often separated as a group equal in rank to the lemurs. These primates, represented by *Loris*, the slender loris of India, and by *Galago*, the "bush baby" of Africa, are specialized in various ways. For instance, the muzzle is reduced, as in higher primates, and the eyes are turned forward as compared with their more lateral direction in the lemurs. Modern lemurs have three premolar teeth, but in the lorises these are reduced to two. The braincase is high. Like the lemurs, the lorises are predominantly noctural animals that prowl through the trees in search of fruit and insects.

HAPLORHINI—THE OMOMYIDS AND TARSIERS

The omomyids appeared during Eocene time, probably as descendants from plesiadapid ancestors, and evolved through the Eocene and Oligocene epochs. They are exemplified by *Tetonius* from North America and *Necrolemur* from Europe, both genera being characterized by enormous orbits that impinge upon the facial region of the skull so that it is short and narrow. Such adaptations evidently were antecedent to those characterizing the modern tarsiers, so it is reasonable to think that the Eocene omomyids were, like their modern relatives, animals of the night. The jaws were short, the canines were enlarged, and the molar teeth were primitive.

A single genus, *Tarsius*, living in the East Indies and the Philippines, is the surviving tarsier of modern times. This is a small animal, no larger than a squirrel, covered with soft fur. The enormous eyes are set close together and look directly forward. Indeed, the eyes of the tarsier are so large that they occupy much of the face, giving the tarsier a weird, nightmarish appearance, and they crowd the nose to such an extent that it is very small and narrow. The external ears are also large, and indicate that the tarsier has an acute sense of hearing, as well as of sight. These animals are completely noctural, and they rely on the large, light gathering lenses of the eyes and the expanded ears that catch faint sounds to guide them through the tops of the trees in the tropical night, where they prey upon insects and other small animals.

The body is compact, but the tail is very long. The hind legs are also long, and the tarsier has a peculiar elongation of the calcaneum and navicular, two proximal bones of the ankle, to make the foot an elongated lever. Elongation of the hind foot usually signifies a running or jumping animal, and the tarsier is noted for its ability to make tremendous leaps through the tops of trees. The fingers and the toes are slender and are supplied with adhe-

Figure 22-3. Some modern primates. A lemur, tarsier, New World monkey *(Cebus)*, Old World monkey (macaque), and an ape, the chimpanzee. Prepared by Lois M. Darling.

sive pads for the purpose of grasping branches very firmly. Such specializations, established at an early date in the phylogenetic history of the tarsiers, have enabled these little primates to survive successfully as alert, quickly moving denizens of high trees, over a span of many millions of years.

PLATYRRHINI—THE NEW WORLD MONKEYS

We now come to a consideration of the advanced primates, often grouped, in the wide sense of the word, as anthropoids. The anthropoids may be divided into lesser groups—notably the New World monkeys, the Old

World monkeys, and finally the great apes and men.

All these higher primates have certain characters in common that set them apart from their primitive prosimian relatives. In all the anthropoids the eyes are large and face foward, and are completely separated from the temporal fossa by a solid bony wall, a specialization beyond the simple bony postorbital bar of the primitive prosimians. There are either three premolar teeth (in the New World forms) or two (in all but the most primitive Old World types). The molar teeth, though low crowned, are generally quadrangular, having lost the primitive tribosphenic pattern. The brain is comparatively large, and the cranial part of the skull is rounded.

Most of the anthropoids are able to sit in an upright position, and thus the hands are freed for manipulating objects. The thumb and the great toe are commonly set apart from the other digits, as in the prosimians, but in many anthropoids this character is advanced to a high degree of perfection.

The New World monkeys, the most primitive of the anthropoids, are now found in Central and South America, where they generally live high in the trees of tropical forests. They are characterized by flat noses (hence the name platyrrhine) in which the well-separated nasal openings are directed outwardly. Furthermore they retain three premolar teeth, as mentioned above, and some of them are characterized particularly by a prehensile tail, with which they are able to grasp branches. Thus they have a sort of "fifth hand" as an aid for climbing. There are two subgroups of modern New World monkeys, the small, rather squirrel-like marmosets, and the larger cebids, or capuchins, spider monkeys, howling monkeys, and their relatives.

The fossil history of the New World monkeys is scanty. In the Oligocene and Miocene sediments of Argentina and Colombia are found a few genera (such as *Dolichocebus, Homunculus* and *Cebupithecia*) related to some of the more primitive modern cebids, but beyond this evidence there is limited knowledge of the primates in South America during the Tertiary period.

There are two explanations of the origin of monkeys in South America. On the one hand, it is quite possible that the ancestors of these monkeys, perhaps the descendants of northern adapids entered the southern continent from North America in very early Tertiary times, before the two continental masses had been separated. If this is so, they lived during early Tertiary times in South American environments, where their bones were not buried and fossilized. The other possibility is that primates came into South America during mid-

dle Tertiary times as waifs, as accidental and unwilling passengers on natural rafts or floating logs, perhaps floating to the west from Africa when the South Atlantic ocean was much narrower than it is today. If this be so, we have yet to find their ancestors.

CATARRHINI—THE OLD WORLD MONKEYS, APES AND MEN

Some very fragmentary jaws representing two genera, *Pondaungia* and *Amphipithecus*, from the Eocene of Burma, may be the earliest representatives of the catarrhine primates. The rami in these two forms are deep and short, and are set with low-crowned teeth, these being features that would seem to indicate their relationships to the higher primates. Beyond this nothing can be said. However, rather abundant fossils of the Old World monkeys or cercopithecoids can be traced from Oligocene times to the present. One of the first and most primitive of the Old World monkeys was the genus *Parapithecus*, found in Egypt in sediments of early Oligocene age. This was a very small monkey, as indicated by a jaw less than two inches in length. The jaw was rather deep, as is characteristic of the higher primates, and the condyle, the joint for articulation of the jaw with the skull, was placed on a high ascending ramus, well above the line of the teeth. There were three premolar teeth, and the canine tooth was comparatively small. All the teeth formed a continuous series, in contact with each other.

Various fossils of monkeys are found in the Old World in deposits of Miocene and later ages, but for the most part they are fragmentary and give us mere glimpses of cercopithecoid history. However, enough material has been found to indicate that the cercopithecoids were widely distributed throughout the Old World in middle and late Cenozoic times. One of the best-known of these fossils is the genus *Mesopithecus*, a Pliocene monkey related to the modern langurs, and known from complete skeletal materials.

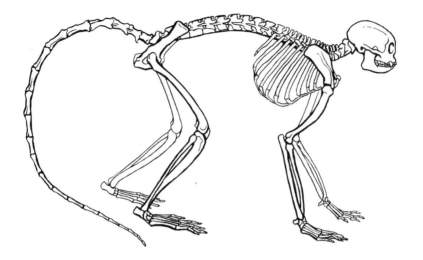

Figure 22-4.   The skeleton of *Miopithecus*, a relatively primitive Old World monkey. About one-fifth natural size.

The more advanced Old World monkeys, as represented by fossil and recent forms, are small to medium-sized mammals. They have only two premolars, a character they share with the highest of the primates, apes and men. The brain is well developed, perhaps more so than in the New World monkeys, and consequently the braincase is large and rounded. The tail is variously developed; it may be long or it may be very short, but it is never prehensile or grasping, as in some of the South American monkeys. In these primates, as in the New World monkeys, the external ears are comparatively small, pressed close to the sides of the head, with the edges rolled over.

In evolving during middle and late Cenozoic times, the cercopithecoids have followed two trends, represented by the cercopithecines and the colobines. The cercopithecines are the various Asiatic and African monkeys, such as the rhesus monkeys, the guenons, and the baboons. These monkeys are gener-

ally omnivorous, eating fruits, berries, insects, lizards, and any other animals they can catch, but some of the highly specialized members of the group are primarily fruit-eaters. They are characterized by cheek pouches, in which they can store considerable quantities of food.

The baboons show an interesting secondary trend in the evolution of the primates away from the basic tree-living life to a life on the ground. In so developing, the baboons have imitated in certain respects some of the ground-living carnivorous mammals, like dogs. This is particularly apparent in the skull. Baby baboons have round heads and short noses, as do most of the higher primates, but in the adults the snout becomes very long and dog-like, and the canine teeth are greatly enlarged. Their elongated jaws and large canine teeth make them aggressive animals, well able to cope with other group-living mammals. This mode of life has proved very successful for the

Figure 22-5.   Skulls of primates: A, B, and C roughly two-thirds natural size, the others roughly one-fourth natural size. **(A)** *Notharctus*, an Eocene lemuroid. **(B)** *Tetonius*, an Eocene tarsioid. **(C)** *Mesopithecus*, a Pliocene cercopithecoid or Old World monkey. **(D)** *Pan*, a modern chimpanzee. **(E)** *Australopithecus*, a Pleistocene man-ape from Africa. **(F)** *Homo erectus*, a primitive Pleistocene man from Asia. **(G)** *Homo sapiens neanderthalensis*, Neanderthal man. **(H)** *Homo sapiens sapiens*, Cro-Magnon man.

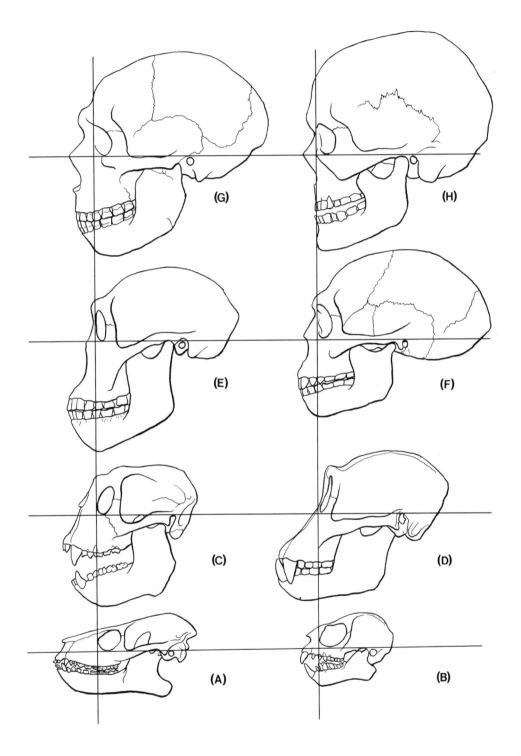

various baboons, now widely distributed throughout the African continent.

The colobine monkeys are the langurs; slender, long-legged, long-tailed, now inhabiting Asia, with one genus in Africa, but living during Pliocene times in Europe. The colobines are consistent tree-dwellers and are herbivorous.

THE APES

Parapithecus, described above, was discovered in the Fayum region of Egypt, approximately sixty miles to the southwest of Cairo, and it is in this same area and from sediments of about the same age that the earliest apes have been found. This fact seems to indicate that in Oligocene times northern Egypt was a tropical forest that harbored the progenitores of the higher primates. From the sediments that yielded *Parapithecus*, the remains of a very primitive ape, *Propliopithecus*, have been excavated. This was a very small animal, the lower jaw being less than three inches in length. Moreover, the jaw was proportionately much deeper than the jaw in *Parapithecus*, and there were only two premolar teeth, as is the case in the higher cercopithecoid monkeys, apes, and men. The teeth formed a continuous series in the jaw, and the lower molars were characterized by five low cusps, in which respect they resembled the lower molars in the later apes. In the several characters, *Propliopithecus* shows strong resemblances to the apes.

Most fortunately there has been discovered in this same Fayum region, an undoubted early ape, represented by a virtually complete skull and lower jaw, with the teeth present. This extraordinary fossil, which has been named *Aegyptopithecus*, although of Oligocene age, comes from a higher and, therefore, a later level than the horizon in which *Propliopithecus* has been found. *Aegyptopithecus*, as might be expected, is more advanced toward the typical ape adaptations than is *Propliopithecus*. *Aegyptopithecus*, evidently about the size of a modern gibbon, had a somewhat expanded

cranium, large, forwardly directed eyes adapted to binocular vision and depth perception, a face that was appreciably longer than in the later apes, a deep jaw, teeth in a continuous series but with the canines of large size, two spatulatelike incisors, two premolars, and three low-crowned molars each with five cusps. The dentition indicates that this early primate was a fruit eater. The skeleton, so far as it is known, suggests that *Aegyptopithecus* was well adapted for life in the trees. Such an ape could very well have been derived from *Propliopithecus* and, in turn, could be nicely antecedent to the Miocene apes, exemplified by the genus *Dryopithecus*.

The characters that so distinguish *Aegyptopithecus* established a base from which developed those evolutionary trends that distinguished the later apes. A few of the more prominent ones may be mentioned. For instance, there has been a strong trend toward a great increase in size, so that almost all the apes have been larger than most of the monkeys, and some of them have been the giants among primates. There has been not only a great increase in body size among the apes during the course of their phylogenetic development but also a marked increase in the size of the brain with, of course, a correlative development in the cranial portion of the skull. Consequently, the head in the apes is large and rounded to a greater degree than it is in the monkeys. The teeth are low crowned and rather generalized, and in the lower molars there are five cusps (already present in *Propliopithecus*) as contrasted with the four cusps in the lower molars of monkeys. The canine teeth are very large in some apes, but this is an adaptation for fighting, not for eating meat. Most of the apes are vegetarians, although some of them may eat a little meat.

Since the apes have been for the most part rather large animals, unable to walk along the branches of trees as do monkeys, they have evolved as brachiating animals, swinging by their arms from branch to branch. This type of locomotion is predominant among the apes, and only in the heavy-bodied giant apes has it been abandoned. Because they have been brachiators, the apes have developed long arms and long fingers, the fingers serving as hooks for hanging on to large tree limbs. The hind legs in the apes are short, and most of these animals are clumsy walkers. And of course the apes have no external tail.

The fossil remains of apes are not common, but they are sufficiently numerous to show that these primates were widely distributed throughout Europe, Asia, and Africa during the middle and later portions of Cenozoic times. Among the earliest apes to appear after *Aegyptopithecus* were *Limnopithecus* and *Proconsul*, from the lower Miocene sediments of Africa. *Proconsul* is probably immediately ancestral to *Dryopithecus*, and from the earliest species of *Dryopithecus*, of middle Miocene age, the later species of this genus evolved, together with a complex of what may be called dryopithecine apes. The dryopithecines, represented by a number of genera, were on the whole generalized apes, in many respects similar to the modern chimpanzee. Probably, most of the apes of Pleistocene and recent times arose from them. From some of the dryopithecines came the evolutionary line leading to the first ancestors of man, as we shall learn.

The modern apes are the comparatively small, long-armed gibbons of Asia (the most skilled brachiators of all primates), the orangs of the East Indies, and the chimpanzees and the gorillas of Africa. The gorillas are the giants among modern primates, far exceeding any men in bulk and strength. However, there seem to have been apes living in Asia during the Pleistocene epoch that were even larger than the gorillas. These gigantic apes, named *Gigantopithecus*, unfortunately known at the present time only from teeth and jaw fragments, were the greatest of all primates, past or present— the nearest things in nature to the giants of fairy tales.

AUSTRALOPITHECINES

Within recent years important discoveries, made in Pleistocene deposits in Africa, have revealed the remains of some advanced preho-

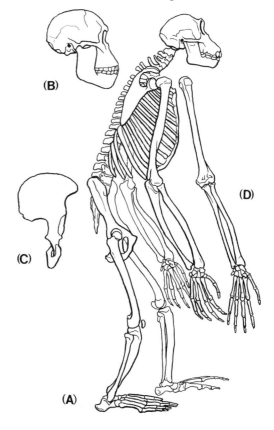

Figure 22-6. (A) Skeleton of the modern chimpanzee, *Pan*. (B) Skull and jaw of *Australopithecus*, a Pleistocene hominid from South Africa. (C) Pelvis of *Australopithecus* (note the expansion of the ilium as an adaptation to an upright posture. (D) Right arm of *Proconsul*, a Pleistocene ape from South Africa. All about one-tenth natural size.

minids that lived during the last great ice age. The first fossil to come to light, the skull of a baby animal, was named *Australopithecus*. Subsequently, the skulls and portions of skeletons of adults were found in South Africa and in East Africa, as well as in Israel, and the Orient and, although they received different names, it seems likely that they may all be included within the genus *Australopithecus*.

It seems probable that the australopithecines were derived from dryopithecine ancestors, this derivation involving some increase in size, enlargement of the brain as indicated by the increased capacity of the cranium, the modification of the dental arch from a U-shaped con-

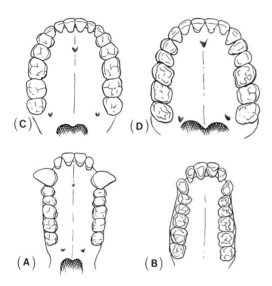

Figure 22-7.   Upper dental arcades of **(A)** *Gorilla;* **(B)** *Sivapithecus,* a Miocene ape from India; **(C)** *Australopithecus,* a Pleistocene hominid from South Africa; **(D)** *Homo,* modern man. Drawn to unit scale.

figuration as is characteristic of fossil and recent apes to a parabolic curve, the diminution of the canine teeth so that they are barely larger than the other anterior teeth and the modification of the skeleton for upright locomotion, as is manifest in the configuration of various skeletal elements, particularly the pelvis.

The oldest australopithecine is *Australopithecus afarensis,* known from rather complete materials discovered in the Afar triangle of eastern Africa. This early man-ape was rather small, but the males were as much as fifty percent and more larger than the females. The arms were long and the hands would seem to have been capable of fairly precise manipulation—possibly for the making of crude tools. The legs were comparatively short. Of particular interest was the discovery a few years ago in this region of fossilized footprints, quite obviously made by *Australopithecus.* Here is dramatic evidence, to be correlated with the structure of the pelvis, for the upright posture that characterized *Australopithecus.*

The combination of an apelike brain with teeth similar to those in men, and the attainment of an upright posture, present a mixture of conservative and progressive characters that indicate the intermediate position of the australopithecines between the dryopithecine apes and the most primitive men. These man-apes approximate closely what the immediate forebears of men may have been like, and thus they give us some insight into the very beginning of human evolution.

The transition from australopithecine ancestors to the first men of the genus *Homo* would seem to have occurred about two million years ago with the appearance of *Homo habilis* in Africa, a man that may have been directly ancestral to *Homo erectus,* to be discussed below.

FACTORS OF HUMAN EVOLUTION

Perhaps human beings may not be descended from the australopithecines as we know them, but it is very possible that man arose from australopithecine-like ancestors. The origin of the human stock from progressive australopithecines probably occurred in late Tertiary times, for man is essentially a Pleistocene animal. Having become differentiated from his primate relatives, man evolved during the Pleistocene period along certain lines that have made him what he is today. The evolutionary development of human beings was not of great magnitude within the course of Pleistocene history; rather it was a matter of the perfection of details that set man apart from all other primates, and from all other animals for that matter.

Four factors have been of prime importance in determining the evolutionary development of man from an apelike primate. They have been the growth and elaboration of the brain, the perfection of the erect posture, a slowing down of postnatal development, and finally the growth in human populations. Let us consider these four factors of human evolution.

Man has been through the course of his evolution a thinking animal. Structurally and physically man was and is far inferior to many of the large animals that have shared his environments; consequently, most of his phenomenal success in adjusting himself to varied environments and in overcoming his enemies has been the result of his superior intelligence. The largest of the modern apes, the gorilla, has a braincase with a capacity of about 500 to 600 cubic centimeters. In the australopithecines the cranial capacity was approximately 600 cubic centimeters, certainly an advance over their dryopithecine ancestors. In the most primitive men of Pleistocene age the cranial capacity was about 900 cubic centimeters, whereas in late Pleistocene and Recent men this figure has ranged from a minimum of about 1200 to a maximum of over 2000 cubic centimeters. The size of the brain is a rough index to mental development, and it is reasonable to think that as the brain got larger through time the intelligence of man increased. Of particular significance is the size of the brain in relation to body size. The most primitive men with cranial capacities of about 900 cubic centimeters were about as large physically as modern men; it is evident therefore that the brain of man has about doubled in its relation to body weight within the limits of Pleistocene time. As we follow the evolution of man we can see the braincase getting increasingly larger as compared with the rest of the skull. Since much of the development of the human brain has been a matter of growth in the frontal region, the cranium has bulged in front, giving to men of late Pleistocene and Recent times a high, broad forehead. Correlatively the face has become increasingly vertical and the jaws proportionately short and small. Because of the shortening of the jaws, the bones containing the teeth have as mentioned changed from a long, U-shaped arch in the ancestral types to a short parabolic curve in the more modern men.

In addition to the physical changes brought about in the skull and jaws as a result of the increase of the brain, there have been the less tangible but very important behavior patterns resulting from the growing intelligence of man as he has evolved. Growth of the intellect has resulted in the ability to communicate ideas by facial gestures and especially by speech. The importance of speech to the quickening evolution of man is so obvious as to need no extended discussion at this place. As a result of his increased intelligence, man became a tool-making animal at an early stage in his history. The manufacture and use of tools, like speech, has been of incalculable effect in hastening the progress of man beyond the status of a mere forest-living animal that competed on more or less equal terms with other animals in wresting a living from the environment.

This brings us to the second important factor in the evolution of man, the development of the upright posture. In this respect man has diverged widely from most of the other primates, which are either completely tree-living animals or ground-living types that walk and run on "all fours" as do baboons and gorillas. The upright posture has freed the hands completely from the necessity of assisting in locomotion and has made them available for handling things, for use in defense, and finally for the manufacture of tools. It is the ability to pick up things and examine them that has helped to make all the higher primates, and especially man, very curious animals that are always experimenting with things and generally learning much as a result of this experimentation.

Physically the upright posture of man has involved a change in the shape of the spinal column. In the ancestors of man, as in the modern apes, the backbone between the skull and the pelvis formed a simple curve, so that the body leaned forward from the hips, and the head was thrust forward from the shoulders. In man the vertebrae are aligned in a complex, S-shaped flexure that throws the body and neck into an upright position, with the head balanced on top of the neck. Naturally this pose is especially efficient for an animal that has freed the forelimbs from locomotor functions, since the entire body from head to heels is aligned along a vertical axis, in line with the force of gravity. Correlative with the development of the upright posture and bipedal locomotion, man has evolved long legs for comparatively fast running over the ground, and arms that are short as contrasted to the arms in most other higher primates.

The third important factor in the evolution of man has been the increase in the process of growing up. A gorilla becomes mature at about the age of ten years, whereas it takes a man about twice as long to reach his full stature and his full powers. This time difference in maturing means that young human beings spend a longer time under the care of parents than other primates, and this longer interval in turn leads to the perfection of family life. This growingup period, as surely as increased intelligence and the freeing of the hands for toolmaking, has been of tremendous import in the advance of man to the high position he now holds in the world.

Finally, the growth of human populations has been of the utmost importance in the evolutionary history of man, particularly in his development during the last few thousand years. From the family group, man progressed to the clan and the tribe, and at length to the complex nations that have developed within the last few millennia. This trend in human evolution, so important to the modern world and the world of the future, has depended upon the social behavior of man, first developed within the family group. It has been made possible by man's capacity for cooperation and by his realization of the necessity for restraint in his behavior. The factor of social relationships, like other physically intangible aspects of evolution, has been of particular importance in making man unique among the animals, a being with a culture.

THE EVOLUTION OF MAN

Perhaps no aspect of evolution has received such intense study as the evolution of man, yet this is a subject concerning which there is much debate, and about which there is much still to be learned. There are various reasons for the incomplete knowledge and controversial nature of human evolution. Fossils are not

common. Because of the importance to modern man of fossil men, every specimen has been subjected to close study, and differences that would be regarded as minor in other animals have here been accorded great importance. But even though the imperfections in the record are great, and there is much disagreement among authorities who work in this field of knowledge, the general outlines of the evolution of man are fairly well defined.

As mentioned above, it is possible that man arose from australopithecinelike ancestors. By early to middle Pleistocene times the pithecanthropoids had evolved. These are known from fossils found in Java and in China, near Peking. The Javanese fossils have been designated as *Pithecanthropus*, those from China as *Sinanthropus*, but it is quite evident to an objective observer that these early men were essentially the same, and therefore should be placed within a single genus. The modern trend among students of early man is to include them in *Homo erectus*. For convenience, those fossils from the East Indies can be called the Java Man, those from China the Peking Man.

These early human beings were definitely more primitive than modern men. The cranium was low, especially in the frontal region, and the bony brow ridges above the eyes were comparatively heavy. The jaws were strong and large, and projected farther forward than the jaws in modern man; hence the pithecanthropoids are said to be prognathous. However the chin was rather receding. The teeth were thoroughly human but they were heavy, and the canine teeth were somewhat longer than the other teeth. Such skeletal evidence that is at hand would indicate that these early men were ground-living, erect-walking primates.

The pithecanthropoids were probably forest men that traveled in small family groups and sought shelter in caves. It seems likely that they knew how to use fire, and that they utilized simple tools and weapons of wood and stone. To our modern eyes they seem like crude people in a low stage of human development, yet they had many advantages over the animals

Neanderthal

Cro–Magnon

Australopithecus   Pithecanthropoid

Figure 22-8.   The man-ape of Africa and early men of Eurasia. Prepared by Lois M. Darling.

that were around them and with which they had to contend. They had made great strides along the path leading to modern men.

In Europe, there are some fragmentary fossils that indicate the presence of men in this region during the first half of the Pleistocene period. One of these fossils is the famous Heidelberg jaw, found in sands of the first interglacial stage near Heidelberg, Germany. It is a very large, heavy jaw, but the teeth are of moderate size and of modern appearance. The Heidelberg jaw may indicate the presence in Europe of a somewhat modernized pithecanthropoid.

In late Pleistocene times there appeared the Neanderthal men, *Homo sapiens neanderthalensis*, known from a considerable number of skulls and skeletons found in Europe, Asia, and Africa. Neanderthal man lived during the third interglacial stage and was a man sufficiently advanced to make tools and weapons and to bury his dead.

Neanderthal man was rather short and stocky, not much more than five feet in height. The shoulders were stooped, the head was thrust forward somewhat, and the knees were slightly bent. The skull showed some traces of pithecanthropoid features. Thus the cranium was somewhat lower than in modern man and the brow ridges were heavy, the face was large and the jaws were prognathous, and the chin was receding.

In spite of his uncouth appearance, Neanderthal man had developed a complex, though primitive, culture. This man was a hunter who used fire. He skillfully fashioned beautiful chipped stone tools and weapons. He lived in a land inhabited by many large aggressive mammals, great cats, bears, wolves, rhinoceroses, and mammoths, and he was able to prevail against them.

Neanderthal man was of the Old Stone Age, the period in the development of human cultures when men relied on chipped flint implements and weapons. One of the last of the Old Stone Age men was the Cro-Magnon man of Europe. Cro-Magnon man was a modern man, belonging to our subspecies, *Homo sapiens* sapiens. He was a tall man, with a large brain and a high, wide forehead. The face was straight, not prognathous like his predecessors', and there was an angular point on the chin. Cro-Magnon man frequented rock shelters and caves, and in the deep recesses of many caves in southern Europe he drew wonderful pictures of the animals that lived around him. He was an accomplished artist, and Cro-Magnon pictures and sculptures rank with some of the finest artistic work ever produced by man. It is possible that Cro-Magnon man interbred with, or displaced and finally exterminated, Neanderthal man in the Old World.

Man was essentially an Old World animal, but with the close of the Pleistocene epoch and during sub-Recent times various men spread over most of the earth's surface. Some twelve to twenty thousand years ago Asiatic men migrated into the New World to populate both North and South America. They were the progenitors of our modern Indians. Other men spread from Asia into the East Indies and the Pacific Islands. As men became proficient with tools they learned how to make clothes and dwellings, and thus were able to escape from the subtropical and tropical lands to which they had originally been limited. They spread north and south as well as laterally, and learned to live comfortably even in the polar regions.

There are four basic stocks of modern men, all belonging to the species *Homo sapiens*. The Australian "blacks" are the most restricted of modern men. The other basic races of modern men are the caucasoids or "whites", the negroids or "blacks", and the mongoloids or "yellow" and "red men"—the Asiatics and the American Indians. Man is a very restless being, made more so in modern times by the perfection of rapid and easy methods of transportation. Consequently the races of men have been mixing with each other for thousands of years, and within the last century this process has been greatly accelerated. Therefore, although the basic racial stocks of modern men remain distinct, their limits are blurred, and are becoming more and more so as the years progress. This is an age of assimilation.

HUMAN CULTURES

A discussion of cultures properly belongs in a textbook of anthropology. However, since the later evolution of man was affected by his cultural development, perhaps it is pertinent to say a few words at this place about the sequence of cultures.

During most of the Pleistocene epoch men made chipped stone implements and weapons. This was the Old Stone Age, and it may be divided into a series of type cultural levels, according to the technique and perfection of the stone work. The Old Stone Age or Paleolithic Age was followed by the New Stone or Neolithic Age, which began perhaps about fifteen thousand years ago. Neolithic men made polished stone implements in addition to their chipped stone tools and weapons. In some parts of the world, as in many Pacific Islands and in the New World, the Neolithic stage of culture continued until modern times; it is only now being displaced.

About five thousand years ago man began to learn the use of metals, and the Bronze Age began. The Bronze Age was followed by the Iron Age, when the great Old World civilizations arose and flourished. We are now in what might be called the Steel Age, a relatively new phase of culture, and are entering with much foreboding the Atomic Age.

What of man's future? This question is in the forefront of much of the thinking and planning of modern men, all over the world. Man has had increasing control over his own fate, even over his evolution during several thousand years; but within the last century, and especially within the last decade or two the degree of this control has increased at a prodigious rate. More than ever before in his history man has assumed crucial responsibilities for himself and for the environment in which he lives, and the manner in which he fulfills these responsibilities will determine to a large degree the course of his history in the years to come.

Gray Squirrel

Rodents and Rabbits

THE EVOLUTIONARY SUCCESS OF THE RODENTS

The criteria of evolutionary success are various. We like to think of ourselves as the most successful of all animals, for in our own way we rule the world. Yet we are but a single species, and the great dominance we now enjoy has been a recent development of the last few thousand years. As contrasted with the evolutionary success of man, or of the order of Primates, or of any other mammalian order, the rodents have been supremely successful during most of Cenozoic times. If the range of adaptive radiation, the numbers of species, and the numbers of individuals within a species are criteria of success in evolution, the rodents far outshine all other mammals.

At the present time the 1700 species of rodents outnumber all other species of mammals combined, and it is probable that this was true throughout most of Cenozoic times. Moreover, most species of rodents are abundantly represented in their respective ranges, so that in individuals they commonly are more numerous than any other mammals. Indeed, it is probable that their biomass, the total mass of all the rodents in the world, exceeds the biomass of the whales, the largest of all mammals.

Several factors have contributed to the evolutionary success of the rodents. In the first place these mammals throughout their history have for the most part remained small. Small size has been advantageous to the rodents, for it has allowed them to exploit numerous environments not available to larger animals, and to build up large populations. The establishment and continuation of large rodent populations are furthered at the present time, as they probably were in the past, by the rapid rate of breeding in these little mammals, which means that they can quickly occupy new ranges and adapt themselves to changing ecological conditions.

The adaptability of most rodents has also stood them in good stead during the millions of years of mammalian dominance. They live on the ground and under the ground, in trees and rocks, in swamps and in marshes, and they range from the equatorial regions to the polar belts. They have been able to compete successfully with other mammals, and frequently to prevail by the sheer weight of their numbers.

All these factors have brought about the long-continued success of the rodents throughout the world. They have persisted where other mammals have failed, and it is quite likely that when human beings decline, at some unforeseeable date in the future, the rodents will still be making their way on the earth with unabated vigor.

In spite of their importance as the most numerous of all mammals, either as species or individuals, rodents cannot be given adequate consideration in this survey of mammalian evolution. Any discussion of the modern rodents involves such a broad range of families and genera and so much detail concerning adaptations to environments that it necessarily becomes very long and complex. Any consideration of the fossil rodents is inadequate, in part because the history of these mammals is incompletely known.

Our unsatisfactory knowledge of the fossil rodents is to a considerable degree a result of the former lack of interest among paleontologists in these mammals. In the earlier collections of fossil mammals rodents are comparatively rare, since the inconspicuous remains of these small mammals were frequently overlooked by collectors who concentrated on the larger animals. Moreover, large mammals have been given priority during past years in collecting programs for the purposes of impressive museum displays.

Within the last several decades, however, many paleontologists have turned their attention to the rodents, so that knowledge of the fossil record of these mammals has increased greatly during this period. Much of this increase is due to new and improved field methods for retrieving microscopic fossils, including rodent teeth, thereby expanding fossil collections far beyond their previous limits. The trend is continuing, and it is to be expected that in the years to come the history of the rodents as revealed by fossils will be greatly augmented.

CHARACTERS OF THE RODENTS

From their beginnings to the present time, the rodents have been gnawing animals, and, in line with this basic adaptation, they have the incisor teeth limited to two pairs of large, sharp-edged chisels, one pair in the skull and an opposing pair in the low jaws. These chisel-like teeth grow from a persistently open pulp cavity, so that the wearing down of their cutting edges is compensated for by the continual growth of the teeth. Along the front edge of each incisor tooth is a broad, longitudinal band of hard enamel, and it is the differential wear between this enamel and the softer dentine composing the remainder of the tooth that brings about the formation and the maintenance of the sharp chisel edge.

The lateral incisor teeth, the canines, and the anterior premolar teeth are suppressed in the rodents, so that there is a long gap between the gnawing incisors and the cheek teeth. Most rodents feed upon plants, although some may eat insects and others are omnivorous. The cheek teeth, which consist of the molars and in some rodents of one or at the most two premolars, commonly take the form of tall prisms; and such teeth, with the grinding surfaces complicated by folding of the enamel, are well suited for grinding hard grain and other plant food. In the more primitive rodents the teeth may be low crowned, with blunt cusps.

The skull is long and low, and the brain is primitive. The articulation between the skull and the lower jaw and the development of the cheek muscles are of such a nature that the mandible can be slid backwards and forwards in relation to the skull, as well as moved up and down and sidewise. This is variously accomplished in the rodents, according to the manner in which the origins of the several layers of the masseter muscle are arranged on the side of the skull. In typical mammals the masseter is a powerful muscle having its origin on the zygomatic arch or cheek bone of the skull, and its insertion on the lower border of the jaw. It serves to close the jaw with great force. The rodents show four variations or patterns of this arrangement.

In the most primitive rodents the so-called protrogomorph pattern is seen, and is characterized by a long portion of the masseter originating beneath the front of the zygomatic arch and running back to the angle of the lower jaw. This muscle, which pulls the jaw forward, lies above the deeper portion of the masseter, which is vertically directed between the zygomatic arch and the lower border of the jaw in the usual fashion.

In those rodents typified by the sciuromorph pattern, a branch of the masseter extends up in front of the orbit onto the side of the face. In the rodents having the hystricomorph pattern, another branch of the masseter has grown up inside the zygomatic arch and extends forwards through a greatly enlarged infraorbital opening in front of the eye (which in most mammals is for the passage of blood vessels and nerves), to expand over the side of the face. In the rodents that have the myomorph pattern of jaw muscles, there is a combination of the latter two patterns, with both branches of the masseter extending forward—one beneath the zygomatic arch and one inside the arch and through the infraorbital opening. These patterns of jaw muscles are shown in figure 23-1.

The skeleton behind the skull is not highly specialized in most rodents. The forelimbs are generally very flexible for climbing, running, and food gathering, and usually all the toes are retained. These toes, like those of the hind feet, usually have claws. The hind legs are frequently more specialized and less flexible than the forelimbs. Some rodents are adapted for hopping; their hind limbs are long and powerful, and their forelimbs are comparatively short.

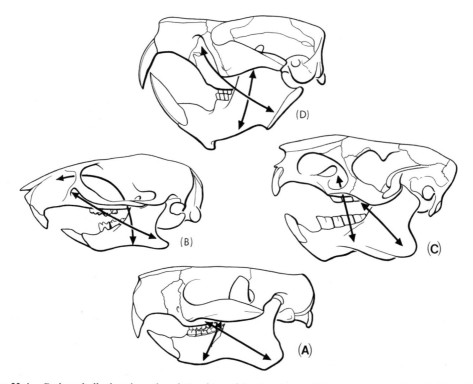

Figure 23-1. Rodent skulls that show the relationships of the deep layers of the masseter muscle, as indicated by arrows. **(A)** *Paramys,* a primitive Paleocene and Eocene rodent, showing the protrogomorphous type of jaw musculature, in which both layers of the muscle reach from the zygomatic arch to the lower part of the mandible. **(B)** *Cricetops,* an Oligocene rodent showing the myomorphous type of jaw musculature, in which the deep layer of the masseter muscle is extended through an enlarged infraorbital foramen on to the face. **(C)** *Neoreomys,* a Miocene rodent showing the hystricomorphous type of jaw musculature, in which the deep part of the masseter muscle is accommodated within a greatly enlarged infraorbital foramen. **(D)** *Palaeocastor,* an Oligocene rodent showing the sciuromorphous type of jaw musculature, in which the middle layer of the masseter muscle originates from an area in front of the zygomatic arch. (A) one-half natural size; (B) twice natural size; (C) one-half natural size; (D) two-thirds natural size.

A CLASSIFICATION OF THE RODENTS

Among the orders of mammals the rodents are the most difficult to classify. For many years the common practice has been to divide the rodents into three large subordinal categories—the sciuromorphs, the myomorphs and the hystricomorphs, based on the specializations of the masseter muscles that have been outlined above.

But it is now thought that the several patterns of specialization within the masseter muscle may have developed independently more than once, and that there are transitions from one pattern to another. Therefore the specializations of the masseter muscle probably are not of prime significance for the classification of the rodents. Furthermore, with the discovery and the study of abundant fossil materials during the past two or three decades, the threefold subdivision of the rodents has become more and more unsatisfactory; hence, modern authorities are reluctant to continue with this system. The problem is what to substitute. Many of the rodents can be included in three large suborders, more or less as outlined above but, when this is done, a considerable residue of superfamilies, families, and genera will not fit into the threefold arrangement of suborders. And any system that includes them will be unsatisfactory in one way or another, especially according to someone who may be reviewing the classification. Dr. A.E. Wood, one of our leading students of fossil rodents, has said, "The current status of rodent phylogeny and classification is such that anyone can point out inconsistencies in anybody's else's classification."

A recently developed classification of the rodents, which with modifications will be adopted here, recognizes two large suborders of these mammals, each suborder divided into several infraorders, as follows:

Suborder Sciurognathi
  Infraorder Protrogomorpha
    Primitive rodents, paramyids and their relatives, aplodonts or sewellels, mylagaulids.
  Infraorder Sciuromorpha
    Squirrels, chipmunks, marmots, prairie dogs.
  Infraorder Castorimorpha
    Beavers and their relatives.
  Infraorder Myomorpha
    Voles, rats and mice, pocket gophers, dormice, kangaroo rats, jerboas, jumping mice.
  Infraorder Theridomorpha
    Extinct European rodents, anomalurids, ctenodactylids.
Suborder Hystricognathi
  Infraorder Hystricomorpha
    Old World porcupines.
  Infraorder Phiomorpha
    Oligocene to Miocene African rodents of diverse types. Cane rats of India and Africa, rock rats and mole rats (bathyergids) of Africa.
  Infraorder Caviomorpha
    South American cavies, capybaras, agoutis, chinchillas, spiny rats, and New World porcupines.

A listing such as this is admittedly very incomplete because it does not include a host of extinct rodents for which there are no common names. But by citing examples of the various modern rodents belonging to the several major groups of these mammals perhaps some impression may be had as to the great variety of rodent evolution.

EVOLUTION OF THE RODENTS

One of the earliest and most primitive of the known rodents is *Paramys*, a protrogomorph from Paleocene and Eocene sediments of North America. *Paramys* was rather like a large squirrel, with clawed feet for grasping and probably for climbing and a long tail for balancing. The cheek teeth were low crowned, with rather blunt cusps in the unworn condition. Yet even though *Paramys* was a primitive rodent, it was none the less a specialized mammal, with no indications of its derivation from more primitive mammals. The skull was elongated and rather low. The incisors were large chisels. There were two premolars and three molars on each side of the skull, one premolar and three molars in each half of the lower jar, the maximum number of cheek teeth in this order of mammals.

From ancestors approximated by *Paramys* the protrogomorphs evolved along numerous trends of adaptive radiation during Cenozoic times. A conservative stock was the Aplodontoidea, of which the sewellel *(Aplodontia)* of Northwestern North America is the only surviving representative. The aplodontoids reached the zenith of their development in middle Tertiary times, when they were represented by several families, among which the Miocene mylagaulids, were particularly abundant. The mylagaulids were burrowing rodents. Some members of this group, such as *Ceratogaulus*, had horns on the skull, the only rodents showing this specialization.

The squirrels, the sciuromorphs, appear in the fossil record in middle Tertiary times, but it seems obvious that these rodents represent a long-continued and probably little-changed line from paramyid ancestors. Most squirrels, like *Sciurus*, have been tree-living types, and this environment has afforded them the safety and the food supply that has insured their long continuation to the present day. However, some members of this group of rodents are specialized for life on the ground and in burrows. These are the various ground squirrels

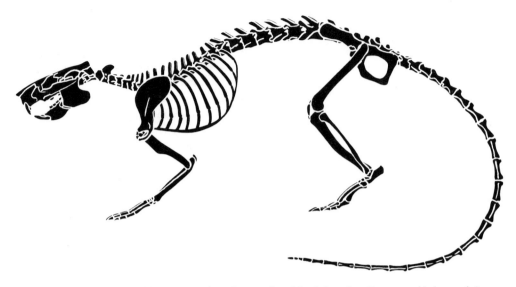

Figure 23-2.   Skeleton of the Eocene rodent, *Paramys,* from North America. About one-third natural size.

and chipmunks, like *Tamias*, the prairie dogs (*Cynomys*), and the marmots or woodchucks, these last being the largest of the squirrel family.

The beavers of Pleistocene and Recent times are the sole survivors of the castorids, a sciuromorph group that made its appearance in Oligocene times. The first castorids, such as *Palaeocastor* and *Steneofiber*, were small, burrowing rodents, and their skeletons have been found associated with the natural casts of their burrows. During the course of time the castorids showed a trend toward adaptations to the water, culminating in the beavers, with which we are familiar. In Pleistocene times there were giant beavers in North America, and of these *Castoroides* was one of the largest of all rodents, attaining a size comparable to that of a small bear.

The eutypomyids, from the Oligocene of North America, have been considered as relatives of the beavers, but this concept probably is erroneous. They are characterized by complexly folded enamel in the cheek teeth and by rather aberrantly specialized feet.

Perhaps the word "rodent" means rat or mouse to most people, since they are the ro-

dents with which we are most familiar, often to our great annoyance. The myomorphs show a considerable range of adaptive radiation, but the central and most abundant types are the various rats and mice. It must not be supposed, however, that the names rats and mice refer only to the unwelcome dwellers in our homes, for there is a tremendous array of rats and mice belonging to numerous genera and species living throughout the world. The rats and mice are at the present time the most successful of the rodents, outnumbering all other rodents in numbers of species and probably in numbers of individuals as well.

The ubiquitous myomorphs may be divided into at least five groups of superfamily status. The muroids are the rats, mice, voles, and lemmings in their varied manifestations—the most numerous of all mammals. Many genera and species, native to the several continental areas, have been components of mammalian faunas in many regions of the world during much of the latter portion of the Cenozoic era. In addition to the muroids, the myomorphs include the long-ranging geomyoids, represented today by such divergent types as the burrowing pocket gophers (*Geomys*) and the hopping, desert-living kangaroo rats (*Dipodomys*). The gliroids or dormice (*Myoxus*) are, in some respects, such as the presence of low-crowned cheek teeth and the retention of a premolar, the most primitive of the myomorphs. Recent evidence would seem to indicate that these rodents originated from European Eocene paramyids and they may form a separate group quite independent of the myomorphs. For the time being they are retained here, in their traditional position. Another group of this infraorder of rodents is the dipodoids, consisting of the Old World jerboas (*Dipus*) and the New World jumping mice (*Zapus*). In these myomorphs the hind limbs are greatly elongated for making long hops, whereas the forelimbs are comparatively small. The development of adaptations for leaping in the jerboas and the jumping mice obviously took place independently by processes of parallel evolution. Fossils of these rodents are not common, but the evidence for them in the geologic record extends back to Oligocene times. Finally, we mention the spalacoids, the Old World burrowing rats, the fossil record of which extends into late Miocene times.

The infraorder Theridomorpha would seem to have had its beginnings in the Eocene of Europe, where primitive rodents known as pseudosciurids may form a link between the ancestral paramyids and the later theridomorphs. Early theridomorphs, small scampering rodents, were the dominant rodents in Europe during early Tertiary times. Theridomorphs survive today as the anomalurids, some of which are known as scaly-tailed squirrels. These handsome rodents are highly adapted as arboreal mammals and some of them show interesting parallels to the flying squirrels of North America, having large, membranous folds of skin between the legs that enable them to glide with great skill between trees. The ctenodactylids or gundis, small, compact, short-tailed rodents of northern Africa may belong in this infraorder.

Brief mention may be made of the pedetids, represented today by the springhaas or "jumping hare" of Africa, a large rodent with long hind legs. Its ancestry extends as far back as the Miocene epoch.

Having briefly considered the sciurognath rodents, we may now turn to the hystricognaths, composed of the two infraorders designated as the Phiomorpha and the Caviomorpha. The Phiomorpha are represented today by the large cane rats of Africa and the strange African mole rats or bathyergids, of which one genus, *Heterocephalus*, is almost devoid of hair and adapted to a colonial, subterranean existence. The phiomorphs are indicated by several families of such rodents in the fossil record of Africa and it would appear that they went through their evolutionary development on that continent.

The hystricids or Old World porcupines may be regarded as probable phiomorphs, although they represent a rather independent line of evolutionary development reaching back into the Miocene epoch. They may have arisen from early Tertiary phiomorphs, and probably have nothing to do with the porcupines of the New World. They evolved in Asia and Africa and are found in those continents today.

The Caviomorpha is a large infraorder of the Hystricognathi, that evolved in South America from Oligocene to Recent times. Although the caviomorph record is predominantly South American it seems likely that these rodents had their beginnings in Middle America, and may have entered South America in late Eocene or early Oligocene times by rafting, that is, by floating across oceanic barriers on logs or masses of vegetation. Possible caviomorph ancestors of the proper geologic age are found in Texas and central Mexico, which may have been centers from which the caviomorphs were then rafted into the then island continent of South America. Alternatively, they might have drifted westward from Africa when the South Atlantic Ocean was much narrower than it is today.

In South America the caviomorphs evolved during the interval from Oligocene to Recent times along varied lines of adaptive radiation.

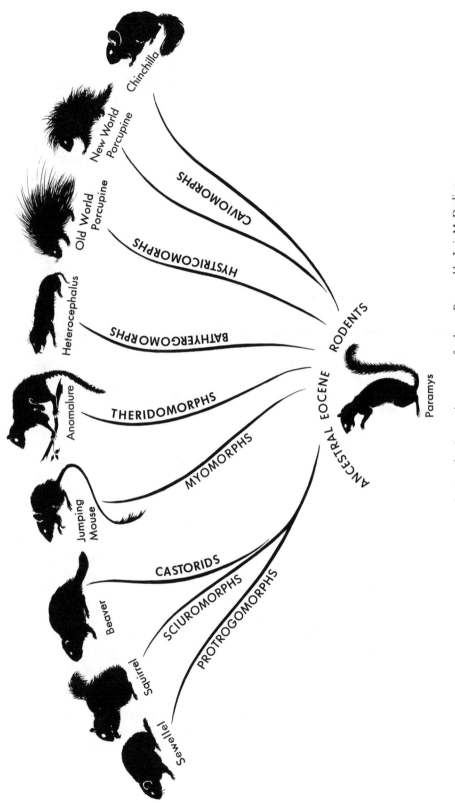

Figure 23-3. Adaptive radiation of various major groups of rodents. Prepared by Lois M. Darling.

One of these lines was that of the New World porcupines, a very successful group of rodents that pushed into North America in Pleistocene times, when the two continents had become reunited. The other caviomorphs of South America, a few scattered members of which invaded the northern continent in the Pleistocene epoch, have been the cavies or guinea pigs *(Cavia),* the capybaras *(Hydrochoerus),* and their relatives, some of which are the largest of modern rodents, the agoutis *(Agouti)* and pacas, the chinchillas, and finally the octodonts, which may be designated as the "spiny rats." Until the final stages of the Pliocene epoch these rodents were isolated on the island continent of South America, completely free from competition with any other rodents. Consequently they evolved along many lines, some of them developing as small, rather mouselike animals, others as running, rabbit-like animals, and others as large, herbivorous rodents. Although the invasion of South America in late Pliocene and Pleistocene times by various North American mammals brought about the extinction of many of the indigenous mammals of ancient ancestry on that continent, it did not so affect the caviomorph rodents, and they continued with vigor into our modern age. Indeed there was little Pleistocene extinction of rodents on any of the continents.

The development of the rodents through geologic time along these several infraordinal lines, so briefly summarized in the foregoing paragraphs, indicates a high degree of parallel evolution among these mammals. Such parallelism, involving many of the anatomical features and some of the behavior patterns that characterize the rodents, has in the past confused our understanding of phylogenetic relationships among these most numerous of all mammals. But with the growing recognition of the important role that parallelism has played in the evolution of the rodents, and with a constantly increased knowledge of the fossil forms (this, the result of improved field techniques and the increased interest, already

mentioned, among paleontologists in these mammals) our knowledge as to the details of rodent evolution should become even clearer as time goes on.

THE HARES, RABBITS, AND PIKAS

It was long the practice to consider the hares, rabbits, and pikas as rodents, placing them in a suborder, the Duplicidentata, so named because of the two incisor teeth on each side in the skull. This suborder was set against the Simplicidentata, the rodents with a single incisor on either side in the skull. The grouping of the rabbits and their relatives with the rodents was based primarily on the general similarity of adaptations for eating in these two large groups of mammals, for in both rabbits and rodents there are enlarged incisor teeth for gnawing, separated by a long gap or diastema from the grinding cheek teeth. In recent years, however, there has been a growing tendency among students of mammals to regard the rabbits as quite independent of the rodents, and to view such similarities as exist between these mammals as a result of convergent evolution. There is good reason to think that this interpretation is correct. Very primitive rodents and rabbits are found in sediments of Paleocene and Eocene age, and these fossils show that the two groups of mammals were quite distinct from each other at that early stage in the history of the mammals. Moreover, when the seemingly similar characters of the rabbits and rodents are critically examined it soon becomes apparent that the resemblances are superficial rather than basic.

For instance, the rabbits have enlarged incisor teeth, as do the rodents, but this is a character that has developed independently many times in various groups of mammals. As for the cheek teeth, there are no real similarities. In the rabbits there are two or three premolar teeth present, as contrasted with a much greater limitation in the rodents. Moreover, the cheek teeth in the hares and their relatives are tall prisms with transversely ridged crowns

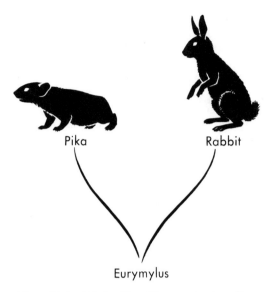

Figure 23–4. Relationships of the two suborders of lago-morphs. Prepared by Lois M. Darling.

for cutting, rather than crushing teeth of the rodent type. The masseter muscle, though powerful in the rabbits, never shows the high degree of specialization so characteristic of the rodents. Moreover the system of mastication in the rabbits is quite different from that of the rodents, being transverse, or from side to side, whereas in the rodents the movement of the jaws is propalinal, or forward and back-ward. In the postcranial skeleton there are few similarities between rabbits and rodents. The rabbits are animals specialized for hopping, and as a result the hind limbs are very long and strong. The tail is reduced to a mere vestige.

Thus it would seem probable that the rab-bits are an independent order of mammals de-scended from an ancient eutherian ancestry. Various basic relationships have been sug-gested for the rabbits and their relatives, rang-ing from an ancestry in common with that of the condylarths (the primitive hoofed mam-mals) to the carnivores to the tree shrews, and recently back to the former idea of an early affinity with the rodents. This last supposi-tion is based upon the discovery of Paleocene mammals known as anagalids that seem to show an early divergence of rabbits and ro-dents from a common ancestry. Whatever may be their ancestry it is apparent that the rabbits and their relatives are sufficiently dis-tinct to be placed in a separate order, the Lagomorpha.

EVOLUTION OF THE LAGOMORPHS

An early indication of rabbits in the fossil record is the genus *Eurymylus,* an anagalid from upper Paleocene sediments of Mongo-lia. Although this ancient mammal must be excluded from the direct ancestry of later lago-morphs, since it had lost the second upper in-cisor, it nevertheless indicates what the very early Tertiary rabbits were like. These animals are sparsely represented in Eocene faunas, but with the advent of the Oligocene epoch they seem to have become abundant, and to have continued this abundance into modern times.

The lagomorphs became divided at an early date into two separate groups of family rank, and have maintained this dichotomy ever since. On the one hand there are the pikas, of which the modern pika or "cony," *Ochotona,* is typical. Through the extent of their history they have been small, compact, short-legged lagomorphs, with short ears. Contrasted with the pikas or Ochotonidae, there are the Lepor-idae, or hares and rabbits, such as *Lepus* and *Sylvilagus.* These lagomorphs have evolved as swift runners that bound over the ground in long hops. The hind limbs are very long to give power and distance to the leap. The front limbs are adapted for taking up the shock of landing. The ears are elongated, particularly the hares', and serve as acute sound-gathering devices.

RODENTS, RABBITS, AND MEN

The development of man as an agricultur-ist and as a civilized human being has been inextricably interwoven with the fortunes of the rodents and rabbits. When man was a primitive hunter it is probable that rodents and rabbits were of little concern to him, except as secondary sources of food; but as he established his dominance over the larger animals of the earth, he came into increasing conflict with the rodents and the rabbits.

When man began to grow vegetable crops, cereals, and grains, he discovered that the rodents and the rabbits were eager to share his food supply. So it is that the history of civilized man has entailed, among other things, a long battle with certain rodents and rabbits, which have persisted in invading his premises, to eat his food and to bring disorder into his household. The battle has been going on for a good many thousand years.

Of course, the common house mouse and the European rats are virtual parasites on man, going wherever he goes, and eating whatever he eats. The damage done by these rodents each year is tremendous. But various rodents of the fields also are and have been for many centuries annoying and costly to man. Likewise the rabbits cause much damage as a result of their depradations, although these animals are also viewed with favor, since they are for us a source of food.

In addition, the rodents and rabbits are of extreme importance and concern to man as carriers of disease. During past ages, rats and lice spread the bubonic plague through Eurasia, with terrifying and disastrous effect, and it is only within comparatively recent times that this great scourge has been conquered. Moreover, it requires eternal vigilance on the part of man in the modern world to prevent the spread of various diseases borne by rodents and rabbits.

All these considerations point up the fact that the rodents and the lagomorphs have been extraordinarily successful animals during the last fifty or sixty million years. From the evolutionary viewpoint these mammals represent in many respects the climax of mammalian success.

Fin-Back Whales

Cetaceans

THE RETURN TO THE SEA

Of all mammals the whales, dolphins, and porpoises, or cetaceans, are certainly the most atypical, and in many ways the most highly specialized in the extent to which they have diverged from their primitive eutherian ancestors. These mammals probably arose in early Cenozoic times from primitive carnivore-like mammals known as mesonychids. Like the bats, the whales (using this term in a general and inclusive sense) appear suddenly in early Tertiary times, fully adapted by profound modifications of the basic mammalian structure for a highly specialized mode of life. Indeed, the whales are even more isolated with relation to other mammals than the bats; they stand quite alone. Therefore, it seems evident that the whales, having separated from their mesonychid forebears at an early date, enjoyed at the outset a series of extraordinarily rapid evolutionary changes that made them by middle Eocene times well adapted for life in the ocean.

The whales returned to the sea to imitate the fishes after a fashion and to take the place of the ichthyosaurs, the fishlike marine tetrapods of Mesozoic times. In becoming adjusted to a life in the open ocean the whales, as descendants of fully land-living ancestors, were faced with the same problems that had been encountered and overcome millions of years previously by the first ichthyosaurs. These problems have already been discussed in the description of the ichthyosaurs, but perhaps they may be repeated briefly at this place.

The problems to be surmounted by the ancestral whales in their new environment were in short those of locomotion, of respiration or breathing, and reproduction. The adaptations to locomotion in the water involved a streamlining of the body, the eventual development of a fishlike tail as the main propeller, and the transformation of the legs into balancing paddles. Like the ichthyosaurs of the Mesozoic, the whales retained the lung breathing that was their heritage from land-living ancestors, and in so doing they evolved adaptations that increased the efficiency of respiration. Of course embryonic development in whales posed no particular problems for these mammals, but there were special adaptations for the survival of the young in the water, from the moment of birth on.

The return of whales to the sea is a fine example of convergence in evolution. In following this evolutionary trend the whales have shown many adaptations that are unusually similar to those of the ichthyosaurs, yet the ancestors of these two groups of tetrapods were quite distinct—early diapsid reptiles on the one hand and primitive eutherian mammals on the other. Convergence such as this illustrates the remarkably similar adaptations by dissimilar animals to an environment that imposes stringent limitations upon its inhabitants.

WHALES AS MARINE VERTEBRATES

As already mentioned, whales are fishlike mammals, highly streamlined for efficient swimming. The body is torpedo shaped, and there is no visible neck distinct from the trunk. There is no hair on the body of the modern whales, and it is likely that this adaptation developed early in the history of these mammals. The smooth skin makes a sleek surface that offers little resistance to the water as the animal moves forward. Since whales are warm blooded, they have developed a heavy layer of fat or blubber beneath the skin, to give the necessary insulation that in most mammals is furnished by the covering of hair.

The vertebrae are much alike and numerous, making for a flexible backbone, but the neck vertebrae are commonly shortened and fused into a single bony mass. Long muscles and tendons attach to the vertebrae and run back to the tail, to furnish the propulsive force that drives the animal through the water. The whales are unlike most marine vertebrates in that the tail terminates in a horizontal fin that moves up and down, rather than in a vertical fin that moves from side to side. The horizontal tail fin of the whale, generally called flukes, is a neomorphic or new structure in these mammals and, although stiff and strong, contains no bones. Likewise almost all whales have a fleshy dorsal fin, another neomorphic structure. This fin is a stabilizer that prevents the animal from rolling, and may be compared in function with the fleshy dorsal fin of the ichthyosaurs or the bony dorsal fins of fishes.

As in the ichthyosaurs, the limbs in whales have been modified as paddles. However, in all known whales, both fossil and recent, the pelvis and hind limbs are reduced to mere vestiges; only the front limbs remain as functional paddles. The arm bones are short and flattened, the wrist bones are flattened discs, and the fingers are commonly greatly elongated by multiplication of the phalanges as supports for the paddle.

In all but the earliest whales the external nostrils are shifted to the top and the back of the skull, to form the "blowhole" so characteristic of these mammals. The nostrils can be closed by valves, and the lungs are highly elastic and extensible, for taking in great quantities of air. Whales are able to remain submerged for long periods of time—as long as an hour in some of the large whales—and some of them can dive to great depths. In line with these remarkable accomplishments there are profound adaptations in the physiology of the whales.

In addition to the modifications in the skull brought about by special adaptations for breathing, the whales show marked specializations of the ear. The tube to the external ear and the eardrum is drastically reduced, and it is obvious that whales do not hear sounds in the same manner as other mammals. They are very sensitive to vibrations in the water, which are transmitted to a heavy, shell-like bone formed by a fusion of the periotic bone and the auditory bulla, and separate from the rest of the skull. Indeed, modern experimental work has shown that the cetaceans, like the bats, have a very sophisticated sonar apparatus by means of which they can locate and identify objects in the water. Moreover, whales

and porpoises communicate with each other by projecting underwater sounds or vibrations of varying frequencies; their system of communication seems to have reached a rather high stage of development. Indeed evidence shows that some toothed whales emit sounds so intense as to stun their prey. In contrast to this highly developed sense, whales have no olfactory sense. Consequently olfactory lobes are lacking, and the brain is large and rounded. The brain is highly developed in these mammals, and recent experiments indicate that whales are remarkably intelligent. The brain and the sensitive auditory apparatus are telescoped into a small space at the back of the skull; the remainder of the skull is made up of very long jaws.

Whales are and have been throughout their history carnivorous, living on large invertebrates, fishes, and even other whales, or on microscopic marine animals known as plankton. In the first category are the toothed whales, ancient and modern. Those in the second category are the whalebone whales, in which the teeth are suppressed and transverse plates of fibrous keratin or baleen hang from the roof of the mouth to strain plankton from the water.

As for reproduction, young whales are very large and well formed when they are born. As soon as they are born, the mother pushes them to the surface so that they can get their first breaths of air. From that time on the baby is able to swim along with its mother. The mammary glands are enclosed in a sort of pocket, so that the young whale can nurse without shipping a lot of sea water.

ACREODI

The Acreodi, here accorded ordinal status, contains a single family of carnivorous land mammals, the Mesonychidae. These mammals, as typified by such Paleocene and Eocene genera as *Dissacus* and *Mesonyx* were long regarded as primitive carnivores known as creodonts, and subsequently as primitive hoofed mammals known as condylarths. Perhaps it is significant that detailed anatomical features of the teeth and the base of the skull show similarities with those of the early cetaceans. Consequently there is now some sentiment among students of these mammals for regarding the Acreodi as belonging to an independent order, intermediate between the primitive creodonts and the primitive cetaceans. Indeed, the genera *Ichthyolestes* and *Gandakasia,* from the middle Eocene beds of Pakistan and originally considered as creodonts, are now variously classified as either mesonychids or cetaceans. Evidently the Acreodi occupy a position close to the ultimate ancestry of the whales.

EARLY WHALES

Recent discoveries in the early and middle Eocene sediments of Pakistan in which are found mesonychids showing cetacean-like features, have yielded some diagnostically significant fossils, named *Pakicetus,* that have teeth resembling not only mesonychid teeth but also the teeth of middle Eocene cetaceans. It is perhaps significant that the otic region of *Pakicetus* lacks the specializations characteristic of whales that have established the highly efficient underwater directional hearing in these mammals. Yet there is good reason for classifying *Pakicetus* as a true cetacean. *Pakicetus* is found in river sediments along with various land-living mammals—an indication that this earliest and most primitive cetacean had not as yet ventured into the open ocean, although it probably did frequent marine embayments into which the rivers flowed.

The first whales, known as archaeocetes, are typified not only by *Pakicetus* but also by other genera of middle Eocene and later ages, of which the upper Eocene genus *Basilosaurus* (commonly designated as *Zeuglodon*) is known from relatively abundant remains from "Zeuglodon Valley," in the Fayum region of Egypt. These early whales were large, and in this respect they indicate an evolutionary trend typical of the whales. Many of these mammals,

Figure 24-1. *Pakicetus,* an ancestral cetacean from the Eocene of Pakistan. **(A)** Skull and Jaw, about one-sixth natural size. **(B)** Restoration of *Pakicetus,* showing possible appearance, in silhouette.

freed from the limiting effects of gravity, have become the giants of the animal world past or present. Even *Basilosaurus* (or *Zeuglodon*) of the Eocene epoch was some sixty feet in length. In this whale the tail was very long and the forelimbs were modified into paddles. Until recently it had been assumed that the hind limbs were suppressed in this Eocene whale. But as a result of recent field explorations in Egypt, a complete hind limb of *Basilosaurus* has been brought to light. Although relatively small the limb is well formed, and includes the foot, containing three toes. The anatomy of the foot indicates its derivation from a mesonychid ancestor.

The skull in the archaeocetes was somewhat more primitive than the skull in later whales. The bones of the facial region were not telescoped into the back of the skull, as they were in the more advanced whales. There were forty-four teeth, as in primitive placentals, of which the incisors and the canines were simple, sharp cones. The cheek teeth were cusped, but with the cusps arranged in a single fore and aft line, and coming to a high point in the middle. Teeth of this form are seen in other mammals, especially certain seals that

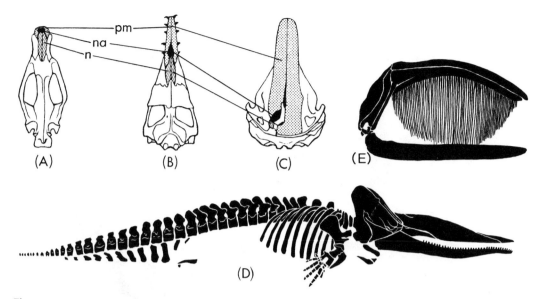

Figure 24-2. **(A–C).** The possible derivation of the cetacean skull from a terrestrial mesonychid condylarth, showing the retreat of the nasal opening or external nares, na, and of the nasal bones, n, to the back of the skull, and the enlargement of the premaxillary bones, pm. Note the asymmetrical development of the premaxillaries and position of the nares in (C). (A) *Apterodon,* a mesonychid from the Oligocene of northern Africa. (B) *Prozeuglodon* an early cetacean from the Eocene of northern Africa. (C) *Aulophyseter,* a cetacean from the Miocene of North America. (Although A is geologically younger than B, it represents a structural type that might have been ancestral to the cetaceans.) A–C not to scale **(D)** The skeleton of the modern sperm whale, *Physeter.* This whale is commonly about sixty feet in length. **(E)** Skull with baleen plates, and jaw, of *Eubalena,* a mysticete whale. Skull about four meters in length.

live on fish. The nostrils in the archaeocetes were placed in a forward position, not on top of the skull as in the later whales.

ADAPTIVE RADIATION OF THE MODERN WHALES

During late Eocene or Oligocene times the modern whales arose as descendants of the archaeocetes, and by Miocene times almost all the families of modern whales had appeared. Two lines of cetacean evolution developed from the archaeocete stem, as already mentioned. One of these was the toothed whales or odontocetes; the other was the whalebone whales or mysticetes.

Most of the modern whales are odontocetes. In late Oligocene times there appeared some comparatively small odontocetes, designated as the squalodonts, much like the modern porpoises in general appearance, but characterized by cusped cheek teeth somewhat similar to the teeth in the early archaeocetes. Evidently the

squalodonts, of which the Miocene genus *Prosqualodon* is typical, were forms intermediate between the archaeocetes and the modern types of whales. In spite of their archaic teeth, the skull was highly advanced, with the nostrils completely dorsal in position and the skull bones modified accordingly. The squalodonts were important whales during the Miocene epoch, but they did not survive long after the beginning of the Pliocene epoch.

Their place in Pliocene, Pleistocene, and Recent times was taken over by the small, toothed whales with which we are familiar, the porpoises and dolphins. These are compact, very swiftly swimming cetaceans that feed upon fish. The teeth are greatly multiplied in number and have the form of simple spikes. A common odontocete of Miocene times was *Kentriodon;* two of the most widely spread modern small cetaceans are *Delphinus,* the common dolphin, and *Phocaena,* the common porpoise of worldwide distribution. Re-

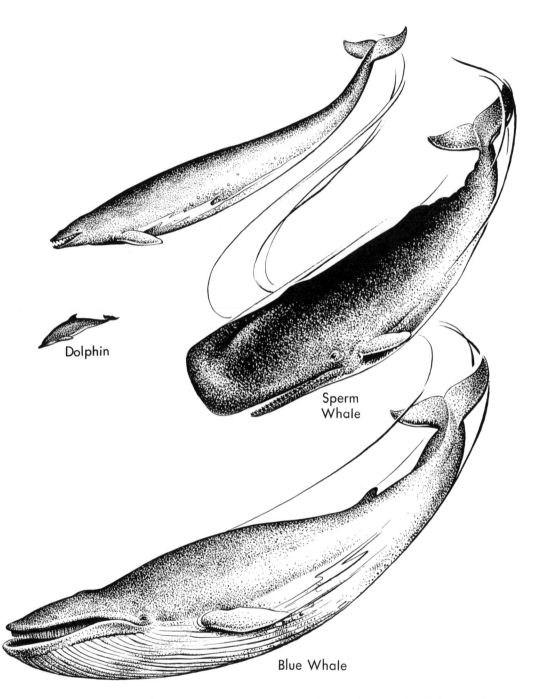

Dolphin

Sperm
Whale

Blue Whale

Figure 24-3.   An Eocene whale (zeuglodont) and three recent cetaceans, drawn to the same scale. The largest of these is an animal ninety feet or more in length. The zeuglodont was an archaeocete; the dolphin and the sperm whale are odon-tocetes or toothed whales; the blue whale is a mysticete or whalebone whale. Prepared by Lois M. Darling.

lated to these cetaceans are the fierce killer whales, the so-called blackfish, and some of the river dolphins. Other small to medium-sized odontocetes are the narwhals, the beaked whales, and the platanistids that live in the Amazon and the Ganges rivers.

There was an early trend toward giantism in the toothed whales, reaching its culmination in the physeterids or sperm whales. The modern sperm whale or cachalot *(Physeter)* is a giant odontocete with a great, square snout filled with gallons of sperm oil, contained within the huge spermaceti organ. Recent studies indicate that this organ serves as a re-verberation chamber whereby bursts of sound are used for long-range echolocation. Peglike teeth are present on the lower jaw, and this whale feeds upon squids. In the great days of whaling the sperm whale was much sought for its oil, and incessant hunting has reduced it from its original vast numbers to the limited herds of the present day. Moby Dick was a white sperm whale.

The largest of the modern whales are the mysticetes or whalebone whales, of much lesser variety than the toothed whales, but none the less highly successful cetaceans. As mentioned above, these whales feed upon plankton, and it may have been the abundance of their food supply that led to the strong trend to giantism among these largest of all vertebrates. The primitive mysticetes were the cetotheres, in which the teeth had been lost and the skull had progressed toward the high degree of modification that is so characteristic of the modern whalebone whales. *Mesocetus*, of Miocene age, was typical of this group.

In the latter portion of Cenozoic times the evolution of the whalebone whales reached its ultimate stage with the development of the skull into a highly arched structure for bearing large plates of baleen, so that the eyes and braincase were limited to a very small poste-

rior region. Some modern representatives of this evolutionary trend are the Greenland whales, the right whales *(Balaena),* the finbacks or rorquals and the titanic blue whales *(Balaenoptera)*. These last may reach lengths of one hundred feet and weights of one hundred and fifty tons. We are wont to look back at some of the giant dinosaurs with awe, yet such modern whales as the blue whale far exceed in size the largest dinosaurs. These are the ultimate extreme of giantism in the evolution of animals.

WHALES AND MAN

A few hundred years ago whales roamed the seas in vast herds. Then there was a demand for whale oil, and the large-scale systematic hunting of whales began, reaching a high point in the last and present centuries. Many species of large whales have been drastically reduced in numbers. Whale oil is not used for lamps as it once was, but the hunting of whales continues on a vast scale, for the oil is used for soaps, the meat is processed for food, and the bone is ground up for fertilizer. Factory ships go out to hunt whales with airplanes, radar, modern whale guns, and other refined apparatus. Consequently many whales are in danger of extermination, and it is only through strict adherence to international treaties that these large mammals can be preserved for posterity. Fortunately strong movements for the preservation of whales have arisen and grown throughout the world in recent years, while at the same time there has been a decrease in the commercial exploitation of whales. It is to be hoped that they will not disappear from the seas, for the giant whales together with their smaller cetacean cousins are marvelous animals that should never be allowed to vanish at the hand of man. There is still much to be learned about these wonderful mammals.

Hesperocyon

Fox

Creodonts and Carnivores

25

ADAPTATIONS OF THE CARNIVOROUS MAMMALS

The early mammals that arose from primitive insectivore ancestors evolved along varied lines of adaptive radiation, to occupy the numerous ecological niches that had been vacated by the wide extinctions of reptiles at the end of the Cretaceous period. Several orders of mammals became adapted for feeding on plants, whereas two orders of land-living placental mammals, the creodonts and carnivores, became dominantly specialized for eating other vertebrate animals. Why was not the killing and eating of prey more widely followed among the orders of mammals? Perhaps the answer to this question is that the carnivores specialized as efficient predators at such an early date, and became so widely distributed, that no other mammals were able to compete with them on their own terms. It is an interesting fact that the successful predators outside the orders Creodonta and Carnivora are those mammals that have lived in regions or in habitats where they have been free from carnivore competition, for instance, the carnivorous marsupials in Australia (and in past ages in South America), the insectivores, the bats, and the predatory whales and porpoises.

Adaptations for a predatory mode of life are on the one hand generally less extreme than those required for a life of plant-feeding, yet on the other hand they involve more "evolutionary risks" in the long battle for survival. The plant-eater must possess complicated teeth and digestive organs to gather and convert bulky plant food into energy, but in general the source of food for animals of this type is abundant and readily available. The meat-eaters, on the other hand, depend largely upon their ability to catch other animals. This ability may not require advanced specializations, although often it has resulted in the evolution of highly modified animals, but it does make the carnivore dependent on a very uncertain and variable source of food.

Consequently there has been intense competition among carnivorous mammals through the ages, either on the level of species or of individuals.

The adaptations that have been characteristic of both orders of the carnivorous mammals from the beginning of the Cenozoic era to present times may be outlined briefly. These mammals usually have had strong incisor teeth for nipping, and enlarged, daggerlike canine teeth for stabbing. In most carnivores the canines have been the principal weapons with which they kill their prey. Certain cheek teeth in the carnivorous mammals have been commonly transformed into blades that act against each other like the blades of a pair of scissors, for cutting and slicing meat into small pieces that can be swallowed easily and assimilated by the digestive system. These cutting teeth in the carnivores are called the *carnassials*. Naturally the carnivores have strong jaws, and strong crests and zygomatic arches on the skull for the attachment of powerful jaw muscles.

The carnivores usually have been very intelligent animals, because a great deal of mental alertness and coordinated action are required if other animals are to be overcome and killed. The sense of smell usually has been highly developed as an aid to hunting, and in many carnivores the eyesight has been very keen. The body and limbs generally have been strong and capable of lithe, powerful movement. There has been little reduction of the toes, which have sharp claws. These animals today are frequently fast runners over short distances, or adept climbers.

THE CREODONTS

Fortunately there is a good fossil record for both orders of carnivorous mammals, the creodonts and the carnivores (in the strict sense of the word), so that we can follow their evolutionary histories in considerable detail. For many years these two groups of meat-eating mammals were placed within a single order, the Carnivora, with the creodonts ranked as a suborder—an arrangement based on the view that the creodonts and the other carnivores were closely related, with the former group occupying an ancestral position to the latter group. The work of recent years makes it apparent, however, that the creodonts had a quite separate evolutionary history from the other land-living carnivorous mammals; that the resemblances between them were the results of parallelisms that grew out of similarities in habits and, consequently, in adaptations. In this way, two groups of mammals became carnivores independently of each other; they have been designated as the Creodonta and the Carnivora.

The creodonts constitute the older and the more primitive of the two groups. These mammals appeared in early Cenozoic time and enjoyed the culmination of their evolutionary development during the early phases of Tertiary history. They were rather archaic in morphological adaptations for the chase and the kill, and as long as the herbivorous mammals on which they preyed retained their primitive adaptations—as long as they were relatively clumsy and slow—the creodonts prevailed as the dominant meat-eaters. But with the appearance of advanced herbivores, roughly during the transition from Eocene to Oligocene times, the creodonts were at a disadvantage and were gradually replaced by the more specialized, more adept, and more intelligent Carnivora. A few of the creodonts persisted into late Tertiary times, but for the most part these carnivorous mammals were no match for the "true" carnivores and therefore were replaced by the evolutionary lines of meat-eating mammals that have continued into our modern world.

The term Creodonta is not being used here in the same sense that it was formerly employed. Certain groups of mammals, notably the arctocyonids long regarded as creodonts, are now placed among the most primitive of the hoofed mammals, the condylarths. If this seems like

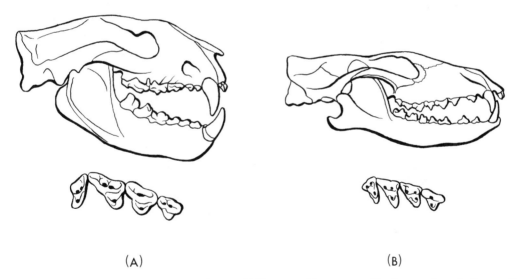

(A)   (B)

Figure 25-1.   The two lines of creodont evolution. **(A)** The oxyaenids, represented by *Oxyaena*, from the Eocene of North America. Skull, lower jaw, the right upper last two premolars, and the first and second molars. Note that the posterior edge of the first molar forms a long cutting blade. **(B)** The hyaenodonts, represented by *Sinopa*, from the Eocene of North America. Skull, lower jaw, right upper fourth premolar, and molars. Note that the posterior edge of the second molar forms a long cutting blade. The development of these upper molars as carnassials is characteristic of the creodonts, and distinguishes them from the later carnivores, in which the fourth upper premolar tooth is the carnassial. (A) About one-third natural size. (B) Somewhat less than natural size.

a drastic shift from primitive carnivore to primitive herbivore, it must be remembered that the very early eutherian mammals were quite generalized in morphology, and probably in habits, also. Thus, the distinction between carnivore and herbivore was a small one. The mesonychids, also formerly regarded as creodonts, are now generally placed in a separate order, the Acreodi, as has been noted in the preceding chapter. By removing the above two groups (generally classified as families) from the creodonts, the order becomes restricted to two remaining groups, the hyaenodonts and the oxyaenids.

As mentioned above, the creodonts were archaic carnivorous mammals. The skull was low with a small braincase; the molar teeth were basically tribosphenic, but with various molars frequently modified to form cutting blades. No ossified auditory bulla surrounded the middle ear, as in the advanced Carnivora. The skeleton was generalized; the limbs were generally rather short and heavy; the tail was long; and the toes terminated in sharp claws.

At an early stage in their evolutionary history the creodonts branched into two phylogenetic lines, the oxyaenids, in which the first upper molar and the second lower molar were the carnassials, or cutting teeth, and the hyaenodontids, in which the second upper and third lower molars were the carnassials. Adaptations were varied among these creodonts. Some were small and slender like the Eocene hyaenodont, *Sinopa*. Others were large and powerfully built, like the Eocene oxyaenids, *Oxyaena* and *Patriofelis*, or the Oligocene hyaenodont, *Hyaenodon*. It is obvious that these creodonts were the arch predators of early Tertiary times and anticipated the divergent evolution that was to take place among the carnivores in later Tertiary times, after the credotons had become extinct.

After the extinction of all other creodonts at the end of the Eocene epoch the hyaenodonts continued into the Oligocene epoch, and from then through the Miocene and into the early phases of the Pliocene epoch. Evidently these particular creodonts were sufficiently well adapted as beasts of prey so that they could compete successfully with the progressive land-living carnivores, often known as fissipeds, that arose at the end of the Eocene times, to blossom into abundance and great variety during subsequent geologic ages.

THE MIACIDS

The various creodonts that have thus far been described shared their function of early Tertiary predators with still another group of carnivores, the miacids. These carnivorous mammals appeared in Paleocene times, and like many of the creodonts they continued through the Eocene epoch to become extinct at the close of that phase of geologic history. They had certain primitive characters, such as the general archaic structure, with a low skull, elongated body and tail, and short limbs. The miacids, however, were progressive in some very important features. For one thing, they seem to have had a proportionately larger and more highly developed brain than the typical creodonts, a feature that would have been of great advantage to them as beasts of prey. Of particular importance is the fact that in these carnivores the carnassial teeth were more anteriorly placed than in any of the other early Tertiary carnivores, for they consisted of the fourth upper premolar and the first lower molar. The molar teeth were tribosphenic in form, and the last upper molar was absent. These are exactly the conditions typical of carnivores, and for this reason the miacids are regarded by many authorities as the most primitive representatives of the Order. However, two primitive characters distinguish the miacids from the later carnivores to which they were ancestral. In the first place there was no ossified tympanic bulla that in later carnivores (and other mammals) forms a chamber that encloses the bones of the middle ear. And secondly the bones of the wrist were all separate, whereas in later carnivores there was a fusion of the scaphoid and lunar bones to form

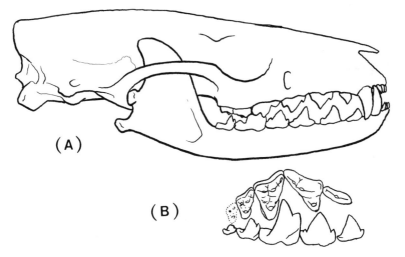

Figure 25-2. *Viverravus*, a Paleocene–Eocene miacid carnivore from North America. **(A)** Skull and lower jaw. **(B)** Right upper third and fourth premolars, and molars, in crown view, and right lower third and fourth premolars, and molars, in lateral view. Skull twice natural size.

a single element. These may seem like small features of carnivore anatomy, but they are important in determining relationships, the details of the basicranium being particularly crucial.

The miacids were small carnivores of weasel-like form. They were probably forest dwellers, preying upon small animals that lived in the dense undergrowth or in trees. *Viverravus* and *Miacis* were characteristic Eocene genera.

THE FISSIPED CARNIVORES

The fissipeds are the modern and familiar land-living beasts of prey that have been dominant from late Eocene and early Oligocene times to the present day. They are the dogs and their relatives, the bears, the raccoons and pandas, the varied mustelids such as the weasels, minks, badgers, wolverines, skunks and otters, the Old World civets, the hyenas and the cats. It is probable that at the time the early fissipeds appeared or soon after, the aquatic pinnipeds, the sealions, seals, and walruses, originated. However, the fossil record of the latter carnivores does not extend back beyond the Oligocene epoch.

The fissiped carnivores may be divided into three major categories of superfamily rank. These, with their constituent families, are:

Superfamily:	Miacoidea.
Family:	Miacidae: the miacids of Paleocene and Eocene age.
Superfamily:	Canoidea (or Arctoidea).
Family:	Canidae: dogs, wolves, foxes, and their relatives.
	Ursidae: bears.
	Ailuridae: pandas.
	Procyonidae: raccoons, coatis, kinkajous.
	Mustelidae: weasels, martens, minks, wolverines, badgers, skunks, otters.
Superfamily:	Feloidea (or Aeluroidea).
Famiy:	Viverridae: Old World civets.
	Hyaenidae: hyenas.
	Felidae: cats.

The twofold division of modern fissipeds into canoids and feloids is based upon various technical details of anatomy, especially the structure of the tympanic bulla that surrounds the middle ear. When fossils are taken into account the distinction between these two groups of carnivores is not very sharp, since some of

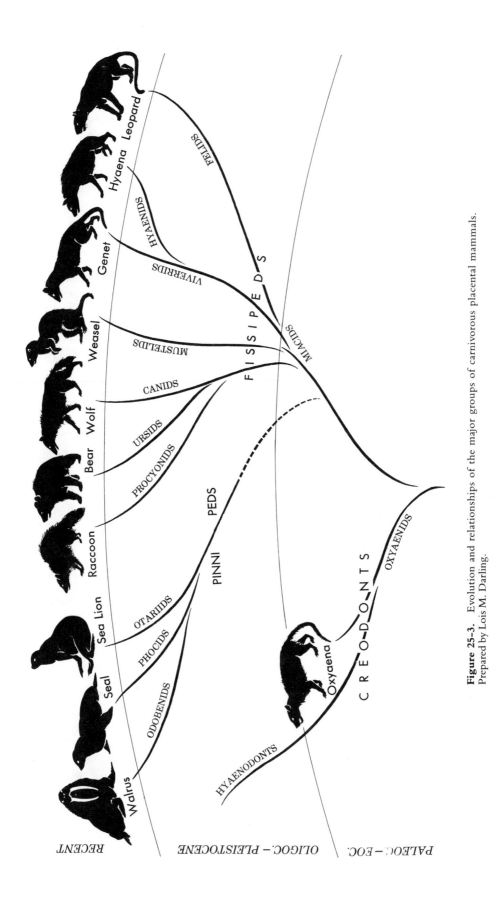

Figure 25-3. Evolution and relationships of the major groups of carnivorous placental mammals. Prepared by Lois M. Darling.

the primitive forms in each superfamily approach each other closely in structure. On the whole, however, this makes a good practical arrangement for grouping the fissiped carnivores beyond the miacids, and it probably expresses their basic relationships with a fair degree of accuracy.

THE CANOID CARNIVORES

The early canoid and feloid carnivores were, like the miacids, probably forest-dwellers that preyed upon the small game they could catch in the undergrowth and in trees. *Cynodictis*, a late Eocene form, and *Pseudocynodictis* (properly known as *Hesperocyon*) of the Oligocene were among the first canids, and although they retained many characters of their miacid ancestors they showed certain features that were to characterize the evolutionary development among the dogs or canids. There was some elongation of the limbs and feet in these early dogs, and the carnassial teeth were more highly specialized as shearing blades than they had been in the miacids. Also, the brain-case was expanded. Thus we see the early dogs embarking along the evolutionary path of long feet and limbs for running, sharp carnassial teeth for cutting meat, and a large brain, an indication of a high degree of intelligence. In these ways the dogs advanced; in other ways they have remained as comparatively primitive carnivores. For instance, there has been little loss of teeth or change in the form and function of the dentition beyond the stage characteristic of the late Eocene or early Oligocene carnivores.

From the small Oligocene *Hesperocyon* the canids progressed to the Miocene genus, *Cynodesmus*, from thence to the Pliocene *Tomarctus*, and finally to the modern dogs like *Canis* of Pleistocene and Recent times. This sequence represents the "main line" of canid evolution, but as so often happens there were several side lines of canids during Miocene and Pliocene times. *Amphicyon* was a very large, heavy, rather clumsy dog with a long tail.

Borophagus was another large dog, with a very deep, heavy skull and robust teeth. *Borophagus* and various related genera constituted an important line of Miocene and Pliocene dogs.

In Pleistocene and Recent times the history of the dogs reached its culminating phases in the differentiation of our modern canids, the wild dogs, wolves, and foxes of the northern hemisphere, and various highly specialized dogs of South America and Africa. These are intelligent animals that hunt and live in family groups and packs. For the most part they run down and kill their prey, often pursuing their victims over many miles of terrain. The foxes, however, are more solitary in habits and frequently hunt by stealth or by cunning strategems, for which they are justly famous in folklore.

The social instincts of the wild dogs have made them ideal as companions, and they were certainly the first animals to be domesticated. Man and dog lived together and worked together as early as Neolithic times, and this relationship has continued ever since, through a period of many thousands of years. It is difficult to make positive statements as to the origin of the domestic dog, *Canis familiaris*, but these friends and companions of home and field are probably in the main of wolf ancestry. Since the dogs are structurally primitive in many respects, they are genetically plastic. The truly astonishing variety of modern breeds of dogs is a proof of this that needs no elaboration in words. Under the guiding hand of man, the domestic dog has indeed departed far from his wild, wolflike ancestor in form and physical appearance, yet psychologically he is still a wolf—an intelligent, friendly canid that likes to run and hunt.

In Miocene times some dogs began to evolve as large, heavy carnivores. From such an ancestry it is probable that the first bears, typified by *Ursavus*, of Miocene age, arose. From *Ursavus* there evolved the Pliocene *Indarctos* and related genera, with massive skulls and robust teeth. In carnivores of this type the carnassial teeth lost their shearing function, and the molar teeth became square in outline, with blunt cusps. At the same time the legs and feet became heavy and the feet short, and the habit of pursuing the prey declined. The tail was reduced to a mere stub. Thus the bears evolved as massive middle and late Cenozoic carnivores. The trend of bear evolution reached its climax in the Pleistocene and recent bears, typified by *Ursus*. In the modern bears, some of them the largest of all land carnivores, the molar teeth are elongated and the crowns are complicated by a wrinkling of the enamel. This is obviously a specialization for an omnivorous diet and a great departure from the predominantly meat-eating habits of the dogs.

Bears are very adaptable animals, as are the dogs for that matter, and they are widely distributed throughout the world. The middle Tertiary origin and the evolution of the bears took place in the northern hemisphere. They entered South America during Pleistocene times, but curiously they never invaded Africa.

From the central canid stock there was another evolutionary trend that also led away from the chase and the kill, to adaptations for climbing and an omnivorous diet. This was the line of the procyonids—the raccoons and their allies.

The procyonids probably diverged from the canids during the Oligocene epoch, for in Miocene times they were well established, as indicated by the genus *Phlaocyon*, a small, climbing carnivore with handlike forepaws, flexible limbs, a dentition in which the carnassials had lost their shearing function, and molar teeth that were square with blunt cusps. The adaptations characteristic of *Phlaocyon* have been continued with little change in the modern *Bassariscus*, the ring-tailed "cat" or cacomistle of Mexico and the southwestern United States. Here we see in effect the structural ancestor of the procyonids, a small carnivore that lives among rocks or in trees, where it eats almost anything it can catch or gather. It is partly a meat-eater, partly a vegetarian.

From *Phlaocyon* the advanced procyonids evolved during late Tertiary times. Many of them have remained relatively small and have been confined to North America, where they originated, or to South America, a region that they invaded. These are the familiar raccoons, *Procyon* and its relatives, the coatis, *Nasua*, and the kinkajou, *Potos*, of South and Central America, and some other forms. All of them are forest-living animals that spend much of their time in the trees or along the banks of streams, where they feed upon a great variety of foods. The catholic diet of the common raccoon of North America is well known to many farmers and fishermen, who have had their fields, their chicken coops, or their fishing grounds raided by these intelligent little carnivores.

The pandas evolved in Eurasia during middle and late Tertiary times. The lesser panda, *Ailurus*, now lives in the Himalayan region, but fossils show that it once extended as far west as England. It looks very much like an enlarged raccoon, even to the ringed tail and the mask on the face. The giant panda, *Ailuropoda*, was rather widely distributed in Asia during Pleistocene times, but is now confined to a comparatively small area in western China. This animal is as large as a bear, and like a bear is heavily built, with a very short tail. The relationships of the giant panda have been long debated. Some authorities have included it with the lesser panda among the procyonids or raccoons. Detailed studies would seem to indicate, however, that it probably should be regarded as an ursid or bear. The giant panda is interesting because it is a carnivore that has turned completely herbivorous. The molar teeth are low crowned, with blunt cusps, and the enamel is wrinkled, making a broad grinding surface. These decorative and popular animals live exclusively upon green bamboo shoots.

Although the dogs, bears, and procyonids are closely related, the mustelids, on the other hand, are set apart from the other canoids, and have been a separate phylogenetic line since the

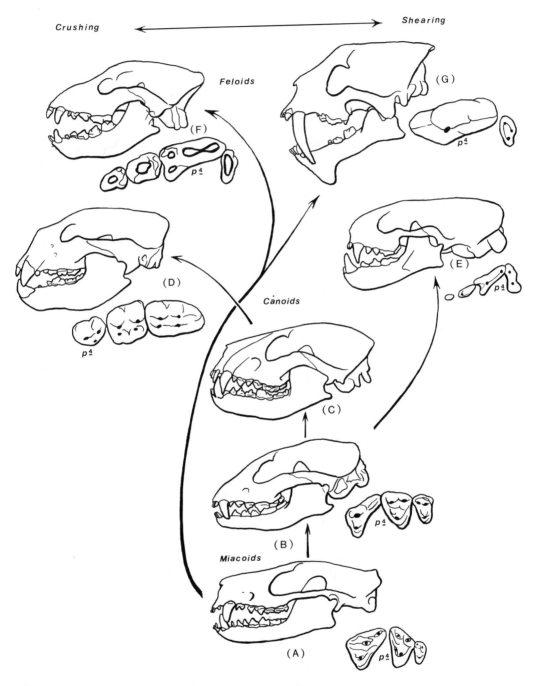

Figure 25-4.   Carnivore skulls, and left upper posterior premolars, and molars, showing evolutionary lines and the dominance in later forms of crushing or shearing actions in the cheek teeth. Not to scale. **(A)** *Vulpavus*, an Eocene miacid. **(B)** *Hesperocyon*, an Oligocene canid. **(C)** *Cynodesmus*, a Miocene canid. **(D)** *Arctodus*, a Pleistocene bear. **(E)** *Mustela*, a Pleistocene weasel. **(F)** *Hyaena*, a Pliocene hyena. **(G)** *Hoplophoneus*, a Oligocene saber-tooth cat. P^4 indicates the fourth premolar, the upper carnassial tooth adapted for shearing in most carnivores, but secondarily modified for crushing in some groups.

time of their origin, at the beginning of the Oligocene epoch. *Plesictis*, one of the first of the mustelids, was a small carnivore of generalized structure, with tribosphenic molars. However, the carnassial teeth were well developed, and the posterior molars were suppressed. The face was short and the braincase long and expanded, characters quite typical of the mustelids. From this ancestry the mustelids evolved with bewildering variety during middle and late Cenozoic times. Their evolution was characterized by the development of several short-lived lines of adaptive radiation, now extinct, which add to the complexity of mustelid phylogeny and make an interpretation of their history particularly difficult. It is not possible at this place to go into the details of mustelid development, but perhaps it may be useful to discuss briefly the modern mustelids, the persisting groups that have emerged from the complex melange of middle and late Tertiary mustelid history.

Generally speaking, there are about five groups of modern mustelids, all of subfamily rank. In the first place there are the primitive mustelids, the mustelines, many of them retaining the characters of their middle Tertiary ancestors. In this group are the weasels, the martens, the minks and their relatives, and the wolverines. These are very active, highly carnivorous animals, living in trees and on the ground in the forests. Some of them, especially the weasels (*Mustela*), are savage out of all proportion to their size.

The second mustelid group is the mellivorines, now represented by the ratel or honey badger (*Mellivora*) of Africa. Specializations here have been toward a ground life and a varied diet. A third group is the melines, the badgers of Eurasia (*Meles*) and North America (*Taxidea*). They are large, heavy mustelids that live in burrows. They are aggressive, but not highly carnivorous.

The fourth modern group of mustelids is the mephitines, the skunks (*Mephitis* and other genera) of North America. These small mustelids are ground-dwellers that burrow, and feed upon a great variety of things— small animals, insects, worms, berries, plants, carrion, and garbage. The skunks are protected by special scent glands that emit a strongly scented liquid, the smell and effects of which need no description for American readers.

Finally there are the lutrines, *Lutra*, and its relatives, the otters. They are aquatic mustelids specialized for catching fish, or even for feeding upon shellfish. Many of them live along the banks of streams, but the sea otters of the Pacific Ocean spend their life almost entirely in the shallow waters along the coasts.

From this brief review it can be seen that the evolution of the mustelids has been highly divergent, and of all the carnivores they certainly show the widest range of adaptive radiation.

THE FELOID CARNIVORES

The most primitive of modern carnivores are some the Old World civets, little modified descendants of the progressive miacids, that may be regarded as essentially late Eocene carnivores living on into modern times. The genet, *Genetta*, now inhabiting the Mediterranean region, is very near to the central stock from which all the civets have evolved. This is a small, forest-living carnivore, with a long body and a very long tail. The limbs are rather short and the feet are provided with claws that can be withdrawn to some extent, like the retractile claws of a cat. The skull is elongated and low, and narrow, the carnassial teeth are sharp, to form efficient shearing blades, and the molars retain the primitive tribosphenic pattern. The last molars are absent. The modern genet has a spotted coat, and it is probable that this is a primitive color pattern that has been retained through the ages. It has specialized scent glands for marking territory and for defense, a characteristic adaptation in the modern civets.

The civets are abundantly represented in considerable variety among the modern faunas of Asia and Africa. From a central, conservative stem, approximated by the genet and its relatives, the viverrids have branched along varied lines of adaptive radiation. One branch is composed of the various African and Oriental palm civets and the binturong or bearcat of Asia, this last one of the largest of the civets. An extreme offshoot of this general line of adaptation is *Eupleres,* the falanouc of Madagascar, in which the teeth have been reduced to relatively simple pegs, as an adaptation to eating ants and insects. Another evolutionary branch is represented by *Cryptoprocta,* the fossa of Madagascar, a very catlike civet. The position of this carnivore, whether a catlike civet or a civetlike cat, has long been debated. It is very possible that the fossa arose from a primitive civet, but near the ancestry of the cats, so that it shares catlike as well as civet characters. Finally, one large branch of the civets is the group of mongooses, small active civets that are famous as predators upon snakes and upon various small mammals.

Civets first appear in sediments of upper Eocene and lower Oligocene age, and are represented by such genera as *Stenoplesictis* and *Palaeoprionodon.* The subsequent epochs of the Cenozoic era reveal very little of the past history of the viverrids, probably because these predominantly tropical, forest-living carnivores were rarely preserved as fossils. The few genera known from Miocene and Pliocene sediments in Eurasia indicate that the viverrids continued as very primitive carnivores during much of Tertiary times. Thus the sequence from *Palaeoprionodon* through middle and late Tertiary forms, of which the Mongolian genus *Tungurictis* is an example, to the generalized modern civets, indicates only a minor amount of evolutionary progress.

In Miocene times one evolutionary branch split from the central civet stock and followed a trend toward increase in size and particularly the development of a heavy skull and very robust teeth. This was the line of hyenas, which share with the bears the distinction of being the youngest among the families of carnivores. Simply stated, the hyenas are very large, heavy descendants of the civets, in which the legs have been elongated for running and the teeth and jaws usually enlarged for cracking bones. The enlargement of the teeth is concentrated especially on the last two cone-shaped premolars, which are used for breaking the bones of large carcasses on which the hyenas feed. The jaws and the jaw muscles are necessarily very strong. The carnassials are highly specialized shearing blades in the hyenas, and the molars behind the carnassials are reduced to mere remnants.

Ictitherium, of late Miocene and early Pliocene age, was the first hyena. This carnivore was truly intermediate between the civets and the hyenas, larger and heavier than the former but much lighter and smaller than the latter. The step from *Ictitherium* to advanced and fully modern hyenas was a quick one, and we find fossil hyenas very similar to the modern animals in sediments of early Pliocene age. In effect the hyenas quickly reached the peak of their adaptational perfection soon after they split from the civets, and they have maintained their specialized form with little change since it was first attained. Hyenas live in Asia and Africa at the present time; but during the Pleistocene epoch they were widely distributed through northern Europe. One modern hyena, *Proteles,* the aardwolf of South Africa, is curiously specialized for eating termites, and the cheek teeth are reduced to small pegs, although the canines remain large.

The cats had an evolutionary history something like that of the hyenas, but it began at an earlier date. Once having split from a viverrid ancestry and, once having departed from the civet stem, the cats very rapidly evolved into fully specialized cats. They have maintained their high degree of specialization without much change for millions of years. The separation of the first members of the cat family from their civet ancestors took place during late Eocene times, and the upper Eocene genus, *Proailurus,* may represent an early step in the evolution of the cats. By early Oligocene times the cats were highly evolved cats, not very different from their modern relatives.

Sabre tooth cat

Cave Bear

Figure 25-5. Two great fissipeds, representing the climax of feliod and canoid evolution during the Pleistocene epoch. Prepared by Lois M. Darling.

Of all land-living creatures, the cats are among the most completely specialized for a life of killing, and for eating meat. They are very muscular, alert, supple carnivores, fully equipped for springing upon and destroying animals as large or larger than themselves. They generally hunt by stealth, and catch their prey with a long bound or a short rush of great speed. The limbs are usually heavy and strong, and the feet are provided with sharp, usually retractile claws that are used for catching and holding their victims. The neck is very heavy to take up the severe shocks imposed by the violent action of the head and the teeth. The teeth are highly specialized for just two functions—stabbing and cutting. The canine

are therefore long and strong, and the
ssials are large, perfected shearing blades;
ther teeth are reduced or completely sup-
sed. The smaller cats are adept tree-
bers, but the larger cats spend most of their
e on the ground.

All cats are constructed pretty much to the
pattern that was established by the cats of early
Oligocene times. However, there seems to have
been a dichotomy in the evolutionary history
of the cats that went back to the time of their
definition and continued until the end of the
Pleistocene epoch. On the one hand, the cats
evolved as active, fast-moving predators, the
normal cats with which we are familiar; on
the other hand, they developed as the comparatively heavy and slower saber-tooth cats.
The ancestry of the normal or "feline" cats is
exemplified by *Dinictis* of Oligocene age; that
of the sabertooth cats by *Hoplophoneus*, also
of Oligocene age.

Both of them were medium-sized cats with
long tails. In *Dinictis* the upper canine teeth
were large and heavy, and the carnassials were
well-developed shearing blades. There were premolar teeth in front of the carnassials, but the
molars behind the shearing teeth were greatly
reduced. In *Hoplophoneus* the upper canine teeth
were elongated sabers, and there was a flange
on the lower jaw to protect these down-pointing
swords when the mouth was closed. The carnassials were specialized cutting blades, and
the other cheek teeth were greatly reduced or
suppressed.

As the feline cats evolved, the canine teeth
became relatively smaller than they had been
in *Dinictis*; but otherwise the dentition changed
very little. As the saber-tooth cats evolved, the
canine teeth remained large, as they had been
in *Hoplophoneus*. Such trends indicate that the
feline cats became increasingly perfected for
catching and killing agile animals, while the
saber-tooth cats became specialized for killing large, heavy animals.

The culmination of saber-tooth evolution
was reached during the Pleistocene epoch, in
the large saber-tooth cat, *Smilodon*. This cat
was as large as a modern lion, and the upper
canines were huge daggers of impressive proportions. Anatomical studies indicate that
Smilodon was able to open the mouth very
wide, thus clearing the way for the large canine sabers to function. In attacking its prey,
Smilodon evidently struck down very hard with
the sabers, using the force of the strong neck
and the weight of the shoulders and body to
give power to the thrust. This was an effective method of hunting, as long as there were
large, comparatively slow animals available. But
as the Pleistocene drew to a close, the large
animals on which the saber-tooth cats had
preyed became extinct, and so did the saber-
tooths. They were unable to compete with their
agile feline cousins in the chase of speedy
animals.

The feline cats, on the other hand, have continued into modern times with great success.
As said above, cats are cut pretty much to one
pattern, yet there is a great variety of modern
cats, differing mainly in size and in the habitats that they frequent. The modern cats show
a distinct division into the typical cats, *Felis*
(and related genera), on the one hand, and the
swift-running cheetah or hunting leopard,
Acinonyx, on the other. As for the typical cats,
they are found throughout the world except
in Australia and on remote islands. There are
many small cats, and from some of these, probably from a mixture of the ancient Egyptian
wild cat and the European wild cat, our modern domestic cat has descended. The large cats
are so familiar as to need no particular
description—lions and leopards in Africa, leopards and tigers in Asia, jaguars and cougars in
the Americas.

Cats did not enter South America until
Pleistocene times, when both feline and saber-
tooth cats invaded that continent. The invasions of South America by the cougar or puma
from the north established one of the widest
known ranges for a single species of mammal,
except man, for this cat extends from the snows
of Canada to the southern tip of South America. A comparable distribution is typical of the

Old World leopard, which ranges from the southern portion of Africa into northern Asia.

THE PINNIPED CARNIVORES

The pinnipeds—the sea lions, walruses and seals—appear in the geologic record during the transition from Oligocene to Miocene time, as exemplified by *Enaliarctos*, known from an almost complete skeleton discovered in California. *Enaliarctos*, although undoubtedly a pinniped, retains numerous anatomical characters that indicate its derivation from terrestrial canoid ancestors. Thus the evidence of this earliest known pinniped clearly points to a monophyletic origin for these carnivores, in contrast to the frequently held view that the pinnipeds are diphyletic, with the sea lions and walruses derived from canoid ancestors while the seals presumably originated from mustelid progenitors.

In making the transition from life on the land to life in the water, the pinnipeds became streamlined for swimming. However, their adaptations along this line have never been as complete as in some of the totally marine tetrapods, like the ichthyosaurs or the whales, for they have retained a flexible neck and have failed to evolve a dorsal fin or a propulsive tail. Perhaps the tail had been reduced to such a point in the ancestors of the pinnipeds that it was, so to speak, never available for transformation into a propeller. Consequently, the pinnipeds have had to rely on the limbs in combination with body movements for propulsion through the water. In these carnivores all four feet are transformed into paddles, with webbing between the toes. The front paddles are used for balancing and steering, as well as for making propulsive thrusts. The back paddles are turned back and function like a sort of caudal fin when these animals are in the water. In the sea lions *(Zalophus)* and walruses *(Odobenus)* the back flippers can be turned forward or back at will, and are used when on land as aids to locomotion. In the seals *(Phoca)* the back flippers are permanently fixed in the

backward direction, so that when seals are on land or on ice floes they have to move along on their bellies by a "humping" motion of the body.

Sea lions are found today along the Pacific coast, while the walruses are found in both the Pacific and Atlantic oceans. The seals are widely distributed along the seacoasts of the world.

The teeth are greatly modified in all the pinnipeds. The incisors are commonly reduced or suppressed, whereas in most pinnipeds the premolars and molars are secondarily simplified to the form of pointed, cone-shaped teeth, all much alike. Such a dentition is useful for catching fish. The walruses have large canine tusks, and the cheek teeth, reduced in numbers, are broadened into crushing mills, with which these carnivores grind up the oysters and clams on which they feed. The sea lions have small external ears; in the other pinnipeds the external ear lobes or pinnae are completely suppressed.

The seals are more highly adapted to an aquatic life than are the sea lions and walruses. The teeth are highly modified from the primitive carnivore pattern, so that in some seals the teeth behind the canines are distinguished by having accessory "cusps" or points on a midline in front of and behind the principal cusp. Such teeth are very efficient for grasping and holding slippery fish.

Perhaps the most remarkable adaptations among the seals are those allowing them to make deep and prolonged dives, in which respect they are second only to the whales. An integrated series of specializations in the lungs, the heart and the circulatory system, provide some of these carnivores with an ability to dive to depths as great as 600 meters, and to stay submerged for more than an hour. It is probable that such adaptations were present in many of the seals of Cenozoic age.

RATES OF EVOLUTION IN THE CARNIVORES

The carnivores are interesting not only because of the wide range of their adaptive

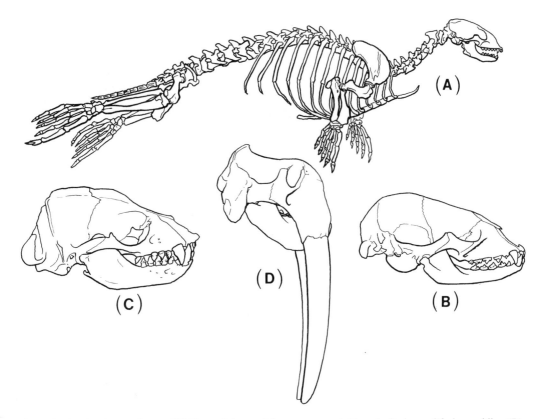

Figure 25-6.   Marine carnivores. **(A)** *Phoca*, skeleton of the common seal. Note the limbs modified as paddles. **(B)** *Phoca*, skull and jaw. **(C)** *Zalophus*, skull and jaw of a male California sea-lion. **(D)** *Odobenus*, the walrus. Note enlargement of the upper canine teeth as tusks. Various reductions.

radiation, but also because of the varying evolutionary rates displayed among the several families of these mamals. For instance, the civets have remained on the whole very primitive through the extent of their history, so that their evolutionary rate may be regarded as having been relatively low. Broadly, the mustelids may show low or medium rates of evolutionary development. The dogs, though generalized in many respects, have nevertheless shown a moderate rate of evolutionary development since the Oligocene epoch. Higher evolutionary rates are seen in the procyonids and in the bears, which branched from canid ancestors. The hyenas and the cats show very high rates of evo-

lutionary development at the beginnings of their histories, when they went through all the steps from viverrid to specialized hyenas and cats in remarkably short periods of time. Then the development of the hyenas and cats remained stationary—the hyenas since lower Pliocene times and the cats since Oligocene times.

As a result of these differing evolutionary rates the carnivores have attained the wide diversity of forms that inhabit our modern world. They have been truly successful mammals during middle and late Cenozoic times, and if they are not unduly persecuted by man they will extend their success far into the future.

Phenacodus

Ancient Hoofed Mammals

26

THE UNGULATES

The term ungulate is a very broad and loosely defined word, used especially by students of modern mammals to indicate the hoofed mammals that feed upon plants. Many paleontologists, however, have a broader concept of the ungulates than is indicated above, their view being based upon the long evolutionary history of the mammals, which reveals relationships between mammalian groups that seem distantly removed from the modern ungulates. Indeed, the word "ungulates" describes not only rather unexpected taxonomic relationships, but also ecological adaptations among various lines of parallel evolution—points that should be kept in mind when the word is used.

It may be helpful to mention here the various groups of so-called "ungulates" that evolved through Cenozoic times. Even before the advent of Cenozoic history the most archaic of the ungulates, the condylarths, made their appearance. They continued through early Cenozoic history and gave rise to several groups of early ungulates—the Dinocerata, and perhaps five orders of South American ungulates, the notoungulates, the litopterns, the astrapotheres, the pyrotheres, and the trigonostylopians. Moreover, quite possibly, the two dominant ungulate orders of late Cenozoic and recent times—the perissodactyls and the artiodactyls—also had condylarth origins. At this point there should be mentioned the tubulidentates or aardvarks which, although not generally considered as ungulates, may nevertheless have had condylarth ancestors. Still another group of mammals, called ungulates in the widest sense of the term, and probably quite independently derived, is composed of the proboscideans—the elephants, mastodonts and their relatives—the hyracoids, the sea cows and desmostylids, and probably the embrithopods.

BASIC ADAPTATIONS OF THE UNGULATES

Within the limitations outlined above, it may be said that adaptations among the ungulates have been most conspicuously developed in the teeth, which have been modified for cropping and grinding plants, in the digestive tract, which has been modified for converting bulky plant material into nourishment, and in the limbs and feet, which have been generally modified for running over hard ground. In addition, many ungulates have defensive weapons on the skull in the form of horns or antlers, or show modifications of some teeth for fighting and defense.

It is common but not universal for these mammals to have closely appressed incisor teeth that bite together along a slightly curved transverse line at the front of the skull. These teeth make an efficient nipping or cropping mechanism for gathering into the mouth leaves of trees and bushes, or grass. Although functional canine teeth are present in some ungulates, it is very common for these mammals to lack canines, or for such teeth, if present, to have lost their caniniform shape and function. In some ungulates the canine teeth join the incisor series, to increase the efficiency of the cropping function. However, the most remarkable adaptations in the dentitions of ungulates are to be seen in the cheek teeth, which in various ways act like grinding mills. Generally the crowns of the molar teeth become square or rectangular, by the strong development of the hypocone or of some other cusp near it in the upper teeth, and by the suppression of the paraconid and the growth of the talonid, so that it is equal in height and area to the trigonid in the lower molars. The primitive sharp tooth cusps become modified to form blunt cones, or crests, or ridges of complexly folded enamel. These modifications increase the area of the tooth crowns. Many ungulates feed upon hard grasses, and in them there has been an increase in height of the crowns of the cheek teeth—the development of hypsodont teeth, as they are called. By the increase in area of the molar crowns and by the increase in their height, the amount of tooth surface available for grinding plants during the life of the animal is enormously multiplied. Finally many ungulates show a "molarization" of the premolar teeth, a process of enlargement and modification whereby the normally small premolars become as large as the molars, to increase the cumulative area of the dental grinding mills.

Naturally it is not possible for us to derive any information on the digestive tract in the extinct ungulates. In the living hoofed mammals, however, there is usually a chamber in the digestive tract within which bacterial action can break down plant cellulose.

A few ungulates have enlarged incisor or canine teeth with which they can defend themselves, but a more common means of defense has been by the development of antlers or horns upon the skull, as already mentioned.

By far the commonest means of defense among the ungulates is flight by fast running. Therefore the hoofed mammals show dominant trends toward elongation of the limbs and feet. This lengthening of the legs increases the stride, enabling these animals to get over the ground rapidly. This adaptation is useful for flight from enemies; it also allows the ungulates to wander widely in search of food.

The ungulates commonly walk on the tips of their toes, a manner of locomotion that is designated as unguligrade. In feet of this sort the wrist and the ankle are far off the ground, as seen in the "knee" of the horse's forelimb and the "hock" of its hind limb. The toes are usually covered with hoofs that protect the feet and take up the shock of running over hard ground. In many progressive ungulates most of the function of walking and running is carried by the middle toes, so that there is a strong trend toward the reduction of the lateral toes. But in some of the ungulates, especially the large, heavy types, the feet remain short and broad, and there is little or no reduction of the toes. Thus the feet are wide, spreading structures that give broad bases for the support of great weight.

THE FIRST UNGULATES

The earliest ungulate, *Protungulatum*, known from upper and lower teeth and the lower jaw,

is found in the late Cretaceous and early Paleocene sediments of North America. It is distinguished by the blunt cusps of the teeth, obviously adapted for crushing and grinding the food upon which this animal subsisted. The upper molars are broad and rather square, owing especially to the development of a strong hypocone, while in the lower molars the talonid forming the back part of the tooth is almost as high as the anterior trigonid. These evolutionary trends in the dentition, slight and subtle as they may seem to be, none the less mark the beginnings of wide adaptations for an herbivorous diet. Otherwise *Protungulatum* and related genera belonging to the arctocyonid condylarths, show many resemblances to carnivores, such as a low and rather elongated skull, a strong sagittal crest, a jaw articulation essentially in line with the occlusal plane of the dentition and large canine teeth. Indeed, the arctocynoids were for many years classified as creodonts, those most primitive of the carnivorous mammals.

From such beginnings the evolutionary history of the ungulates, like that of the land-living carnivores, shows two general phases of development. There was an early phase during Paleocene and Eocene times, when primitive ungulates in great variety spread over the

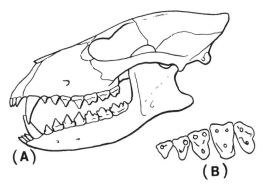

Figure 26-1. *Protoungulatum*, the earliest known ungulate, from upper Cretaceous and lower Paleocene sediments in North America. **(A)** Reconstructed skull and lower jaw in lateral view. **(B)** Last two upper premolars and molars in crown view. The lower jaw as preserved is about four centimeters in length.

face of the earth. The ancient ungulates began to decline during Eocene times, although
a few of them persisted briefly into the
Oligocene epoch. At the same time the modern ungulates were arising, to evolve in ever-
increasing diversity and complexity from
about the beginning of the Eocene epoch to
present times. However, this two-phase history of the ungulates is complicated by the fact
that in South America there was a long continuation of peculiar ungulates descended
from the primitive condylarths and unlike any
hoofed mammals in the other continents.
These South American ungulates lasted until
that continent was reunited with North America at the end of the Tertiary period, when they
quickly disappeared before the influx of invading mammals from the north.

THE CONDYLARTHS

The most primitive of the ungulates, the
condylarths, (of which *Protungulatum* is the
earliest known representative) appear in sediments of late Cretaceous and early Paleocene
age, and thus provide proof of the very early
differentiation of the herbivorous mammals.
Some of these ancient condylarths may not
have been far removed from their insectivore
ancestors, for they were small and had comparatively primitive teeth, and clawed feet. Indeed, as a result of intensive studies during the
past few years, it has become apparent that
various fossil mammals, for many years placed
among the creodonts, more properly should
be considered as condylarths. They are particularly the Arctocyonidae (already mentioned). The fact that they were for so long
regarded as primitive carnivorous mammals,
but are now, with good reason, regarded as
primitive herbivorous mammals, is an indication of the generalized nature of the mammals at the beginning of Cenozoic history. In
short, the differences between meat-eaters and
plant-eaters in those ancient days were small.

The arctocyonids, were generally small; the
skull was long and low; all of the teeth were

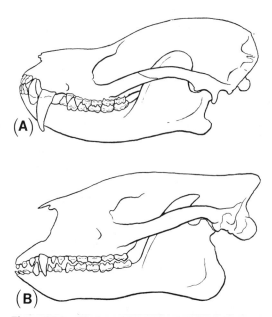

Figure 26-2. Skulls and jaws of ancient hoofed mam-
mals. **(A)** *Deltatherium*, a primitive condylarth from the
Paleocene of North America. One-half natural size. **(B)**
Phenacodus, an advanced condylarth from the Eocene of
North America. About one-fourth natural size.

present; the molars retained much of the original tribosphenic pattern; the back was supple; the limbs were relatively short; the feet
had claws; and the tail was long. *Tricentes* of
middle and late Paleocene age, and *Chriacus*,
extending from the beginning through the final phases of Paleocene history, represent
these primitive persistent types. However,
some arctocyonids evolved into large mammals during the Paleocene epoch. Among
them were *Claenodon* from the middle Paleocene sediments of North America, and
Arctocyon, from the upper Paleocene of Europe;
they were clumsy animals as large as small
bears, with blunt teeth that were possibly an
adaptation to a rather omnivorous diet.

From these early beginnings the more specialized condylarths radiated in various directions during the Paleocene and Eocene
epochs. In some, like *Meniscotherium*, of late
Paleocene and Eocene age, there was a distinct
advance in the teeth, which became almost se-

lenodont with crescentic rather than cone-shaped cusps; yet the feet remained primitive. In others, like the Paleocene genus *Periptychus*, there was a great increase in size, and peculiar specializations in some of the premolar teeth, which became very large. In middle and late Paleocene times there appeared *Tetraclaenodon*, with low-crowned but "squared" cheek teeth and with very broad claws on the ends of the toes. This type was probably directly ancestral to *Phenacodus*, one of the most completely known of the primitive ungulates, an animal that lived during the latter portion of the Paleocene and the early part of the Eocene epochs.

Phenacodus, known from complete skeletons, gives us a very good idea of what an ancestral ungulate was like. It was a fair-sized animal as large as a sheep, or even larger. In many ways it did not look as much like an ungulate as like some sort of primitive carnivore, for the skull was long and low, the tail was very long, the limbs were comparatively short and heavy, and the feet were short with all the toes present. The canine teeth were rather large. However the cheek teeth formed an almost continuous series, and the molars had square crowns, with a well-developed hypocone in the upper molars and a high talonid in the lower molars. The clavicle or collar bone was absent, as it is generally in the hoofed mammals, and the toes terminated in hoofs rather than claws. Evidently *Phenacodus* was a plant-eater that lived in forests or savannas, where it may have wandered widely in search of food. It was probably a clumsy runner.

LARGE UNGULATES OF EARLY TERTIARY TIMES

At an early stage in their evolutionary history certain groups of ungulates evolved toward large size. Two of the early orders of large ungulates were the Pantodonta and the Dinocerata, the latter often designated as uintatheres. Although not abundantly represented by genera and species, they none the less constitute very important segments of the mammalian faunas of Paleocene and Eocene times.

One of the earliest of the large hoofed mammals was *Pantolambda*, of middle Paleocene age, a pantodont about as large as a sheep. It had a rather long, low skull, in which the canines were large and the upper molars were triangular, with crescentic-shaped cusps. The limbs were rather heavy, and the feet were comparatively short, with all the toes present.

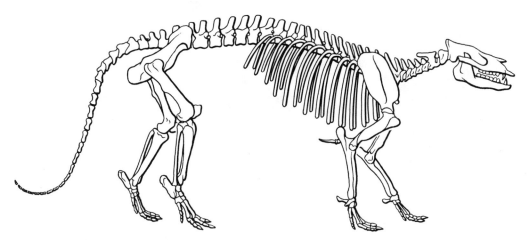

Figure 26–3. *Phenacodus*, a primitive ungulate from the Eocene of North America. The skeleton of this condylarth is about six feet in length.

Uintatherium

Coryphodon

Figure 26-4.   Two ancient hoofed mammals of Eocene age, both drawn to the same scale. *Uintatherium*, one of the Dinocerata, was as large as the modern white rhinoceros of Africa. *Coryphodon* was a pantodont. Prepared by Lois M. Darling.

Evidently these toes terminated in small hoofs. *Pantolambda* must have been a slow-moving ungulate that probably browsed upon the leaves of trees.

The evolution of the pantodonts to large size progressed rapidly through late Paleocene times, as illustrated by the genus *Barylambda*. This was a large animal, standing four feet or more in height at the shoulder. The entire skeleton of this pantodont was extraordinarily heavy, giving the impression of great stolidity and strength, obviously a difficult beast for the early creodonts to pull down and kill. In spite of its large size *Barylambda* had a comparatively small skull and a primitive ungulate dentition.

Perhaps the best known of the pantodonts is the lower Eocene form, *Coryphodon*, an animal as large as a tapir, having a heavy skeleton, with strong limbs and broad, spreading feet. In this early ungulate, as in many large hoofed mammals, the upper limb elements were long as compared with the lower limb elements and the feet. This type of limb gives great strength for the support of a heavy body but is not adapted for fast running. The tail was short, as is common in the hoofed mammals. The skull of *Coryphodon* was large, and the jaws were armed with elongated, saberlike canine teeth, which seems strange in a plant-eating mammal, but nevertheless was not uncommon in some of the early ungulates. The molar teeth had advanced beyond the primitive condition seen in *Pantolambda*, and on each molar crown were two prominent cross-crests, indicating that *Coryphodon* was an advanced browser, like the tapirs of modern times.

The pantodonts continued through the Eocene epoch, and in Asia at least survived into Oligocene times, after which they became extinct.

While the pantodonts were evolving, there was a parallel development of the Dinocerata or uintatheres, perhaps the largest of all early mammals. These animals, known as fossils only from North America and Asia, had

their beginnings during Paleocene times in such genera as *Bathyopsoides* and *Prodinoceras*. As in other large, heavy ungulates, the bones were massive, the limbs were heavy, the upper limb elements were long, and the lower limb elements and feet were short. The feet were broad and spreading. *Bathyopsoides* had a low skull, provided with a tremendously elongated canine tooth on either side. The front of the lower jaw was deeply flanged, making a protection for the canine saber when the mouth was closed.

The culmination of this line of ungulate development was reached in the upper Eocene genus, *Uintatherium*, an animal as large as a big rhinoceros, with an elongated skull grotesquely provided with six horns on top—two small ones on the nose, two above the canine teeth, and two at the back of the head. In this animal the upper canine teeth were very large, and on the molar crowns were cross ridges. The large uintatheres of late Eocene times were the last of the Dinocerata; by the advent of the Oligocene epoch these queer giants of early Tertiary times had become extinct.

AARDVARKS

The aardvark, *Orycteropus*, is a sturdy animal, about the size of a small pig, and it lives in Africa. It is almost hairless, and its skin is a dull-gray color. It has a compact body and very strong legs, terminating in long toes equipped with sharp, flat nails. It has a long head with a tubular snout, and ears that are very long and slender. The tail is heavy. The aardvark burrows in the ground (hence the name, which means "earth-pig" in Afrikaans), and it feeds upon termites, tearing open their nests with its strong clawed feet, and licking up the insects with a long, protrusible tongue.

As in so many ant-eating or termite-eating mammals, the teeth of the aardvark are greatly reduced and modified. There are no incisors or canines, only some columnar-shaped cheek teeth, which when viewed under a microscope are seen to consist of closely appressed

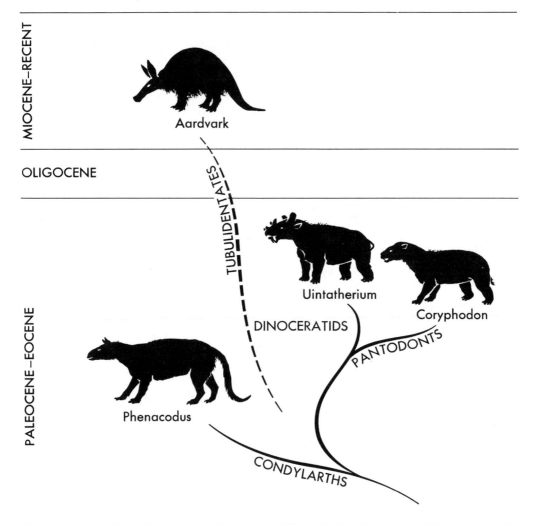

Figure 26-5. The relationships of the primitive ungulates. The aardvark is shown as a possible descendant of the condylarths. Prepared by Lois M. Darling.

tubes of dentine. This character has given the name Tubulidentata to the mammalian order of which the aardvark is the single living representative.

The fossil history of the tubulidentates does not extend back in time with certainty beyond the latter part of the Miocene epoch. In sediments of early Pliocene age fossil aardvarks, similar to the modern form, except for minor differences in size and proportions, are found in India and on islands at the eastern end of

the Mediterranean Sea. Evidently the aardvarks were more widely distributed in late Tertiary times than they now are, and obviously they and their ancestors must have been living somewhere in pre-Pliocene times. Yet to date, nothing is known of their early fossil history.

It has been common practice to regard these animals as related to the edentates, a conclusion drawn from their diet of termites and from the reduction of their teeth. We now re-

Aardvark

Figure 26-6.   The aardvark, *Orycteropus*, is a modern African mammal, highly specialized for feeding upon termites and for burrowing in the ground. This animal, about the size of a pig, is the sole living representative of the order Tubulidentata. It may have descended from condylarth ancestors. Prepared by Lois M. Darling.

alize, however, that various mammals have turned to an ant-eating diet and as a consequence have suffered a loss of teeth. If the skeleton of the aardvark is compared with the skeleton of the condylarths, a rather interesting series of similarities are apparent, which would suggest that perhaps the aardvark is of condylarth ancestry. Perhaps we can think of this peculiar mammal as a condylarth in which the head and the feet have been highly modified as adaptations to a very special diet and to burrowing in the ground.

Macrauchenia

27

South American Ungulates

THE INVASION OF SOUTH AMERICA BY PRIMITIVE UNGULATES

The record of ungulate mammals in South America may possibly begin in late Cretaceous time, where in Peru a single lower jaw with two teeth was found in Cretaceous sediments and described as *Perutherium altiplanense*. The relationships of this fossil have been debated, but there is some reason to think that it may be a very early notoungulate, the notoungulates being a large and varied order of hoofed mammals peculiar to the Cenozoic of South America. Possible Cretaceous notoungulates are reported from Bolivia also.

Whatever the status of Cretaceous ungulates in South America may prove to be, it is evident that some of the first placental mammals to reach this continent were primitive condylarths, obviously derived from the ancient condylarths of North America. These early ungulates, known as didolodonts, appears in the Paleocene deposits of South America, to continue in this region through early Cenozoic time, becoming extinct during the Miocene epoch. Although these condylarths did not continue beyond Miocene times they appear to have been ancestral to three of the ungulate orders that were charcteristic of the continent throughout most of Cenozoic history, namely the Notoungulata (already mentioned, and possibly derived from their condylarth progenitors during Cretaceous time), the Astrapotheria and the Litopterna, including the Trigonostylopia, which some authorities consider as a separate order. The other order of South American ungulates, namely the Pyrotheria (including the related Xenungulata) seemingly has an origin elsewhere within the placental mammals, possibly from the Dinocerata (the uintatheres) or ancestors of this order, so typical of North American early Cenozoic history.

By far the most numerous among these orders of South American hoofed mammals were the notoungulates which are represented in the known fossil record by about twice as many genera as are contained within the other three orders. It is interesting to see how these hoofed mammals divided the South American island continent between them, and how they paralleled the hoofed mammals in other parts of the world, from which they were completely isolated during the long interval between very early and very late Tertiary times. The comparison of South American ungulates with hoofed mammals in other parts of the world nicely illustrates the close correlation between animals and their environments, and indicates how similar environmental conditions will lead, by genetic processes, to the evolution of remarkably similar animals, quite unrelated except through their very remote ancestors.

THE NOTOUNGULATES

Palaeostylops, from the Paleocene of Mongolia and *Arctostylops*, from the Eocene of North America have long been considered as notoungulates, to be interpreted either as ancestral forms, the descendants of which migrated into South America in early Cenozoic time when that continent was still connected to its North American counterpart, there to found a numerous and varied mammalian dynasty, or as ancient notoungulates that abandoned their South American home to invade Laurasia. Recently doubt has been cast upon the notoungulate relationships of the two genera named above. Perhaps, therefore, the notoungulates were limited to South America from the beginning, being derived from condylarths immigrants, as has already been indicated. Certainly the notoungulates were abundantly established in the southern part of South America by early Eocene times.

The first notoungulates were small, primitive, hoofed mammals, with triangular-shaped upper molars characterized by two diagonal crests, the protoloph and the metaloph. The lower molars were likewise crested. *Notostylops* was typical of the early notoungulates, and the comparative abundance of fossils of this animal indicate that it must have lived in large populations throughout Patagonia during early Eocene times.

From such primitive members of the order, designated as notioprogonians, the notoungulates expanded along several lines during Tertiary times. As was common among the hoofed mammals, there was a general trend toward increase in size, reaching its climax in late Tertiary times, when some of these animals became as large as rhinoceroses. From the first, the notoungulates had crested or *lophodont* molar teeth, and as they evolved many of them developed high-crowned, rather prismatic molars, obviously well suited to eating hard grasses and other vegetation that cause a great amount of wear of the tooth crowns. A very common feature of dental evolution in the notoungulates was the development of rather uniform-sized teeth from front to back, without any appreciable gaps or diastemata between them. In many of these hoofed mammals the canine tooth lost its primitive shape, to become one in the continuous series of teeth that extended from the incisors to the last molars. The skull and the jaws were frequently rather deep, and the cheek-bone or zygomatic arch was commonly very heavy. There was never a bony bar or postorbital process, separating the eye from the temporal region in which the jaw muscles were lodged. In primitive notoungulates there were five toes on each foot, whereas in some of the advanced types these were reduced to three. The toes terminated in hoofs in most notoungulates, but in some there were claws on the feet.

Thomashuxleya was a lower Eocene notoungulate, belonging to the suborder of these mammals known as the toxodonts. In this animal, as large as a sheep, the skeleton was rather robust, the limbs were strong, the feet were comparatively short and retained all the toes, and the tail was reduced in length. The long, low skull, large in comparison with the size of the body, was provided with a relatively unspecialized dentition, in which the heavy incisor teeth retained their primitive form.

Thomashuxleya may be compared in a general way with some of the slow, clumsy pantodonts that were evolving at the same time in North America.

The toxodonts evolved in great profusion during late Eocene, Oligocene, Miocene and Pliocene times, after which they began to decline, although they continued into the Pleistocene epoch before they became extinct. By the Oligocene epoch some toxodonts had become quite large, like *Scarrittia*, for example, an animal about the size of a horse, with a continuous dentition, heavy limbs, short feet, and a very short tail.

One line of toxodont evolution developed in a manner broadly similar to the development of the rhinoceroses of other continental areas. *Nesodon* of Miocene age, had a deep skull, tall, prismatic cheek teeth, and feet that that were short and broad with three functional toes, the axis of the foot being through the middle toe. This arrangement is similar in function but of course not in origin to the feet in rhinoceroses. The culmination of this evolutionary line was attained in the Pleistocene genus, *Toxodon*, a very large, heavy mammal, standing five or six feet high at the shoulder, with a large and capacious body evidently as an adaptation for feeding upon and storing quantities of plant food. In this late notoungulate there had been a departure from the continuous series of teeth so characteristic of the middle Tertiary forms, and the large, nipping incisors were separated by a considerable gap from the high-crowned grinding cheek teeth. It is incidentally a matter of some interest that *Toxodon* was discovered by Charles Darwin, when as a young man he went to South America as Naturalist on the survey ship, the "Beagle." Darwin excavated a partial skeleton of this animal from the bank of a creek in the Argentine pampas, and took it to England, where it was described by Sir Richard Owen.

In one group of toxodonts, the homalodotheres, the feet were provided with claws rather than with hoofs. *Homalodotherium*, of Miocene

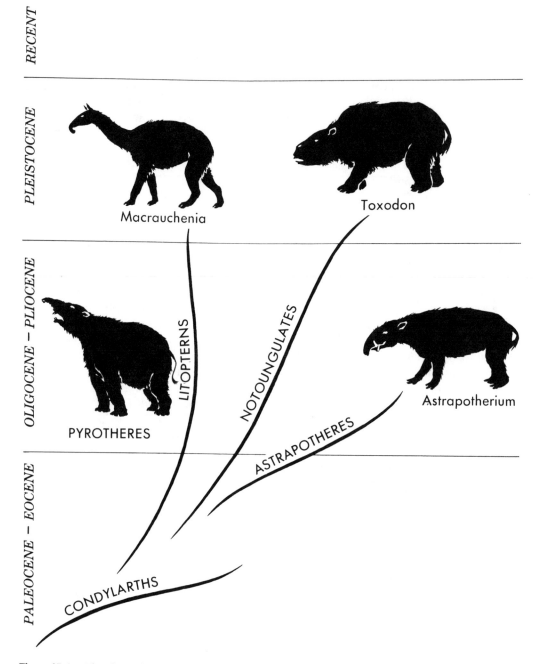

Figure 27-1.   The relationships of certain South American ungulates, most of which were derived from a condylarth ancestry. Prepared by Lois M. Darling.

age, was a heavy animal some six feet in length, and it may be compared with the chalicotheres, which as we shall see were clawed ungulates living in the northern hemisphere during Cenozoic times.

As contrasted with the toxodonts, which developed many parallels to the large hoofed mammals of North America, Eurasia, and Africa, the typotheres and hegetotheres were small notoungulates that have been compared in a very general way with the rabbits or rodents of the north. Many of these notoungulates had lightly constructed skeletons, long limbs, and rather long feet, and evidently were rapid runners. In primitive forms, like *Protypotherium* of Miocene age, the teeth formed a continuous series as in many other notoungulates; but in the specialized types, such as the Oligocene to Pliocene hegetothere *Pachyrukhos*, there were enlarged central incisor teeth, obviously adapted for gnawing, while the canines and premolars were reduced so that there was a gap between the front teeth and the grinding cheek teeth. The typotheres and the hegetotheres, like the toxodonts, reached their greatest diversity and abundance in middle Tertiary times, but some of them continued into the Pleistocene epoch.

This incomplete and very abbreviated summary of notoungulate evolution does scant justice to the most numerous and varied of the South American ungulates. Perhaps it does give some inkling of the wide degree of adaptive radiation in these interesting mammals, which if properly described would require a book to themselves. In brief it might be said that the notoungulates during their evolution ranged from small to very large mammals and that in their ecological adaptations they varied from rodentlike animals through sheeplike animals to large, rhinoceroslike animals. It is really difficult to make any valid comparison of adaptations and probable habits between these South American ungulates and the mammals with which we are familiar, and the approximate resemblances that must be cited are likely to be misleading. Suffice it to say that

the notoungulates were highly successful mammals as long as South America remained disconnected from the rest of the world, but when the isthmian link was reestablished near the close of the Tertiary period they soon disappeared before the impact of progressive invaders from the north. Even so, a few notoungulates were able to hold on well into the Pleistocene, as relics in the modernized fauna of South America.

THE LITOPTERNS

The litopterns, although never as numerous or as varied as the notoungulates, were nevertheless an important group of South American ungulates, first appearing in that region in sediments of Paleocene age, and continuing into the Pleistocene epoch. There are no known early litopterns outside South America, in which respect this order of mammals may differ from the notoungulates; and it is probable that these animals arose on the southern continent as descendants of early condylarths. Indeed, the gap is small between the didolodonts, the early South American condylarths, and some of the primitive litopterns, an indication of the indigenous ancestry of the group now under consideration.

In a way the litopterns are easier for us to comprehend than are the notoungulates for they are more directly comparable to the hoofed mammals with which we are familiar. To put it another way, there were close parallelisms between the litopterns and some of the northern ungulates, parallelisms that make the litopterns seem to us like reasonably orthodox hoofed mammals. The litopterns evolved along two distinct lines of adaptive radiation, each beginning in the Paleocene epoch and continuing through the Tertiary period—one line, the proterotheres, to become extinct during the Pliocene epoch, the other, the macrauchenids, to continue into Pleistocene times.

The proterotheres were the "horses" among the South American ungulates. They never became very large, but some of them evolved

Toxodon

Protypotherium

Thomashuxleya

Figure 27-2. Notoungulates, drawn to the same scale. *Protypotherium* was a small, middle Tertiary typothere. *Thomashuxleya* was an Eocene toxodont. *Toxodon*, an animal as large as a rhinoceros, was the last of the toxodonts, and lived well into Pleistocene times. Prepared by Lois M. Darling.

in ways that were remarkably similar to horses, especially in the adaptations of the feet for running. This evolutionary trend reached its culmination in Miocene and Pliocene times, as exemplified by such genera as *Dia-* *diaphorus* and *Thoatherium*. In these litopterns the skull was elongated and rather low, and there was a bony postorbital bar separating the eye from the temporal region, just as in the horses. The incisor teeth were rather chisel-like, and the cheek teeth were *selenodont*, with crescentic cusps, in which respect they paralleled to some degree the cheek teeth in horses of the same age. It is interesting that in these litopterns there was a molarization of the premolars as in the horses, thereby increasing the grinding area of the dentition.

The backbone was straight and the limbs were slender, an indication of rapid running. The feet were elongated, and the hind feet were especially horselike. In *Diadiaphorus*, a three-toed form, the middle toe was greatly enlarged, terminating in a strong hoof, and the lateral toes were reduced to very small appendages. The upper articulating surface of the as-

tragalus or ankle bone was shaped like a pulley, an adaptation similar to that seen in the astragalus of horses. In *Thoatherium* the evolution of the hind foot had progressed so far that the side toes were reduced to a greater degree than in any of the horses. It seems reasonable to suppose, therefore, that the habits and the mode of life of these litopterns were similar to those of middle Tertiary horses in North America. The proterotheres continued into Pliocene times and then became extinct, at about the time that true horses invaded South America.

The other line of litoptern evolution, the macrauchenids, can be compared in a general

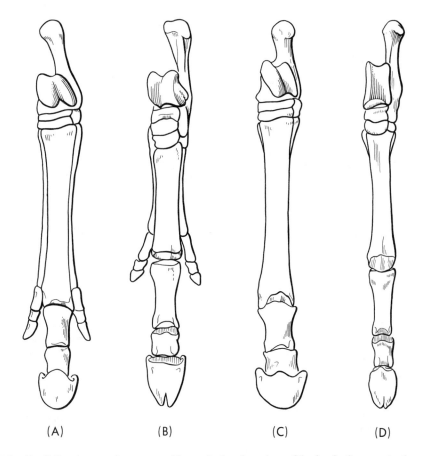

(A)   (B)   (C)   (D)

Figure 27-3. Parallelism between litopterns and horses in the adaptations of the feet for fast, sustained running over hard ground. The total length of the left hind foot is drawn to a unit scale, for the sake of easy comparison. **(A)** *Protohippus*, a three-toed horse. **(B)** *Diadiaphorus*, a three-toed litoptern. **(C)** *Equus*, a single-toed horse. **(D)** *Thoatherium*, a single-toed litoptern.

way with the camels of North America. In these litopterns the skeleton was rather lightly constructed, the back was straight, and the neck and limbs were long. The feet had three functional toes terminating in hoofs, and the axis of the foot passed through the middle toe. The long skull and jaws were provided with a continuous series of teeth that were rather high-crowned in comparison with the teeth in other litopterns. An interesting adaptation in the macrauchenids was the recession of the nasal opening far back on the face, and in the advanced forms even to the top of the skull. In this respect the skull in these litopterns may be compared with the skull in modern tapirs, and it is reasonable to think that the macrauchenids, like the tapirs, had a short proboscis, a flexible extension of the nose. *Theosodon* was a characteristic Miocene macrauchenid. *Promacrauchenia* continued the line into Pliocene times, and *Macrauchenia* lived in the Pleistocene epoch.

Macrauchenia, like *Toxodon*, lived alongside the northern invaders of South America after most of the indigenous hoofed mammals had become extinct. Its final demise during Pleistocene times may have been the result of competition from the new mammals from the north, but, in view of its considerable success after the invasion of South America, we are justified in thinking that this litoptern had become adjusted to the changed conditions so that it was able to hold its own among the immigrants from the north. However, like so many of the large mammals of the Pleistocene, it disappeared before the close of that geologic epoch—for what reason it is difficult to say.

THE ASTRAPOTHERES

Another South American order, probably arising on that continent from the ancient condylarths, is the Astrapotheria, a group that appeared in the Eocene epoch and continued into Miocene times. Among these mammals there was an early trend to gigantism, already apparent in lower Eocene members of the order.

This trend toward large size was accompanied by a series of adaptations that are difficult to interpret.

The Oligocene and Miocene genus, *Astrapotherium*, was a heavy mammal that stood five feet or more in height at the shoulders. This was quite clearly an animal adapted for pushing through the forests or wandering boldly over the plains, as do the modern elephants.

The skull and jaws in *Astrapotherium*, however, were strangely modified. The front of the skull was much abbreviated, with the nasal bones small and retracted in position; the upper incisor teeth had been lost. On the other hand, the upper canines were greatly enlarged, to form downwardly directed daggers of considerable length and strength. Instead of being shortened to match the skull, the lower jaw was long, with well-developed incisors and with enlarged canines. Very probably a long, tough upper lip extended forward from the retracted front portion of the skull, to meet the front of the lower jaw and cover it. Perhaps this constituted some sort of a cropping mechanism. Perhaps the nose extended beyond this upper lip as a sort of flexible proboscis. The posterior premolar teeth were very small, but the last two molar teeth were enormously enlarged to form long, high-crowned grinding mills.

The parallel adaptations to mammals in other parts of the world are not clear. Perhaps the astrapotheres may be compared after a fashion with the large uintatheres of North America that lived in late Eocene times. Perhaps they may be compared with some of the ancient proboscideans, the early mastodonts, in which the trunk was short, the upper tusks downturned, and the lower tusks protruded forward.

A group of South American ungulates known as trigonostylopids has been variously regarded through the years as it may or may not be related to the astrapotheres. Earlier authorities included the trigonostylopids within the Order Astrapotheria, but subsequently

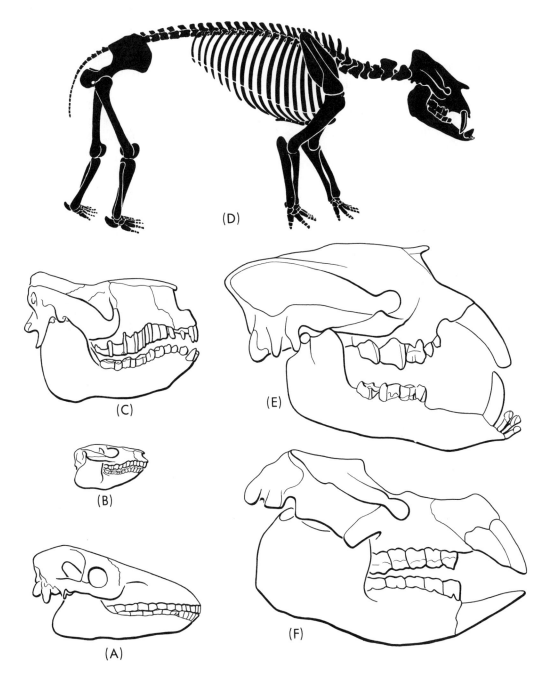

Figure 27-4. South American ungulates. **(A)** *Macrauchenia*, a Pliocene and Pleistocene litoptern. **(B)** *Protypotherium*, a Miocene typothere. **(C)** *Nesodon*, a Miocene toxodont. **(D)** *Astrapotherium*, an Oligocene astrapothere, an animal about nine feet in length. **(E)** *Astrapotherium*. **(F)** *Pyrotherium*, an Oligocene pyrothere. All skulls about one-tenth natural size, except B, which is about one-fifth natural size.

these ungulates were separated as a distinct order, particularly on the evidence of the skull of the early Cenozoic genus, *Trigonostylops*. It is smaller and much less specialized than is the astrapothere skull and does not show the various structural modifications that make the astrapothere skull so distinctive. The teeth are low crowned and on the whole have a generalized pattern. Yet upon the basis of recent studies that take account of the postcranial bones as well as the skull, it has been proposed that the close relation of trigonostylopids to astrapotheres, as originally suggested, is valid.

THE PYROTHERES

The pyrotheres, named from the Oligocene genus *Pyrotherium*, a small and isolated group of South American ungulates confined to early Tertiary sediments, have been regarded, as mentioned above, as of possible uintathere relationships. However, the definitive relationships of the pyrotheres are quite unknown. Probably the pyrotheres entered South America at a very early date, and likewise at a very early date became distinctly specialized.

These mammals were like the proboscideans, the elephants and their relatives, in showing a very early growth to gigantic size. Also the pyrotheres showed various specializations in the skull and dentition that were remarkably similar to those of some proboscideans. For instance, the skull was very large, and the nasal opening was retracted, which would indicate the presence of a proboscis or a trunk of sorts. The zygomatic arches or cheekbones were heavy, and the back portion of the lower jaw, the ascending ramus, was large, as in the proboscideans. Two of the upper incisors on each side were greatly enlarged to form tusks, and in the lower jaw there was a single enlarged, tusklike incisor on each side. The cheek teeth, separated from the incisors by a gap, were low crowned, and each tooth crown consisted of two sharp cross-crests very similar in general appearance to the crested teeth in the dinotheres, a group of proboscideans.

Because of these resemblances the pyrotheres were regarded by earlier students as relatives of the proboscideans. However, the resemblances to the proboscideans that are so strikingly developed in the pyrotheres are probably all the result of parallel evolution.

The xenungulates, largely known from the single genus, *Carodnia*, found in the Paleocene sediments of Brazil and Patagonia, have been considered by some authorities as representing a separate order of South American mammals. *Carodnia* was a large mammal, with a broad, five-toed foot and with rather short limbs. There was a complete set of teeth, of which the incisors were chisel shaped, the canines strong, the premolars generally rather pointed, and the molars strongly cross-crested. At the present time the fossils of this interesting animal are not sufficiently complete to give us information on the shape of the skull or other points about the anatomy. The structure of the third molars in *Carodnia* suggests that this genus, like the pyrotheres proper, may be related to the North American Dinocerata. At this place the xenungulates are placed within the Pyrotheria.

END OF THE SOUTH AMERICAN UNGULATES

The history of the South American ungulates is an interesting and instructive story of parallelism, of long success under favorable conditions, and of sudden extinction when those conditions changed.

The astrapotheres and pyrotheres, restricted to early and middle Tertiary times, filled certain ecological positions for a while but failed to continue, even though protected by the isolation of South America from the rest of the world. The notoungulates and the litopterns, on the other hand, had long evolutionary histories that ran from very early Tertiary times through most of the Cenozoic era. These must be considered as very successful mammals, flourishing along many lines of evolutionary development, the constituent members of which showed adaptations to an astonishing range of ecological habitats. As long as South America remained an island continent the notoungulates and the litopterns prospered, with broad plains and extensive forests in which to browse and graze, and with no enemies except the carnivorous marsupials. It is true that the carnivorous marsupials became adapted in many ways as predators, but still they were marsupials. They probably lacked the intelligence and the cunning that have made our placental carnivore such efficient hunters.

When, toward the end of the Pliocene epoch, the isthmian link between North and South America emerged from the ocean, and the great influx of mammals from the north began, most of the notoungulates and the litopterns quickly disappeared. Only a few large specialized forms, such as *Toxodon* and *Macrauchenia*, were able to continue as competitors of the mammals from the north. And of course they too eventually succumbed to pressures from the northern invaders. The disappearance of the notoungulates and the litopterns before the wave of immigrant northern mammals was brought about by two factors of importance.

In the first place the indigenous hoofed mammals suffered from direct onslaughts by the intelligent, progressive predators that came in from North America. It was one thing for them to defend themselves against the attacks from carnivorous marsupials; it was something else to become the prey of large wild dogs, foxes, bears, and various cats, including mountain lions and jaguars. This in itself was a major factor in hastening the end of many notoungulates and litopterns.

The other factor was competition for living space and food from the invading northern ungulates. With the emergence of the isthmus there came into South America tapirs, horses, deer, llamas, and mastodonts. It seems likely that these animals were more efficient browsers and grazers than the notoungulates and the litopterns, and consequently they literally took the land away from its original inhabitants. This second factor of competition, added to the other factor of direct attack from aggressive predators, was the final blow that terminated the long reign of the South American ungulates.

Eohippus

Perissodactyls

28

PERISSODACTYL CHARACTERS

The perissodactyls are those hoofed mammals (persisting at the present time as the horses, zebras, and asses, the tapirs, and the rhinoceroses) in which there is usually an odd number of toes, and in which the axis of the foot passes through the middle toe. In all the perissodactyls the inner toe, the thumb in the forefoot, and the large toe in the hind foot, has been suppressed, and the same is true of the fifth digit in the hind foot. In most perissodactyls the fifth digit in the forefoot also has been suppressed, but in some of the more primitive forms this finger has remained. Thus, in the hands and feet of perissodactyls

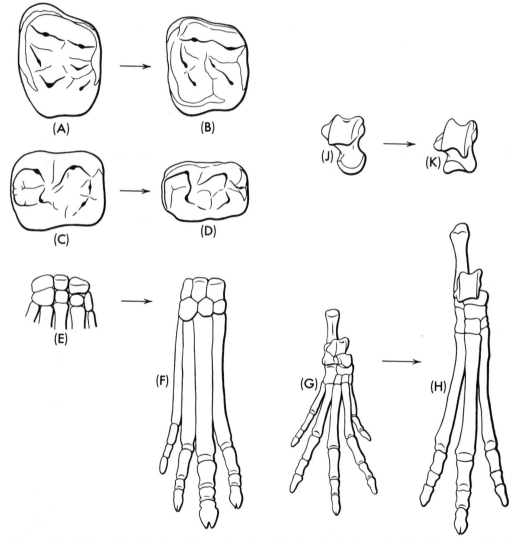

Figure 28-1. The possible derivation of the molar teeth and the feet of a primitive perissodactyl from a condylarth. **(A)** Upper left molar, **(C)** lower right molar, **(E)** right front foot, **(G)** right hind foot, **(J)** right astragalus of *Tetraclaenodon*, a condylarth. **(B,D,F,H,K)** The same elements of *Hyracotherium*, a primitive horse. The teeth are about two and one-half times natural size; the feet are about two-fifths natural size.

there are generally three functional toes or, as in the progressive horses, one.

In the perissodactyl ankle the astragalus has a doubly keeled, pulley- shaped surface for articulation with the tibia, whereas its distal surface, which articulates with certain other bones of the ankle, is flat. The femur, the upper bone of the leg, is characterized by having a prominent process, the third trochanter, on the outer side of its shaft.

In the perissodactyls a full set of incisor teeth is commonly but not invariably present, both above and below, to form an efficient cropping mechanism for biting plants. There is usually a gap between these teeth and the cheek teeth, and canines may or may not be present in this gap, frequently (if present) separated from the incisors in front and from the premolars behind. But perhaps the most characteristic feature of the perissodactyl dentition is the molarization of the premolar teeth. In the primitive perissodactyls this process has not progressed far, but in the more advanced members of the order it reaches such a degree of perfection that all the premolar teeth except the first of the series are completely molariform. This development has greatly added to the grinding surface of the dentition, thus increasing the efficiency of the teeth as mills for crushing hard plants.

Other striking perissodactyl characters, such as elongation of the limbs, growth to large size, the development of horns on the skull, and the like, are typical of the various families within the order and need not be discussed here.

ORIGIN OF THE PERISSODACTYLS

Modern paleontological evidence indicates that the condylarths, which were ancestral to so many groups of hoofed mammals, were probably the progenitors of the perissodactyls. The characters of the teeth and of the feet in the earliest perissodactyls, so basic in determining the directions of perissodactyl evolution, may be derived quite readily (probably through some intermediate "protoperissodactyl") from the teeth and the feet of certain condylarths, particularly as exemplified by the North American Paleocene genus, *Tetraclaenodon,* a close relative of *Phenacodus.* In *Tetraclaenodon* the upper molar teeth were quadrate, with six well- developed, low bunodont cusps forming the crown. This pattern, somewhat modified, is seen in the Eocene genus *Hyracotherium,* the most primitive known perissodactyl. By the simple evolutionary step of a junction between the anterior intermediate and inner cusps to form an oblique ridge or protoloph, and similarly by a junction between the posterior intermediate and inner cusps to form an oblique metaloph, and with a junction of the two outer cusps by a fore and aft ridge to form an ectoloph, the pattern of the typical primitive perissodactyl molar would have been established. In a similar fashion, the cusps of the lower molars in a mammal like *Tetraclaenodon* might have been transformed into the transverse ridges or lophids so characteristic of the primitive perissodactyls and already adumbrated in the lower molars of *Hyracotherium.*

Perhaps a significant morphological alteration between *Tetraclaenodon* and *Hyracotherium* may be seen in the structure of the feet. In *Tetraclaenodon* the bones of the wrist were somewhat rounded and arranged in a serial fashion, with an upper row above a lower row, while in the ankle the lower articulating surface of the astralagus was rounded, as is frequently the case in carnivorous mammals. These features gave the feet great flexibility. In *Hyracotherium* the bones of the wrist were alternately arranged so that they interlocked, and also the lower articulating surface of the astragalus was relatively flat. These features made the feet of *Hyracotherium* comparatively more rigid, and less liable to lateral movements than the feet of the condylarth. Moreover, the feet in *Hyracotherium* had greatly elongated digits as compared with the feet of the condylarth.

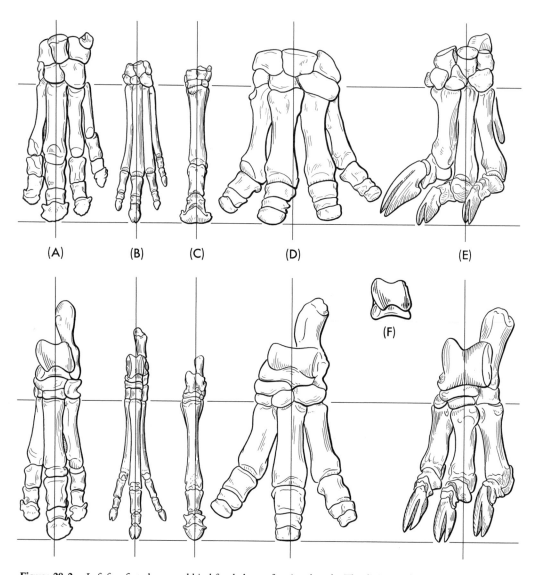

Figure 28-2. Left fore feet above, and hind feet below, of perissodactyls. The digits are drawn to approximate unit lengths, to facilitate a comparison of proportions of the feet in the several groups of odd-toed ungulates. The vertical lines pass through the axes of the feet. **(A)** *Tapirus,* a Pleistocene and Recent tapir. **(B)** *Hyracotherium,* an Eocene horse, a generalized perissodactyl. **(C)** *Equus,* a Pleistocene and Recent horse. **(D)** *Brontotherium,* an Oligocene titanothere. **(E)** *Moropus,* a Miocene chalicothere, with clawed feet. **(F)** The perissodactyl astragalus, showing the proximal pulley for articulation with the tibia, and the flat distal surface for articulation with the distal bones of the ankle.

These changes from *Tetraclaenodon* to *Hyracotherium* point to a shift in adaptations, with emphasis on greater efficiency for browsing and, particularly, for running over firm ground. Therefore, *Hyracotherium,* the most primitive known perissodactyl, was well-equipped to feed on the vegetation that surrounded it, and especially to flee quickly from the attacks of predators. It probably was this adaptation for fast running as a response to pressure from aggressive predators that spelled the success of the earliest perissodactyls; and, conversely, it was perhaps the lack of such adaptations in the condylarths that eventually caused their extinction. With the transition from Eocene to Oligocene times, when numerous carnivorous fissipeds were well established and efficient hunters, the condylarths succumbed. But their well-adapted descendants, the perissodactyls, prospered to follow many lines of varied evolutionary development, as we shall learn.

THE FIRST PERISSODACTYLS

Hyracotherium (commonly designated as *Eohippus*), although classified as the earliest primitive horse, is as characteristically primitive as any of the early perissodactyls, and can be used as a good example of the prototype for this order of mammals. This was a small animal, about the size of a fox. It was lightly built, with limbs clearly adapted for running, a moderately curved back, a somewhat shortened tail and a long, low skull. *Hyracotherium* had nineteen ribs, and behind these about five vertebrae free of ribs, as was common among the perissodactyls. The spines of the vertebrae in the shoulder region were somewhat elongated, for the attachment of strong back muscles. The limbs were slender, and the feet were elongated, with the wrist and ankle raised far off the ground so that the digits were approximately vertical in position. There were four toes in the front foot of this primitive perissodactyl and three toes in the hind foot. In a functional sense all the feet were three toed, and each toe terminated in a small hoof.

The elongated skull had a comparatively small braincase, and the orbit was open behind, there being no bony bar separating the eye opening from the temporal opening as in later horses. The incisors were small, with rather chisel-shaped crowns, and small canine teeth were present. The cheek teeth were bunodont, which means that they had very low crowns, with rounded, conelike cusps. The premolar teeth were not as yet molariform, and the last two upper premolars were trian-

gular in shape. However, the upper molars were quadrangular, with four large cusps, protocone, paracone, metacone, and hypocone. There were also two small, intermediate accessory cusps, the protoconule and the metaconule, which were connected to the two inner cusps by low, oblique ridges—the protoloph and metaloph. In the lower molars the heel, or talnoid, was as high as the fore part of the tooth. The anterior internal cusp, the paraconid, was much reduced, and the two anterior cusps, protoconid and metaconid, and the two posterior cusps, hypoconid and entoconid, were connected by transverse crosscrests or ridges. Such was the dental development and tooth pattern from which evolved the complex and varied teeth of the several lines of perissodactyls through Cenozoic time.

BASIC CLASSIFICATION OF THE PERISSODACTYLS

From an ancestral form approximated by *Hyracotherium* the perissodactyls developed along various paths of adaptive radiation, reaching the height of their evolutionary history in middle Tertiary times when they were the dominant ungulates in most of the world. Since then they have declined, and now must be regarded as an order of mammals that, in spite of the high specializations among some of its members, is on its way toward extinction.

From their primitive beginning the perissodactyls evolved along three distinct lines. One of them, the suborder Hippomorpha, contained the extinct palaeotheres and titanotheres, and the horses. A second line, the Ceratomorpha, contained the tapirs and the rhinoceroses. The third line, the Ancylopoda, consisted of the strange, clawed chalicotheres and their ancestors.

EVOLUTION OF THE HORSES

Of all evolutionary histories, probably none is so widely known as that of the horses. There are several reasons for this, but perhaps the most cogent one is that the fossil record

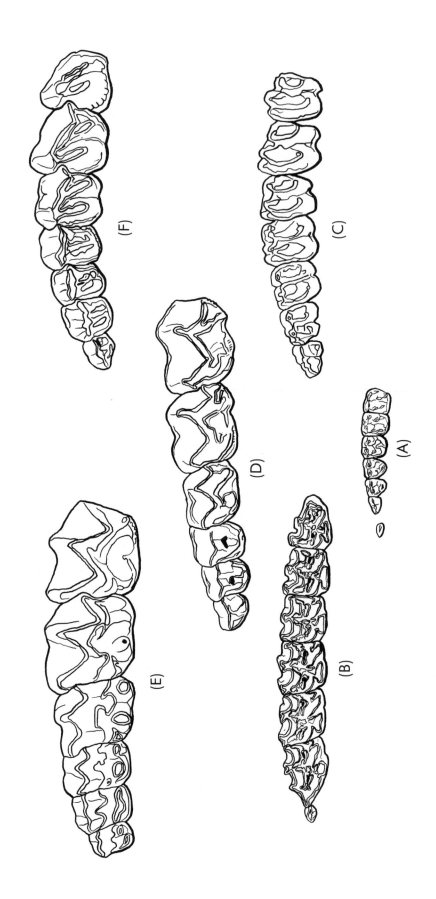

(A)

(B)

(C)

(D)

(E)

(F)

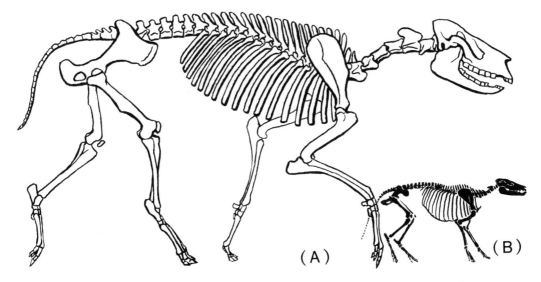

Figure 28-4. **(A)** The skeleton of *Hyracodon*, a lightly built rhinoceros from the Oligocene of North America, showing the structure of a generalized, modern-type ungulate. About five feet in length. **(B)** The skeleton of *Hyracotherium*, an Eocene horse, drawn to the same scale as (A).

of the horses is remarkably complete and well understood. Because of their grazing habits, their life on savannas and plains, and their tendency to live in large herds, horses have been buried and fossilized in great numbers since the early stages of their phylogenetic history. It so happens that in North America, where the entire evolutionary sequence of horse history is recorded, there is an almost complete series of sedimentary deposits from early Eocene times to the present, containing the fossils of horses. Naturally this excellent array of fossil horizons gives a remarkable record through time of the progressive evolution of the horses.

The first horses, belonging to the lower Eocene genus *Hyracotherium* have already been described, and that description need not be repeated here. It is enough to say that *Hyracotherium* was a small animal, with a primitive skull, a generalized perissodactyl dentition in which the cheek teeth were low crowned and bunodont, and slender limbs and feet, with four toes on the front foot and three on the hind. This early Eocene horse was widely spread throughout North America and Europe. With the close of early Eocene times *Hyracotherium* became extinct in the Old World, and from then on the evolution of horses was limited to the North American continent. All the horses that appeared subsequently in other regions, in Eurasia, Africa, and South America, were emigrants from North America.

The progressive trends that characterized the evolution of horses through the Cenozoic era may be listed as follows:

1. Increase in size.
2. Lengthening of legs and feet.

Figure 28-3. Left upper premolar and molar teeth of perissodactyls, showing the various adaptations for increasing the crown areas and enamel lengths in these herbivores. **(A)** *Hyracotherium*, an Eocene equid. Natural size. **(B)** *Equus*, a Pleistocene horse. About one-half natural size. **(C)** *Tapirus*, a recent tapir. About three-fourths natural size. **(D)** *Moropus*, a Miocene chalicothere. About one-half natural size. **(E)** *Brontotherium*, an Oligocene titanothere. About one-fourth natural size. **(F)** *Trigonias*, a Miocene rhinoceros. About one half natural size.

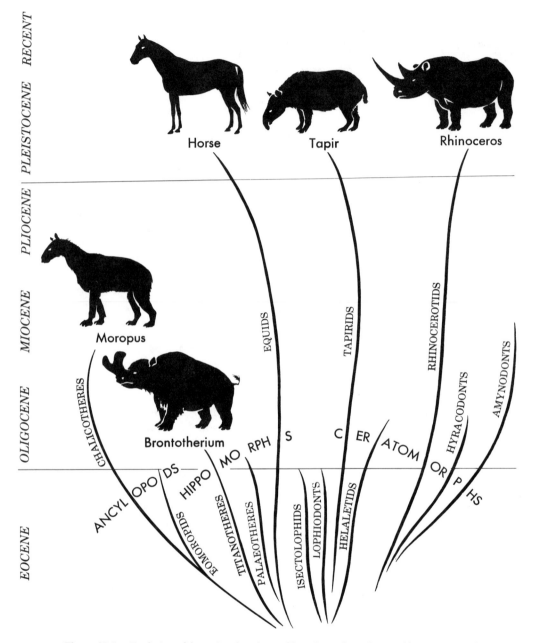

Figure 28-5.   Evolution of the perissodactyls or odd-toed ungulates. Prepared by Lois M. Darling.

3. Reduction of lateral toes, with emphasis on the middle toe.
4. Straightening and stiffening of the back.
5. Widening of the incisor teeth.
6. Molarization of the premolars.

7. Increase in height of the crowns of the cheek teeth.
8. Progressive complication of the crown patterns.
9. Deepening of the front portion of the skull and of the lower jaws, to accommodate the high-crowned cheek teeth.

10. Lengthening of the face in front of the eye, also to accommodate the high- crowned cheek teeth.
11. Increase in size and complexity of the brain.

These changes, initiated in Eocene times, continued to the end of the Cenozoic era, so that in some horses there was a fairly consistent increase in size from the beginning to the end of their phylogenetic history, accompanied by the progressive molarization of the premolars, the deepening of the cheek teeth, the reduction of the lateral toes, and so on. However, when the horses are considered in their entirety no such picture of uniform evolution emerges. The horses are often cited as an outstanding example of "straight-line evolution" or of "orthogenesis," and it is frequently maintained that these animals evolved with little deviation, along a straight path from the little Eocene *Hyracotherium* or eohippus to the modern horse, *Equus*. It is true that most of the progressive changes listed above can be followed through time from *Hyracotherium* to the modern horses, but in middle and late Tertiary times there were various lateral branches of horses that were progressive in some features and conservative in others. When all fossils are taken into account the history of horses in North America is seen to be anything but a simple progression along a single line of development.

At the beginning of their history, however, the horses were confined to a single line of progressive development, as follows:

Middle and upper Oligocene	*Miohippus*
Lower Oligocene	↑ *Mesohippus*
Upper Eocene	↑ *Epihippus*
Middle Eocene	↑ *Orohippus*
Lower Eocene	↑ *Hyracotherium*

During their evolution from *Hyracotherium* to *Miohippus* the horses increased from small, terrier-sized animals to animals as large as sheep. The legs increased in length, as did the feet. The fifth digit on the hand was suppressed, so

that all the feet became three-toed, with the middle toe much larger than the side toes. However, all the toes were functional. The back became progressively straighter and stiffer. The last three premolar teeth became molariform, and the crowns of all the cheek teeth became strongly crested. In the upper cheek teeth an outer W-shaped wall or crest developed, the ectoloph, and from it two oblique crests extended to the inner side of each tooth, an anterior protoloph and a posterior metaloph. On the crowns of the lower cheek teeth were two V-shaped crests with their points directed outward. These crested upper and lower cheek teeth were evidently very efficient choppers for cutting up leaves. However, they were still low crowned. In general the skull remained rather primitive, although there was some lengthening of the facial portion.

By the end of the Oligocene epoch the horses had through these changes attained the status of advanced browsers, capable of eating leaves and soft plants and able to run fairly rapidly and for sustained periods over hard ground. With the advent of Miocene times there was a branching out of horses along several lines of development, probably as a response to an increase in the variety of environments available to them, and especially because of the spread of early grasses and other flowering ground plants.

Archaeohippus, of Miocene age, remained conservative and showed very little increase in size or progressive development of the skull, teeth, and feet beyond the *Miohippus* stage. Another line of horses, the Miocene and Pliocene anchitheres, increased in size so that in the end they were as large as modern horses, but they retained the conservative *Miohippus*-like teeth and functional three-toed feet. These were probably forest-living horses that browsed in the deep woodlands, much as deer do at the present time. During late Miocene and early Pliocene times *Anchitherium* migrated into the Old World, where it spread widely, while in North America a large genus, *Hypohippus,* evolved from an *Anchitherium* ancestry.

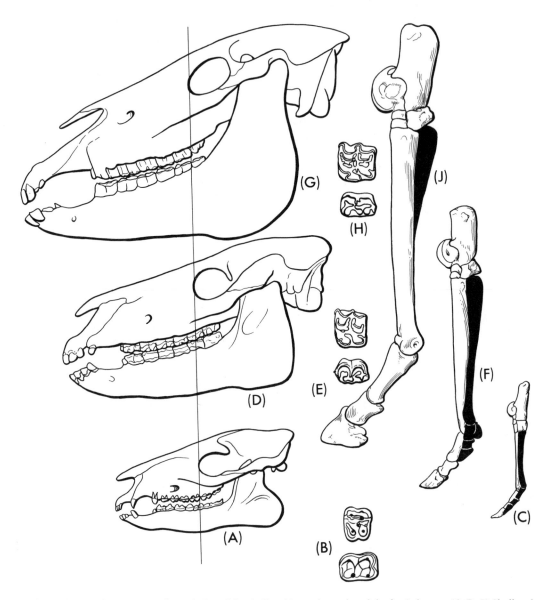

Figure 28-6. Three stages in the evolution of the skull and jaws, the teeth and the feet in horses. **(A,B,C)** Skull and jaws, upper and lower molars and hind foot of *Hyracotherium,* a primitive Eocene horse. **(D)** Skull and jaws of *Parahippus,* **(E,F)** upper and lower molars and hind foot of *Merychippus,* Miocene horses. **(G,H,J)** Skull and lower jaws, upper and lower molars and hind foot of *Equus,* a Pleistocene and recent horse. This figure shows the progressive increase in the length of the facial or preorbital portion of the skull, and the deepening of the skull and lower jaw to accomodate the increasingly higher cheek teeth; the progressive complexity of the molar crowns; and the reduction of the side toes. D and G, one-sixth natural size; A,C,F,J, about one-third natural size; B, about two-thirds natural size; E,H, about one-fourth natural size.

As contrasted with these conservative horses, the main line of horse evolution was continued during Miocene times by *Merychippus*. This horse was as large as a small pony. The feet retained three toes, but the lateral toes were now reduced so that they were of little use, and the animal walked on the single middle toe, the end of which was clad in a rounded hoof. The face was elongated and rather deep, and the lower jaw was also deepened. In *Merychippus* the teeth were definitely high crowned and covered with cement, the crests had become variously conjoined, and the enamel of these crests was folded, so that when the tooth was worn the complex enamel bands projected slightly above the softer dentine and cement. These tooth crowns, so difficult to describe, were efficient grinding mills for breaking down hard plant fibers and seeds, in order that they might be more readily digested. In *Merychippus* there was a postorbital bar behind the eye opening, separating the orbit from the temporal region—a character typical of the later horses.

At the close of Miocene times two groups of horses arose from a *Merychippus* ancestry. One of them was an assemblage of closely related genera, of which *Hipparion* was typical; the other was centered around the genus *Pliohippus*. The hipparions were lightly built horses, progressive in the development of the skull and the very high-crowned teeth with complexly folded enamel, but conservative in the retention of three-toed feet. These horses erupted from their North American center of origin at the beginning of Pliocene times, to spread into all the continents except South America. In fact, *Hipparion* was so characteristic of lower Pliocene mammalian faunas that they are often called the *"Hipparion* faunas." These horses continued through the Pliocene epoch, and a few stragglers held on into the Pleistocene epoch, when they became extinct.

Pliohippus was also a progressive horse, not only in the development of the skull and the teeth but also in the feet, for this horse became a single-toed equid. The side toes were reduced to mere splints that were concealed be-

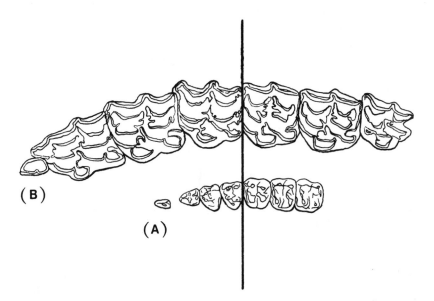

Figure 28-7. Left upper cheek teeth of (**A**) *Hyracotherium*, an Eocene horse, natural size, and (**B**) *Equus*, a Pleistocene-Recent horse, one-half natural size, to show the molarization of the premolars in these perissodactyls. The vertical line denotes the boundary between the premolar and molarteeth. Note the increase in size and the complete molar pattern of the last three premolar teeth in *Equus*.

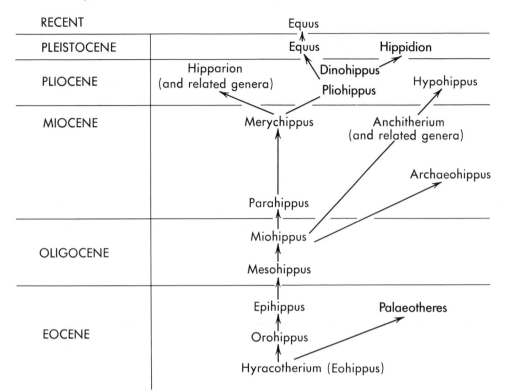

RECENT	Equus
PLEISTOCENE	Equus Hippidion
PLIOCENE	Hipparion (and related genera) Dinohippus Hypohippus Pliohippus
MIOCENE	Merychippus Anchitherium (and related genera)

Archaeohippus

Parahippus

Miohippus

| OLIGOCENE | Mesohippus |

Epihippus   Palaeotheres

| EOCENE | Orohippus |

Hyracotherium (Eohippus)

neath the skin of the upper portion of the foot, a condition that continues in the feet of modern horses.

From an ancestry involving *Pliohippus* two groups of horses arose at the end of the Pliocene epoch. One of them was embodied in the genus *Hippidion*, which originated in South America from *Pliohippus* ancestors that entered the southern continent when the isthmian bridge was re-established. *Hippidion* was a large horse, with rather short legs and feet. It inhabited South America during Pleistocene times, but became extinct before the close of the Ice Age.

The other descendant from *Pliohippus* was our modern group of horses belonging to the genus *Equus*, horses in which the *Pliohippus* trends were carried to a logical conclusion. *Equus* arose in North America and lived in this continent through the Pleistocene epoch, to become extinct a few thousand years ago. In the meantime, at the beginning of Pleistocene times, *Equus* had migrated into the other continents to become a horse of worldwide distribution. It has survived into modern times in the Old World, where it is found as a number of species that we designate as horses, zebras, and asses. (Modern horses in the New World were introduced from Europe by sixteenth-century settlers.) These animals are so familiar as to need no description at this place.

This account, brief as it is, may give some

Figure 28-8. Crown view of worn left upper and right lower molars of perissodactyls, not to scale. (**A**) *Hyracotherium,* an Eocene horse and a generalized perissodactyl. (**B**) *Palaeotherium,* an Eocene palaeothere. (**C**) *Miotapirus,* a Miocene tapir. (**D**) *Equus,* a Pleistocene and Recent horse. (**E**) *Moropus,* a Miocene chalicothere. (**F**) *Brontotherium,* an Oligocene titanothere. (**G**) *Caenopus,* a Miocene rhinoceros. From the ancestral type (A), perissodactyl molars evolved as (D) high-crowned teeth with complexly folded enamel, adapted for grazing on hard grasses; as (E) and (F) low-crowned teeth with simple crescents and cones, for feeding upon comparatively soft vegetation; as (C) cross-crested teeth for browsing in jungles; and as (G) medium to high-crowned crested teeth, for a combination of browsing and grazing.

(D)

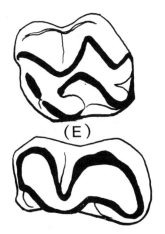

(E)

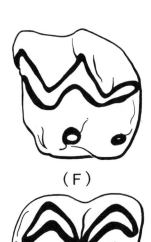

(F)

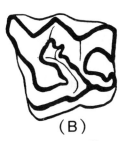

(B)

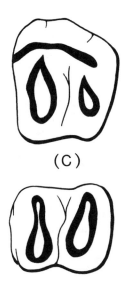

(C)

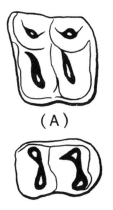

(A)

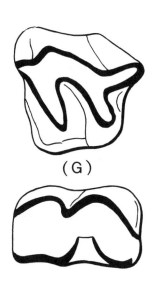

(G)

idea of the rather involved history of the horses during Cenozoic times. Perhaps it may be summarized by the following diagram.

HORSES AND MEN

The history of man has been closely linked with the late history of the horses through many thousands of years. From paintings that were made by Stone Age men in the caves of Europe we know that some of our ancestors were hunters of horses. As man grew out of his Stone Age culture into the metal ages, he learned that horses were useful beasts of burden. Therefore he gave up eating horses and adopted them as riding animals and as work animals that pulled his wagons and his plows.

The history of ancient civilizations in the Old World is a story of man's use of horses and of asses. (Zebras were never successfully domesticated.) Armies traveled and fought on horseback, and with horses whole populations moved from one region to another. Much of man's progress to his modern state has depended on horses, and it is only within the last few decades that these useful ungulates have at last been supplanted by something new—the internal- combustion engine, attached to wheels and implements.

THE PALAEOTHERES

During Eocene and early Oligocene times the palaeotheres were evolving in Europe in a manner somewhat parallel to the way the horses were evolving at the same time in North America. In the palaeotheres, of which *Palaeotherium* was typical, there was a rapid increase in size, so that by late Eocene times these animals were as large as small rhinoceroses. The legs were rather heavy, and all the feet had three toes. The skull was tapirlike in that the nasal bones were retracted, an indication that *Palaeotherium* may have had a short proboscis. The teeth, although low crowned, were progressive in the molarization of the premolars and especially in the patterns of the crowns, which resembled the crown patterns of Miocene horses rather than the Eocene *Hyracotherium*, to which *Palaeotherium* was closely related. It was such horse-like characters that led Huxley to select *Palaeotherium* as one of the early ancestors of the horses, a view that was seen to be erroneous when the abundant fossils record of equid evolution in North America became known. Palaeotheres continued into the Oligocene epoch in Europe and then became extinct.

EVOLUTION OF THE TITANOTHERES

Among the largest of the perissodactyls were the titanotheres or brontotheres, which first appeared during early Eocene times as small, eohippuslike animals, and which attained the peak of their evolutionary development in middle Oligocene times as massive, gigantic beasts seven or eight feet in height at the shoulder. The titanotheres grew up quickly into giants, and once having become giants they soon died out. Theirs was a short, but dramatic phylogenetic history.

One of the first of the titanotheres was a lower Eocene genus known as *Lambdotherium*. In its general aspects *Lambdotherium* was like *Hyracotherium*, and indeed this first titanothere must have been rather closely related to the first horse. It was similar to *Hyracotherium* in size, perhaps a little larger, and like the ancestral horse it was lightly constructed, with a back that was somewhat curved, with slender limbs and elongated feet for running, with

---

Figure 28-9.   Skulls of perissodactyls. **(A)** *Hyracotherium*, an Eocene horse that is in many respects a typical primitive perissodactyl, one-fourth natural size. **(B)** *Equus*, a Pleistocene horse, one-eighth natural size. **(C)** *Miotapirus*, a Miocene tapir, one-sixth natural size. **(D)** *Moropus*, a Miocene chalicothere, one-eighth natural size. **(E)** *Brontotherium*, an Oligocene titanothere, one-twelfth natural size. **(F)** *Rhinoceros*, a Pleistocene rhinoceros, one-eighth natural size. Notice the central position of the orbit, the front border of which is indicated by a vertical line, in the ancestral type A, and the different proportions of facial or preorbital length to cranial or postorbital length in the specialized perissodactyls.

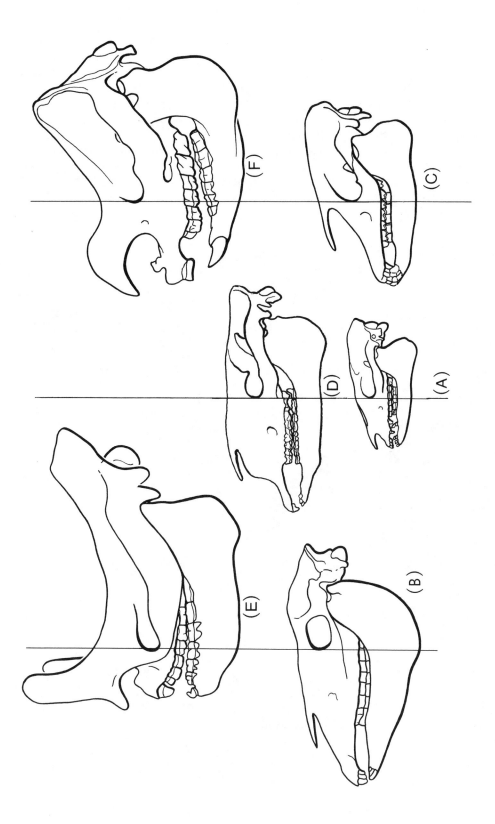

a shortened tail and with a long, low skull. There were four toes on the front foot, of which the outer digit was not reduced, and the hind foot had three toes and the characteristic perissodactyl ankle structure. The skull was primitive and the opening for the eye was confluent with the temporal opening. There were small incisors, separated by a short gap from sharp canines; they in turn were separated by a rather long gap from the cheek teeth. The cheek teeth were low crowned, and there was minimal molarization of the premo-

lars. The upper molar teeth were square and had four prominent cusps. The two outer cusps were joined by a W-shaped outer wall or ectoloph. In the lower molars the outer and inner cusps were alternately arranged, and joined by ridges that made a W-shaped pattern.

Two dominant trends marked the evolutionary history of the titanotheres—a quick phylogenetic growth to large and even huge size, and the development of large horns on the skull. The increase in size preceded the

Brontotherium

Eotitanops

Figure 28-10. A small Eocene titanothere, similar to other generalized perissodactyls, and a gigantic specialized Oligocene titanothere that stood about eight feet in height. Prepared by Lois M. Darling.

development of horns. While these two evolutionary advancements were taking place the titanotheres remained comparatively primitive in other respects, particularly in the dentition and the feet.

Of course, as they grew into large animals and then into veritable giants, the titanotheres showed those modifications of the skeleton that we have seen in other large, heavy ungulates. The elongated legs became very heavy, with the upper limb elements, the humerus and the femur, long in comparison with the lower limb elements. The feet developed as short, broad structures, well suited to support the great weight they had to bear. The body grew very capacious, partly as an adaptation for storing large amounts of plant food, with heavy ribs to enclose the thorax and the abdomen. The tail was comparatively short. In the shoulder region there were long spines on the vertebrae, to afford attachments for strong muscles that held up the heavy head.

The changes that took place in the feet during this process of growing large were mainly of proportion. Even in the latest titanotheres there were still four toes in the front foot as in the first members of the group, and in all of the feet the lateral toes were large. The molars in the later titanotheres became very large as contrasted with the relatively small premolars, and the teeth in front of the premolars were reduced or often suppressed. These changes in proportions of the teeth were correlated with the changes in skull proportions as the titanotheres evolved, because as the titanotheres increased in size through time, that portion of the skull behind the eye became ever longer, while the facial region in front of the eye became shorter. In the last titanotheres, of early Oligocene age, the face was remarkably short, so that the eyes were located far forward in the skull.

The various evolutionary changes that have been outlined above began in early Eocene times. *Eotitanops,* not much later in age than *Lambdotherium,* already showed a considerable increase in size over its predecessor, although in most respects it retained the ancestral perissodactyl characters. Growth in size with consequent changes in the skeleton and the accompanying proportional changes in the skull and teeth can be traced through middle and upper Eocene titanotheres, such as *Palaeosyops* and *Manteoceras.*

Several lines of titanotheres were evolving parallel to each other during middle and late Eocene times. *Manteoceras* represents what might be called the "main line" of titanothere development that led to the gigantic horned titanotheres of Oligocene times. A lateral branch of middle and late Eocene age was that of the dolichorhines, characterized by *Dolichorhinus,* in which the skull was very long but never developed horns. Another side branch was the telmatheres, hornless titanotheres in which the canine teeth became rather large, perhaps as a compensation for the lack of defensive horns.

In *Manteoceras* there were small outgrowths on the frontal and nasal bones, side by side. They increased greatly in the Oligocene titanotheres, such as the gigantic *Brontops,* to form large, heavy horn cores, which in life probably were covered with tough skin or with horns.

During most of their history the titanotheres were restricted to North America, but in late Eocene and early Oligocene times they migrated into Asia, and some of them pushed as far west as eastern Europe. Most of them were advanced, horned titanotheres. *Embolotherium,* a Mongolian form of Oligocene age, had a pair of huge, conjoined horns on the front of the skull that made a sort of battering ram with which this big titanothere could hammer its adversaries.

Why did the titanotheres become extinct during the middle of the Oligocene epoch, after having successfully evolved at such a rapid rate to huge animals, much larger than modern rhinoceroses? It is very probable that their failure was in part because of the lack of progressive development in the teeth. The cheek teeth of titanotheres were always low crowned, adequate for the soft vegetation that grew in very early Tertiary times, but not suited for

Moropus

Figure 28-11.   A Miocene chalicothere, a perissodactyl as large as a modern horse, with large claws on its feet. Prepared by Lois M. Darling.

the harder grasses that were spreading during the middle reaches of the Cenozoic era. Their food supply was changing, and they seem to have been unable to adapt themselves to this change. This factor, together with a comparatively primitive brain, probably had much to do with bringing an end to the titanotheres.

THE CHALICOTHERES

The chalicotheres, perhaps related to the titanotheres, yet sufficiently distinct to be placed in a separate suborder, were successful perissodactyls in that their phylogenetic life extended from Eocene times into the Pleistocene epoch. They seemingly were never very numerous animals. They were unique among the perissodactyls, in that the advanced genera had large claws on the feet, rather than

hoofs, and it is possible that these beasts lived in small groups along streams, where they could dig up roots on which to feed, instead of browsing or grazing across the plains in large herds.

The first chalicotheres, typified by the upper Eocene genera, *Eomoropus* of North America and *Grangeria* of Asia, were generally similar to other primitive members of this order of ungulates. From these ancestors the chalicotheres evolved rapidly through the Oligocene epoch, reaching the full stature of their phylogenetic development by early Miocene times. From the Miocene into the Pleistocene epoch the chalicotheres continued without much evolutionary advancement.

The chalicotheres followed the general perissodactyl trend of size increase, so that the

Miocene and later chalicotheres, like *Moropus* in North America and *Macrotherium* in Eurasia, were as large as large horses. In some respects the later chalicotheres had a horselike appearance, for the skull had a long, deep face, the body was compact, and the limbs were elongated. But here the similarity ends. The teeth were essentially similar to those of the titanotheres, with low crowns, large molars, and small premolars. As in the titanotheres, the teeth in front of the premolars were variously reduced or completely suppressed. In the advanced chalicotheres the front legs were longer than the hind limbs, so that there was a slope back from the shoulders to the hips, somewhat as in modern giraffes. The feet were short, with three functional toes in each foot and, as already mentioned, there were claws on all the toes, these claws being larger in the front feet than in the hind feet. Of the claws on the front feet, the inner one was the largest.

Some large Miocene chalicotheres of North America, belonging to the genus *Tylocephalonyx*, are remarkable in the presence of a large, hollow, bony dome on the top of the skull above the braincase. The function of this structure is puzzling; perhaps it was used in low-impact butting, as seen in modern giraffes, and perhaps it was for visual display.

The chalicotheres managed very well with their strange way of life until sometime during the Pleistocene epoch, and then they became extinct. However, their disappearance probably cannot be blamed upon any inadequacies in their adaptations; their continuation during most of Cenozoic times shows that they were well suited for a very particular mode of life. They finally vanished during the wide extinction of large mammals that took place in late Pleistocene times, when many of the spectacular animals that had graced the Ice Age landscapes of the world disappeared.

EVOLUTION OF THE RHINOCEROSES

Having reviewed the perissodactyls that can be grouped in the suborders Hippomorpha,

namely the horses and titanotheres, and Ancylopoda or chalicotheres, we now come to the other suborder of perissodactyls, the Ceratomorpha, which contains the rhinoceroses and the tapirs.

At the present time, rhinoceroses are a very restricted group of ungulates, represented by two species in Africa and three in Asia—the latter being among the rarest of modern mammals. It can safely be said that the rhinoceroses are well on their way toward extinction, and it is possible that one or two species of rhinoceroses, thanks in large part to the activities of modern man, will become extinct within the next few decades. In Tertiary times, however, the rhinoceroses were numerous and varied, belonging to several lines that were evolving parallel to each other. This parallelism makes the past history of the rhinoceroses difficult to interpret.

The first known rhinoceroses appear in sediments of Eocene age, and like many other very early odd-toed ungulates they show the primitive characters of the stem perissodactyls that have already been described. It is likely that the earliest rhinoceroses were closely related to the primitive tapirs; indeed, the perissodactyls known as hyrachyids (long considered as very primitive rhinoceroses) are now placed among the Helaletidae, a family of early tapirs.

With this assignment of the hyrachyids to the tapirs, the rhinoceroses assume a threefold division represented by three families: the Hyracodontidae, the Amynodontidae, and the Rhinocerotidae. Among them, the hyracodonts, or running rhinoceroses, were the most generalized.

These rhinoceroses appeared in middle and late Eocene times and reached the culmination of their evolutionary development during the Oligocene epoch. *Hyracodon,* from the Oligocene sediments of North America, was a characteristic hyracodont. This rhinoceros was rather small, lightly built, with slender limbs and long feet that were adapted for rapid running, and although the feet were function-

Hipparion

Teleoceras

Figure 28-12. Two Pliocene perissodactyls of North America. *Hipparion* was a three-toed horse, *Teleoceras* a very heavy, short-limbed rhinoceros. Prepared by Lois M. Darling.

ally tridactyl, there were four toes on the front feet. The skull was rather low, the eye was centrally located, and the orbit was confluent with the temporal opening. All of the incisor teeth were present, and immediately behind them were small canines, in effect, forming a part of this series of teeth. These teeth were separated from the cheek teeth on each side by a diastema or gap. The last premolars were molariform in structure; the molars were crested in the manner that was to be characteristic for rhinoceroses throughout their evolutionary history; in brief, each upper molar had a strong outer crest or ectoloph, to which

at slightly oblique angles were two cross-crests or lophs, while the lower molars were strongly cross-crested. These tooth patterns indicate an early specialization for browsing. The hyracodonts, which enjoyed the height of their evolutionary development during Oligocene history, barely survived into Miocene times, and then became extinct.

The second branch of rhinoceros evolution, the amynodonts, arose during late Eocene times, when the hyracodonts were still in the early phases of their evolutionary history. The amynodont rhinoceroses are characteristically represented in the upper Eocene by

Amynodon, in the Oligocene by *Metamynodon.* From the first the amynodonts were large, heavy rhinoceroses, with strong limbs and short, broad feet. They are frequently found in stream channel deposits, and it seems logical to think that they may have been water-lovers with hippopotamuslike habits. The skull was heavy, the incisor teeth and the anterior premolars were greatly reduced, and the canines and the molars were greatly enlarged, the former as large daggers, possibly for fighting, and the latter as long, cutting teeth. For a time the amynodonts were successful, spreading from North America, which would seem to have been the center of origin, into Asia and Europe. They became extinct soon after the close of the Oligocene epoch.

We now come to the central stock of rhinoceros evolution. *Caenopus,* an Oligocene member of the evolutionary line, probably arose from some of the running rhinoceroses. *Caenopus* was a large rhinoceros, standing some four or five feet in height at the shoulders, and it illustrates the early growth to large size that was so typical of most lines of rhinoceros evolution. It was a fairly heavy rhinoceros, seemingly hornless, and it showed some molarization of the premolars. From this point on, the later rhinoceroses evolved in several different directions. Their adaptive radiation included a general increase in size (although the members of a few lines remained comparatively small), the development of broad, three-toed feet for supporting the weight of the strong limbs and the heavy body, the molarization of the premolars, the lengthening of the crowns in the cheek teeth and the continuation of the pattern of strong crests established in the primitive rhinoceroses, and finally the development of horns on the skull.

The horns of rhinoceroses, known mainly from the modern forms, are unique among mammals in that they are formed completely of coalesced hair and have no bony cores. As might be expected, horns such as these were rarely fossilized, but their presence is clearly indicated by roughened areas on the skull, where the bases of the horns were attached.

The baluchitheres were the largest of the rhinoceroses, and the largest of all land mammals, past or present. *Baluchiterium,* which lived in Asia during Oligocene and early Miocene times, was an animal that stood sixteen or eighteen feet in height at the shoulders, and in life it must have weighed many tons. This giant rhinoceros was hornless. It probably browsed upon the leaves of trees.

The Miocene diceratheres were small rhinoceroses with a pair of horns, side by side, on the nose.

The Miocene and Pliocene teleocerines, as typified by *Teleoceras,* were heavy rhinoceroses with remarkably short limbs and feet. They carried a single horn on the nose.

The single-horned rhinocerines appeared in late Miocene times and have continued to the present day as the one-horned rhinoceroses, *Rhinoceros* of India and Java.

Various two-horned rhinoceroses, with one horn in front of the other, appeared during late Cenozoic times. One of these survives as the modern two- horned Sumatran rhinoceros. The well-known wooly rhinoceros of the Ice Age, depicted in many European caves by Stone Age men, was probably related to this modern two-horned type. On a somewhat separate branch are the modern two-horned rhinoceroses of Africa, *Diceros,* the black rhinoceros, and *Ceratotherium,* the white rhinoceros.

The elasmotheres were giant rhinoceroses that lived in Eurasia during Pleistocene times. They had a single large horn on the forehead, not on the nose as in modern one-horned rhinoceroses, and were characterized by tall cheek teeth with very complicated enamel patterns.

From the above it can be seen that the rhinoceroses were numerous in most continental regions during late Cenozoic times. Many evolutionary lines coexisted side by side, but as the Cenozoic era came to a close most of these lines disappeared, one by one. Rhinoceroses became extinct in North America during the Pliocene epoch. At the same time various lines of rhinoceroses died out in Eurasia, but other Old World lines continued into the Pleistocene. Finally, at the close of Pleistocene times some of these disappeared, to leave the five species of rhinoceroses known to us as living animals. And as said above, they represent a vanishing group of ungulates. The heyday of rhinoceroses is long since past.

THE TAPIRS

The modern tapirs of South America and Malaya are in some respects the most primitive of living perissodactyls. They retain the four toes on the front feet and the three toes on the hind feet, seen in various early Eocene members of the order. The body is heavy, the back is curved, and the limbs and feet are stocky and short. On the other hand, the tapir skull is specialized, in that the nasal bones are retracted. These mammals have a very flexible nose, a short proboscis that they can wrap around stems of plants or other objects. All the incisor teeth are present, and the canines are sharp and separated by a gap from the low-crowned cheek teeth. There is strong molarization of the last three premolars, and all the cheek teeth have crested crowns that are obviously related to the crested cheek teeth of the rhinoceroses. In the tapirs, however, the cheek teeth are perhaps more strongly cross-crested than in any other perissodactyls. The modern tapirs have a curious distribution, but it is obvious from the fossil history that they are remnants of a group that was once widely distributed over much of the earth.

Primitive tapirs appear in Eocene sediments. It is evident that during the Eocene epoch the tapirs were evolving rather complexly, along several parallel lines. These are indicated by the several families of Eocene tapirs that have been recognized—the isectolophids, helaletids, lophialetids, deperetellids, and lophiodonts. In general, however, these early tapirs may be compared with other primitive perissodactyls, already described. They were for the most part small, and were characterized by the usual primitive perisso-

dactyl features, including a lack of molarization in the premolars. However, the teeth were strongly cross-crested. *Heptodon*, a lophiodont, was an Eocene genus that may be considered as approximately ancestral to the later tapirs.

In Oligocene times *Protapirus* appeared as a probable descendant of *Heptodon*. This tapir showed most of the characters that have been outlined above in the discussion of modern tapirs, except that there was less retraction of the nasal bones and, therefore, probably not so much of a flexible nose or proboscis as is seen in living tapirs. From here on, the evolution of the tapirs was comparatively simple, and involved mainly a certain degree of size increase. *Miotapirus* of the Miocene had strongly retracted nasals, as in modern tapirs, of which it was without much doubt the direct ancestor. *Tapirus* appeared in the Pliocene epoch and has continued to the present day. In Pleistocene times there was a giant tapir living in China that has been recognized as a separate genus, *Megatapirus*, but except for size this tapir was little different from the surviving tapirs.

Tapirs roamed widely during Pleistocene times, and were prominent in the Ice Age faunas of North America and Eurasia. But as the Pleistocene epoch came to a close the tapirs disappeared in northern continental regions, and have continued only in the East Indies, and in South America, a region they invaded after the isthmian link between the two Americas had emerged above the sea.

In looking at modern tapirs we get in some ways a glimpse of the primitive perissodactyls of early Tertiary times. We are carried back in our minds to a time when the perissodactyls were beginning the long and varied evolutionary history that carried them through the Cenozoic era. It is a history that has passed its culminating point. Several million years ago the decline of the perissodactyls began, as these hoofed mammals gave way to the dominant ungulates of modern times, the artiodactyls.

Mountain Sheep.

Artiodactyls

29

ARTIODACTYLS IN THE MODERN WORLD

Most of the hoofed mammals of the present day are artiodactyls. They are the ungulates having an even number of toes (except for the extinct three-toed anoplotheres), either four or two on each foot, with the axis of the foot passing between the third and fourth toes. The modern artiodactyls are the pigs and peccaries, the hippopotamuses, the camels and llamas, the small tragulids, the deer, the giraffes, the pronghorns of North America, and the sheep, goats, muskoxen, antelopes, and cattle. This roster is in itself some indication of the great variety of artiodactyls that now inhabit the world, and we need think only of the whitetailed deer in eastern North America, the bison on the western plains before the coming of white men, or the vast herds of antelopes on the African veldt, to realize to what extent the artiodactyls are the dominant plant-eating mammals of our modern world. Of course artiodactyls are now threatened by the spread of modern civilization, but this is a very recent phenomenon and the restriction of the even-toed hoofed mammals under the impact of burgeoning human populations, firearms, automobiles, airplanes, and tractors cannot be regarded as evidence of any lack of vitality on the part of the harassed ungulates.

It must be remembered that certain species of artiodactyls have been much utilized by man for many thousands of years, and they will continue to be used in the future, for transportation, as food, as sources of wool or hair, and the like. These are the pigs, the camels and llamas, certain deer and various sheep, goats, and cattle. Moreover, man has been a hunter of artiodactyls of all sizes for thousands of years. So it can be said that the artiodactyls have been very important animals to ourselves and to our ancestors for a long time.

ARTIODACTYL CHARACTERS

In spite of their varied adaptations, the artiodactyls can be united by numerous common characters throughout the skeleton and the soft anatomy. The basic arrangement of the toes, which gives to the order its name, has already been mentioned. To repeat, there are generally two or four toes in each foot, with the axis of the foot passing between the third and the fourth digits. The first digits are almost never present, even in the most primitive artiodactyls. In the ankle the astragalus is distinctive, consisting of two pulleys, one above to articulate with the tibia and one below for articulation with certain other bones of the ankle, quite different from the single pulley astragalus of the perissodactyls, described in the preceding chapter. This double-pulley astragalus makes possible a great degree of flexion and extension of the hind limb; thus, the artiodactyls generally have remarkable abilities for leaping. In addition to these points of difference that distinguish the artiodactyls from the perissodactyls, the ungulate order now under consideration also is characterized by absence of a third trochanter on the shaft of the femur. In the more advanced artiodactyls the radius and ulna may be fused and the fibula may be reduced to a splint, coalesced with the tibia. Fusion also commonly unites the long bones or metapodials of the third and fourth digits into a single bone, often designated as the "cannon bone."

As in the perissodactyls, the artiodactyls generally have a capacious body, for accommodating the complex digestive tract and the large lungs. The back is strong, and most artiodactyls have strong back muscles that work with the muscles of the hind limbs, to give power to the propulsive threat of the legs.

The primitive artiodactyls have a full complement of teeth, but during the evolution of these ungulates there has been a strong trend toward reduction of the upper incisors. This has resulted in their complete suppression in many artiodactyls, to be replaced by a tough pad against which the lower incisor teeth bite, making a very efficient cropping mechanism. In these artiodactyls the lower canine commonly assumes the shape of an incisor and takes a position in series with the incisor teeth, so that there are eight lower cropping teeth. In some artiodactyls the caninces are large, daggerlike teeth, used for fighting and for defense, whereas in many artiodactyls the canines may be various reduced or suppressed.

The cheek teeth of artiodactyls, usually separated by a gap from the anterior teeth, rarely show much molarization of the premolars. In the primitive artiodactyls the cheek teeth are bunodont and low crowned, but in many advanced forms they become selenodont, with crescentic cusps, and are high crowned. The upper molars in all but the most primitive artiodactyls have square-shaped crowns; but instead of the back inner corner of the tooth's being formed by the hypocone, as in the perissodactyls, this section of the tooth is generally formed by an enlarged metaconule, the intermediate cusp commonly between the metacone and the hypocone. In the advanced artiodactyls that have this tooth structure there is no hypocone.

As might be expected, the artiodactyl skull shows changes in proportions and adaptations that are correlative with the specializations of the teeth, and in some groups with the development of antlers and horns. In the advanced forms the face generally is long and deep, and the bones of the back portion of the skull are frequently much compressed. This is particularly true of the horned artiodactyls.

Much of the success of the artiodactyls in late Cenozoic times can be attributed to the complex digestive system that characterizes a large section of this mammalian order. In the modern artiodactyls known as the ruminants the stomach is divided into three or four chambers. The plant food, after being cropped, passes into the first two chambers of the stomach, the rumen and the reticulum. Here it is broken down by bacterial action and reduced to a pulp. This pulpy mass is then regurgitated into the mouth in small masses and is thoroughly chewed; this is the chewing of the cud, so characteristic of the ruminants. The cud, after being chewed, is swallowed and passes

into the other two stomach chambers, the omasum and abomasum (the omasum is missing in the tragulids) for continued digestion. Of particular importance is the thorough extraction of nutrients from ingested plants, made possible by the complex ruminant stomach. This complicated process also enables ruminants to ingest large quantities of plant food in a hurry, after which the food can be thoroughly chewed and digested at leisure, when the animal is in a safe place, free from attack by enemies. The adaptation for eating in a hurry and digesting at leisure certainly gave such artiodactyls a great advantage over the perissodactyls in later Cenozoic times, when predatory carnivores were becoming ever larger and more efficient as hunters of big game.

The evolutionary trends within the artiodactyls, as these trends are reflected by the basic characters as discussed above, may be summarized in the following manner:

Teeth
    Bunodont molars ⟶ Selenodont molars
    Upper incisors   ⟶ Upper incisors
       present   absent
    Lower canine normal⟶Lower canine
                        incisiform
Feet
    Tetradactyl   ⟶ Didactyl
    Digits separate   ⟶ Third and fourth
                       metapodials fused
Skull
    Frontal bones   ⟶ Frontal bones may have
       smooth   antlers or horns
Digestive system
    Non-ruminant   ⟶Ruminant

Although the generally more primitive characters, as listed on the left side, above, are typical of the Palaeodonta, Suina and Ancodonta, while the more advanced characters, as shown on the right side, above, typify the Tylopoda and Ruminantia, the distinctions are not always so simple and clear cut as indicated. For example some of the ancodonts may have selenodont molars, while in others the molars may be intermediate, thus to be designated

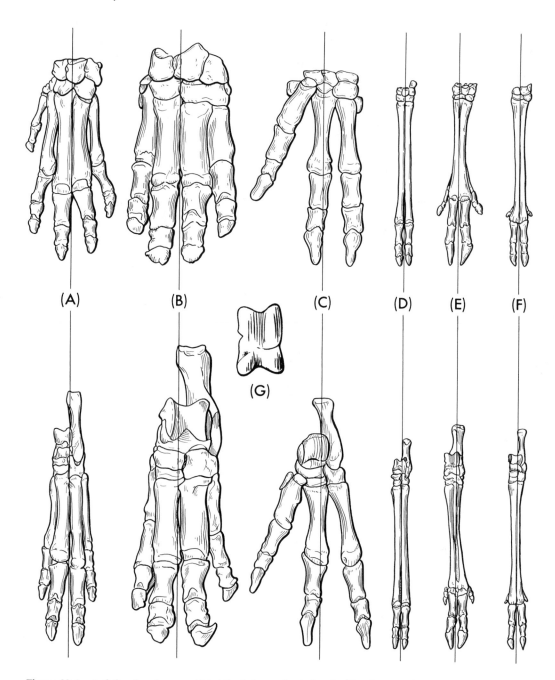

Figure 29-1. Left fore feet above, and hind feet below, of artiodactyls. The digits are drawn to approximate unit lengths, to facilitate the comparison of proportions of the feet in several types of even- toed ungulates. The vertical lines pass through the axes of the feet. **(A)** *Bothriodon*, an Oligocene anthracothere. **(B)** *Hippopotamus*, a Pleistocene and Recent hippopotamus. **(C)** *Diplobune*, an Oligocene anoplothere in which the second digit was reduced so that the foot was functionally three-toed. **(D)** *Oxydactylus*, a Miocene camel. **(E)** *Hippocamelus*, a Pleistocene deer. **(F)** *Merycodus*, a Pliocene antilocaprid. **(G)** The artiodactyl astragalus, showing the double-pulley structure for articulation with the tibia and with the distal bones of the ankle.

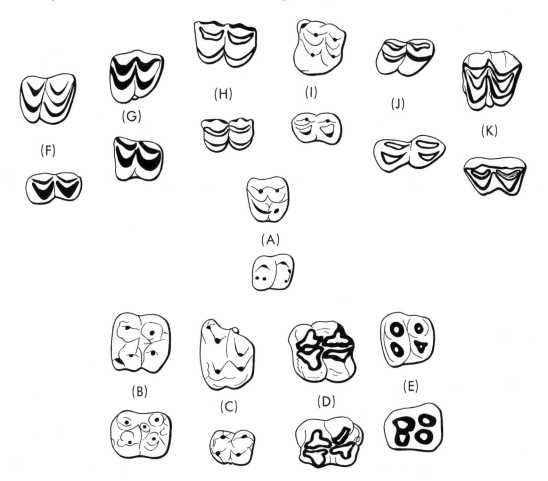

Figure 29-2. Skull and jaw of *Homacodon*, an Eocene artiodactyl from North America. Natural size.

as bunoselenodont. Furthermore, certain attributes that cannot be seen in fossil materials, such as intelligence, indicate that in modern artiodactyls certain Suina, for example the pigs, are more advanced than the ruminants.

BASIC CLASSIFICATION OF THE ARTIODACTYLS

Next to the rodents the artiodactyls are probably the most difficult mammals to clas-

Figure 29-3. Crown views of worn left upper and right lower molars of artiodactyls, not to scale. **(A)** *Diacodexis,* an Eocene dichobunoid and a generalized artiodactyl. **(B)** *Palaeochoerus,* a Miocene pig. **(C)** *Heptacodon,* an Oligocene anthracothere. **(D)** *Hippopotamus,* a Pleistocene hippopotamus. **(E)** *Dinohyus,* a Miocene entelodont. **(F)** *Cainotherium,* an Oligocene cainothere. **(G)** *Merycoidodon,* an Oligocene oreodont. **(H)** *Oxydactylus,* a Miocene camel. **(I)** *Archaeomeryx,* an Eocene hypertragulid. **(J)** *Odocoileus,* a Pleistocene deer. **(K)** *Selenoportax,* a Pliocene antelope. The teeth of *Diacodexis* (A) approximate the central type from which the varied teeth of more specialized artiodactyls evolved. In most of the Suina the teeth are bunodont, with low, conical cusps, or bunoselenodont, with low crescentic cusps. Such teeth are shown in the lower row of this figure. In some of the Suina and in the Tylopoda and the Ruminantia the teeth are selenodont, with highly developed crescentic cusps. These teeth are shown in the upper row of this figure.

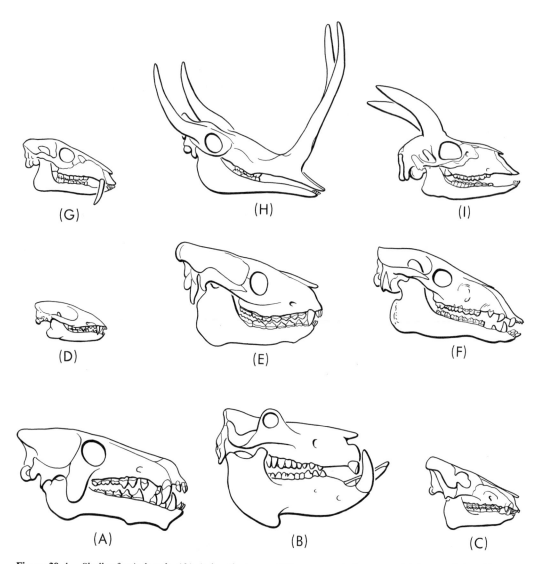

Figure 29-4.   Skulls of artiodactyls. **(A)** *Archaeotherium,* an Oligocene entelodont, one-eighth natural size. **(B)** *Hippopotamus,* a Pleistocene hippopotamus, one-sixteenth natural size. **(C)** *Perchoerus,* an Oligocene peccary, one-eighth natural size. **(D)** *Homacodon,* an Eocene dichobunoid, one of the earliest and most generalized of artiodactyls, one-fourth natural size. **(E)** *Merycoidodon,* an Oligocene oreodont, one-fourth natural size. **(F)** *Oxydactylus,* a Miocene camel, one-eighth natural size. **(G)** *Parablastomeryx,* a primitive Miocene deer, one-fifth natural size. **(H)** *Synthetoceras,* a Pliocene protoceratid, one-sixteenth natural size. **(I)** *Gazella,* a Pliocene antelope, one-fifth natural size. *Homacodon* (D) was an early artiodactyl with a generalized skull and dentition. From an ancestry approximated by this genus, the various groups of artiodactyls evolved along widely divergent paths, as indicated to some degree by the very different skulls of specialized types shown in this figure.

sify. If we consider only the living artiodactyls, the arrangement of these mammals into higher categories is comparatively simple, although the interrelationships of some genera and species, particularly among the numerous and diverse antelopes, poses problems about which there is much difference of opinion. But when we take into account the host of extinct artiodactyls (and the fossil record for this order is an abundant one) difficulties are immediately encountered. Numerous paleontologists have wrestled with the intricacies of artiodactyl classification, and the solutions have been as numerous as the people who have worked on the problem. Indeed, the statement that was quoted on a previous page in connection with the classification of the rodents is equally apposite when applied to the artiodactyls: "Anyone can point out inconsistencies in anybody else's classification."

The living artiodactyls may be very simply grouped within eight distinct families—the pigs, the peccaries, the hippopotamuses, the camels, the tragulids, the deer, the giraffes, and the bovids (sheep, goats, antelope, and cattle, as well as the pronghorns of North America). When the numerous families of extinct artiodactyls are added to this list, it quickly becomes apparent that a series of suborders is needed to group the many families into some sort of logical and usable arrangement.

Various systems have been proposed. One is to subdivide the artiodactyls into the "nonruminants" and the "ruminants"—those that do not and those that do chew the cud. (This is not always an easy criterion to apply to extinct artiodactyls.) Another plan would recognize a threefold division, separating the camels from the ruminants, even though they chew the cud, since these modern denizens the desert have been distinct artiodactyls from late Eocene times. Another threefold subdivision of the artiodactyls would keep the camels within the ruminants, but would set apart the very primitive members of the order (of Eocene age) as a distinct suborder. Other systems would establish four and five and even more suborders within the artiodactyls.

Here, we propose to recognize, in a general way, a fivefold division of the artiodactyls that is composed of three groups of nonruminants, of the tylopods or camels, and of the ruminants or traguloids, deer, giraffes, and bovoids. The last two of these large categories are natural groups, but the three nonruminant groups constitute a diverse assemblage. For simplic-

ity, if not complete logic, these three groups—the palaeodonts or primitive artiodactyls, the Suina or pigs and peccaries, anoplotheres, anthracotheres and hippopotamuses, and finally the ancodonts or oreodonts and cainotheres—are here listed as suborders. Therefore, on a formal basis, the artiodactyls might be viewed as having five suborders that can be listed in the sequence discussed above as the Palaeodonta, Suina, Ancodonta, Tylopoda, and Ruminantia.

PALAEODONTA

Dichobunoids and Entelodonts

The first artiodactyls, like the first perissodactyls, appear in sediments of early Eocene age. Some of these early even-toed ungulates were indeed very archaic, perhaps not far removed from some of the primitive condylarths, and it is largely the presence of an artiodactyl-type astragalus, with the characteristic double pulley arrangement, that justifies the inclusion of such types within the order of mammals now under consideration. *Diacodexis,* from the lower Eocene of North America, is a characteristic early artiodactyl. This was a small animal, as might be expected, with slender limbs, having four functional toes on each foot. The skull was low, and the orbit was centrally placed and confluent with the temporal opening behind it. The full dentition was present, and the canine teeth were rather well developed. The cheek teeth were low crowned, and the upper molars retained the primitive triangular shape, with an essentially tricuspid crown.

An animal like *Diacodexis* might have been ancestral to almost any of the later artiodactyls. In fact, *Diacodexis* was an early member of the dichobunoids, the most primitive of the palaeodonts. Dichobunoids were widely distributed through North America, Asia, and Europe during the Eocene epoch, when they shared with the early perissodactyls the role of hoofed plant-eaters. In the latter part of the Eocene epoch some of the dichobunoids be-

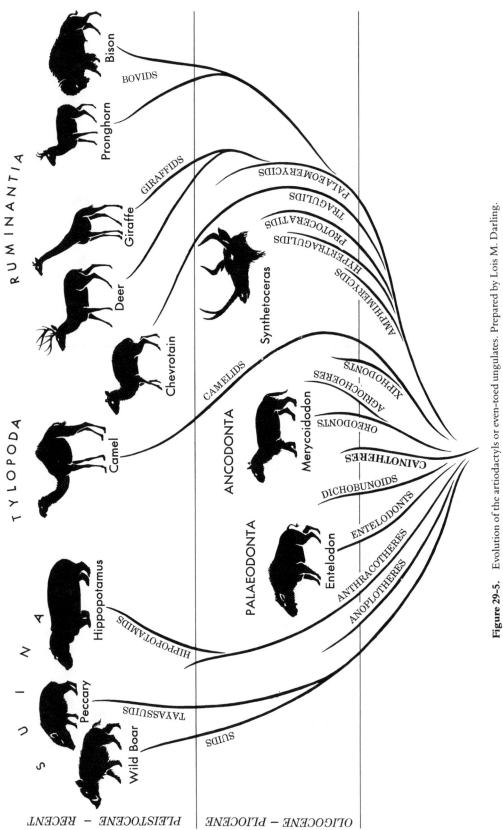

Figure 29-5. Evolution of the artiodactyls or even-toed ungulates. Prepared by Lois M. Darling.

came moderately specialized, and a few of them persisted into Oligocene times before they finally became extinct.

It is very probable that the early dichobunoids were ancestral to other nonruminant groups that arose during Eocene or early Oligocene times. One of these groups was the entelodonts, showing definite relationships to the dichobunoids, and ranging in geologic age from the late Eocene into the Miocene epoch.

The entelodonts showed an early growth toward large size, so that during Oligocene times some of these artiodactyls became as large as modern boars, whereas in early Miocene times they grew into giants, as large as bison. These animals were rather piglike in some features, and for this reason they have often been called "giant pigs," but they were not pigs and such a designation is misleading. Two evolutionary trends especially characteristic of the entelodonts were the development of long legs and feet and a straight back for running, and the growth of the skull and teeth to truly enormous proportions. As the legs became elongated the side toes were reduced, so that at an early stage in their history the entelodonts became two-toed ungulates. These trends are exemplified by *Archaeotherium* of Oligocene age and by the giant *Dinohyus,* of early Miocene age, both found in North America.

The remarkable increase in the size of the head is illustrated by *Dinohyus,* an animal about the size of a bison, in which the skull was more than a meter in length. The braincase, however, remained relatively small. The face was elongated, and the posteriorly placed eye was completely circled by bone, so that it was set off from the temporal fossa in which the powerful jaw muscles were located. On the cheek bone or zygomatic arch and on the border of the lower jaw were large flanges or knobs, possibly for muscle attachments. All the teeth were present, and the canines were of enormous size. The cheek teeth were bunodont, with conical cusps, and the molars above and below were quadrate, each with four cusps.

During Oligocene and early Miocene times

the entelodonts were prominent members of the mammalian faunas of the northern hemisphere. Then they became extinct, perhaps because of competition from the possibly more intelligent pigs and peccaries.

SUINA

Pigs and Peccaries

The first pigs and peccaries were of early Oligocene age, the former appearing in the Old World, the latter in North America. Throughout their phylogenetic histories these artiodactyls, although clearly related to each other, retained essentially separate ranges in the Old and New Worlds, and thus had parallel histories.

Propalaeochoerus, one of the first pigs, lived in Europe during early Oligocene times. This early pig, although considerably larger than the primitive Eocene dichobunoids, was still a fairly small artiodactyl. It had legs of medium length, four-toed feet, and a moderately long, low skull. The canines were well developed, and the low-crowned teeth were bunodont, with conical cusps. The molar teeth were somewhat elongated, and in each tooth there were four main cusps.

The evolution of the pigs from this beginning was marked by a moderate increase in size, a great lengthening of the skull, especially of the face, and likewise of the teeth, the complication of the molar crowns by wrinkling of the enamel, the development of the canines as large, outwardly curved tusks, and the retention of the four-toed feet, with emphasis on the two middle toes. As the pigs evolved through middle and late Cenozoic times they branched along numerous lines of adaptive radiation that made theirs a varied group of artidactyls. Yet in spite of their variety, it would seem that all the pigs have had similar habits, being primarily animals of the forests, where they spend much of their time rooting in the ground for all kinds of food. Pigs are very intelligent mammals, and for this reason have been able to hold their own very well in a highly competitive world.

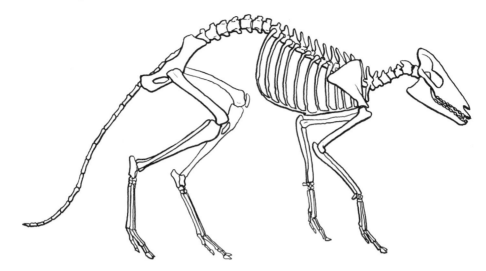

Figure 29-6. *Diacodexis,* an early Eocene artiodactyl from North America. This artiodactyl, about the size of a rabbit, shows strong adaptations for rapid running, as can be seen in the elongated limbs.

Archaeotherium

Figure 29-7.   An entelodont of Oligocene age. More than three feet in height. Prepared by Lois M. Darling.

Among modern swine *Sus* is the domestic pig with which we are familiar, in its wild state typified by the great wild boar of Europe, a rangy, belligerent pig, with a very long skull and large, wicked, canine tusks. Other modern representatives of the pigs are the river hog, *Potamochoerus,* the giant forest hog, *Hylochoerus,* and the highly specialized wart hog, *Phacochoerus,* all of Africa. Pigs are common throughout Asia, and in Celebes is the strange babirusa, in which the upper tusks grow in long curves over the top of the skull.

To go back in time some two million years, it is interesting to see that there was a truly gigantic warthog, *Notochoerus,* with jaws almost as massive as those of a rhinoceros, living in the Olduvai Gorge of eastern Africa as a contemporary of some of the earliest australopithecine hominids.

Pigs were domesticated at an early date in the development of Old World civilizations, and during the last several thousand years these beasts have been taken by men to all parts of the world, even to some of the most remote Pacific islands. There is good reason to think that the progenitors of our domestic pigs were Asiatic pigs.

Perchoerus, one of the early peccaries of Oligocene age, lived in North America at the same time that *Propalaeochoerus* lived in Europe. Although the resemblances between the first known peccaries and the first known pigs are very striking, these North American animals followed evolutionary trends that were increasingly divergent from those that characterized the pigs. The peccaries showed a moderate increase in size, but not so much as did the pigs. In the peccaries there was a greater emphasis on running, so that these animals became long-legged, and the side toes were reduced to mere remnants. The skull in most of the peccaries never became so extremely elongated as it did in the pigs, instead it was usually comparatively short and deep. The canine teeth were always directed straight down and up, to work against each other like sharp shears. Finally, the molar teeth of peccaries have usually been simple as compared with the long, complexly wrinkled tooth crowns of pigs.

From *Perchoerus* the peccaries evolved in North America through the Miocene and Pliocene epochs along two broad lines of adaptive radiation. One line, typified by the Pliocene genus *Prosthennops* and the Pleistocene *Mylohyus,* was characterized by a certain amount of elongation of the face and the frequent development of large bony expansions on the cheek bones. This evolutionary line

was successful for a time, but died out during the Pleistocene epoch. The other line, typified by the Miocene *Hesperhys,* the Pleistocene *Platygonus,* and the recent *Dicotyles* and *Catagonus,* has been characterized by rather deep skulls and short molars, often strongly cross-crested. *(Catagonus,* a large, long-snouted peccary related to *Platygonus* has long been known from Pleistocene specimens found in South America. In 1975 a population of living *Catagonus* was discovered in an isolated scrub-thorn and grass refuge in Paraguay. Interestingly these "living fossils" are identified as *Catagonus wagneri,* a Pleistocene species that was described in 1948). This line survives today as the peccaries that live from southern United States, through Mexico and Central America, into South America. The fossil evidence shows that peccaries entered South America in Pleistocene times, as did so many other northern mammals.

Anthracotheres and Hippopotamuses

In middle and late Eocene times there arose some artiodactyls known as anthracotheres, which were to have a wide distribution, and a long history extending into Pleistocene times. These artiodactyls quickly spread throughout Eurasia during much of the Tertiary period, and in the Oligocene epoch some of them invaded North America, where they continued briefly into Miocene times. These were in general piglike animals, with legs of only moderate length, and with four-toed feet. The skull was long and low, and there was a full complement of teeth. In the early anthracotheres the cheek teeth were bunodont, with conical cusps, but as these animals evolved there was a strong trend toward the development of selenodont or crescentic-shaped tooth cusps. The earliest anthracotheres were very close indeed to some of the primitive palaeodonts, an indication as to their ancestry, but as they evolved they evidently became specialized for life in streams and along river banks. Indeed, some of the last of the anthracotheres, such as *Merycopotamus* of Pleistocene age, were like hippopotamuses in the configuration of the skull and the development of the teeth.

Perhaps it is significant that there are no traces of hippopotamuses in the fossil record until late Miocene or early Pliocene times. Because of the resemblances between the hippopotamuses and certain advanced anthracotheres, it is therefore reasonable to think that the "hippos" diverged from some of the late anthracotheres—that they are in a sense specialized surviving anthracotheres. Their characters are so well known as to need little elucidation at this place. Briefly, they are very large, heavy artiodactyls with short limbs and broad, four-toed feet. The skull is enormous, with elevated orbits, and the lower jaw is very deep. The incisor and canine teeth are enlarged; each molar crown has four trefoil-shaped cusps that may very well have been derived from the selenodont cusps of anthracothere teeth.

During Pleistocene times the hippopotamuses were widely spread throughout Eurasia and Africa; now they are much restricted in their distribution.

Anoplotheres

Some strange artiodactyls were the anoplotheres, confined in time to the Eocene and part of the Oligocene epochs, and in space to Europe. The anoplotheres were about like pigs in size. The skull was generally primitive, long and low, and the dentition was superficially like that in some of the South American ungulates, in that there was a continuous row of teeth grading in form and function from the incisors to the last molars. The later anoplotheres were peculiar artiodactyls because the feet were three-toed, a result of the retention of the second toe as a large digit and the suppression of the fifth toe.

ANCODONTA

Cainotheres

The cainotheres were very small, highly specialized artiodactyls that lived in Europe during late Eocene and Oligocene times. They are so distinct in their adaptations that it is not easy to fit them into the larger artiodactyl categories. It is possible, however, that they can be regarded as broadly related to the oreodonts, or to the anoplotheres, but separately derived from an early Cenozoic ancestor not unlike *Diacodexis*.

Cainotherium was no larger than a rabbit, and was remarkably like the hares in some of its adaptations. The back was strongly curved, and the hind limbs were long, an indication that this little artiodactyl ran with a bounding or hopping gait. The legs were slender; the side toes were short, and only the middle toes were functional. The large eye was closed behind by a bony bar. The teeth formed a nearly continuous series, and the canines were reduced and like the incisors in shape. These animals had inflated auditory bullae, possibly an indication of an acute sense of hearing.

Perhaps the cainotheres lived much as do the rabbits, and they may have competed with some of the early lagomorphs. If so, the cainotheres were not strikingly successful, in spite of the high perfection of their adaptations, for they had a short phylogenetic history, limited to a restricted part of one continent.

Oreodonts

The oreodonts were strictly North American artiodactyls. They arose in late Eocene times and continued into the Pliocene epoch, when they became extinct. They were distinctive in their adaptations, and it has not been possible to relate them with certainty to any other artiodactyl groups. They have been variously considered as allied to the anthracotheres, to the camels or to the ruminants.

Certainly the oreodonts were anthracothere-like in generaly body form. They ranged in size from small animals to beasts comparable in size to very large pigs. The body was generally long and the legs short. The feet were four-toed. In the primitive oreodonts the skull was low, but as these animals evolved the skull

Figure 29-8.   Oligocene oreodonts. These animals were about the size of sheep. Prepared by Lois M. Darling.

became fairly deep. The eye was centrally located, and was closed behind by a bony bar that separated it from the temporal fossa. The teeth formed an almost continuous series, with the upper canine enlarged as a sort of daggerlike tooth. The lower canine, however, was small and in series with the lower incisors, whereas the first lower premolar often took over the function of a lower canine. The molar teeth were selenodont, with crescentic cusps.

There was a dichotomy in the evolution of the oreodonts that began in late Eocene times, with the first appearance of these artiodactyls, and extended through their entire phylogenetic history. *Protoreodon* of the Eocene epoch

probably was the direct ancestor of the agriochoeres, which flourished through the Oligocene, represented by the genus *Agriochoerus*, and into the beginning of Miocene times. These were aberrant oreodonts, in which the thumb was retained on the front foot while all the toes terminated in claws, rather than hoofs. It has been suggested that the agriochoeres were able to scramble around in trees.

Protoreodon, mentioned above, also probably was the direct ancestor of the merycoidodonts, the most abundant and varied of the oreodonts. During Oligocene times these oreodonts were remarkably numerous, and they must have roamed across the landscape in vast herds. Their fossils, especially those of *Merycoidodon,* are the commonest mammals found in the White River badlands of South Dakota, and in other Oligocene exposures. The merycoidodonts were rather sheeplike animals that continued into Pliocene times. *Leptauchenia* of the Oligocene epoch and its relatives were small, lightly built oreodonts. *Ticholeptus,* a Miocene oreodont, typifies a branch of the oreodonts in which the skull and jaws were deep and the teeth high crowned. *Merycochoerus* represents a line of oreodont evolution in which the animals became large and heavy, and in some of the more specialized genera seemingly evolved elongated noses or short trunks. These were probably water-living forms, with hippopotamuslike habits.

As can be seen from this brief outline, the oreodonts were very successful artiodactyls in North America during middle and late Tertiary times. But as the Tertiary period drew to a close, the fortunes of the oreodonts declined. Perhaps they succumbed to competition from the more advanced artiodactyls, specifically the ruminants, that were evolving and spreading at rapid rates during late Tertiary times.

TYLOPODA

The Llamas and the Camels

We think of camels as beasts of the Orient or of the great Sahara, but in fact these ungulates, like the horses, were of North American origin, and they went through the major part of their evolutionary development on this continent. The present distribution of camels in the Old World and of llamas in South America represents emigrations of these animals out of their homeland during the Pleistocene epoch, whereas their present absence in North America is the result of extinction at the very close of the Ice Age. Why the camels should have failed in North America at the end of Pleistocene times is one of those unaccountable evolutionary mysteries, comparable to the extinction of the horse on this continent.

Although camels chew the cud, like the ruminants, they may be conveniently kept apart from the other cud-chewing artiodactyls because of their long, separate history. The first camels appeared in late Eocene times as indicated by *Poëbrodon,* known from some upper teeth and a partial lower jaw from North America. Adequate knowledge of an ancestral camel may be obtained from the Oligocene genus, *Poëbrotherium,* of which complete skeletons are known from North America. This was a llama-like camel, in which the skull was elongated, the cheek teeth laterally compressed, the neck and limbs elongated, and the lateral toes suppressed.

As the camels evolved through Cenozoic times most of them increased in size to become medium-sized or even very large artiodactyls. However, in a few lines of camelid evolution the animals remained small. There was a very rapid reduction and loss of the side toes in the early camels, so that as mentioned even by Oligocene times the feet were two-toed, the lateral toes having completely disappeared. As the camels continued to evolve there was a fusion of the two long bones of the feet, the metacarpals and metatarsals, to form in each foot a single "cannon bone" having separate articulations at the lower end for the two toes. The legs became very long in these animals as an adaptation for rapid running, and in the later camels the hoofed feet were transformed into spreading, padded feet, adapted for walking through soft sand. The neck increased in length.

During evolution the camelid dentition underwent definite changes. In the advanced camels the first two upper incisors were repressed, leaving a single, pointed third incisor that bit down behind the lower incisors. In place of the upper front teeth was a horny pad, that opposed the spoon-shaped lower incisors, an excellent mechanism for cropping plants. The canines remained as pointed teeth and were separated by a gap from the cheek teeth, among which the anterior premolars were suppressed. The molar teeth became

Alticamelus

Figure 29-9.   A small Miocene camel, *Stenomylus*, and a very large camel of Pliocene age, to the same scale. The head of *Alticamelus* was ten feet or more above the ground. Prepared by Lois M. Darling.

high-crowned and elongated, and each tooth had four crescentic cusps.

From *Poëbrodon* of the Eocene the central line of camel evolution proceeded through *Poëbrotherium* of Oligocene age into *Procamelus* and *Pliauchenia* of the Miocene and Pliocene

epochs, and respectively to *Camelus*, the camels, and *Lama*, the llamas of modern times. The llamas represent camels in essentially a Miocene stage of evolution, and the large camels represent the culminating phase of the central camelid line of development. These camels show certain specializations that probably developed within the latter portion of Cenozoic time—long hair in the llamas for protection against cold, and humps in the camels for the storage of fat for nourishment. The camels also have water-storage chambers in the rumen of the stomach.

During middle and late Tertiary times there were several lateral branches of camelid evolution that died out before the advent of the Pleistocene epoch. The stenomylines, exemplified by *Stenomylus,* were small, lightly built, long-legged camels of Miocene and Pliocene age. Evidently they were rapid-running animals, something like the modern gazelles in their habits. *Alticamelus* and its relatives of the Miocene and Pliocene epochs were very large camels with greatly elongated, stiltlike limbs and long necks. They were evidently adapted for browsing in tall trees, and because of these specializations they are often designated as the "giraffe-camels."

In late Eocene and early Oligocene times some artiodactyls, possibly related to the early camels, lived in Europe. They were the xiphodonts, which in the Old World paralleled the early development of the camels in the New World. But they failed to live on, and thus are relatively unimportant in the history of the artiodactyls.

RUMINANTIA

Advent of the Ruminants

We come now to a consideration of the ruminants, the most varied and numerous of modern-day artiodactyls. The ruminants very conveniently may be divided into two large groups, the traguloids or primitive ruminants, and the pecorans or advanced ruminants. The traguloids are represented today by the diminutive chevrotains of Africa and Asia—small, delicate deerlike mammals of primitive form, usually about a foot in height at the shoulder, with a curved back and no antlers on the skull, but with large, daggerlike upper canine teeth. There are no upper incisors but, as in other ruminants, there is a full set of lower incisors that are augmented on each side by the incisiform canine tooth. These teeth are opposed above by the hard pad that is characteristic of the ruminants. The limbs are long and slender, with the third and fourth digits enlarged as the functional toes, but with the lateral second and fourth toes still complete. The large metacarpals of the front foot are separate bones; in the "higher" ruminants these bones are fused to form a single cannon bone as, indeed, is the case in the hind feet of the chevrotains. These little ruminants chew the cud, but the stomach is less complex than in the more specialized ruminants. Generally, the chevrotains, which are secretive dwellers in the African and Asian forests, give us a good idea of the ancestral ruminants.

The pecorans are composed of the ruminants with which we are familiar—the deer, the giraffes and the great group of sheep, antelopes and cattle. The remarkable specializations in the digestive system of these mammals have already been briefly outlined. In addition, the pecorans show various adaptations in the skeleton for browsing and grazing, and for running, that place them on the evolutionary scale far beyond their traguloid cousins. There commonly has been an increase in size; the limbs are elongated; and the fusion of the third and fourth metapodials into cannon bones is complete. In addition, there is considerable fusion in the wrist and ankle complexes, and the ulna in the forelimb and the fibula in the hind limb are much reduced. Thus, the limbs are very strictly specialized for the rectilinear fore and aft motion that is necessary for fast, prolonged running. Finally, the pecorans generally have antlers or horns on the head (at least, in the males) for defense and for sexual combat.

Traguloids

The ancestry of the ruminants may be approximated by various traguloids, notably *Gelocus* from the Eocene of Europe, *Archaeo-*

meryx from the Eocene of Asia, *Hypertragulus* from the Oligocene of North America, and the recent *Tragulus* (the chevrotain) from Asia. Although *Tragulus,* truly a "living fossil", in some respects most nearly resembles what a ruminant ancestor should be like, *Archaeomeryx* is here presented as a generalized, primitive ruminant. This was a small animal about the size of a chevrotain or "mouse-deer," which means that it was scarcely larger than a big jack rabbit. In *Archaeomeryx* the limbs were long and the back was curved. It had a long tail, a primitive feature that is lost in most of the ruminants. The feet were somewhat elongated, and, although the four long bones of the feet, the metapodials, were separate elements, emphasis was on the two middle toes in each foot, and the lateral toes were of secondary importance in locomotion. In the skull the eye was centrally located, about halfway between the front and the back of the head, as in so many primitive mammals, but it was closed behind by a bony bar. This character continued through all the ruminants. All the teeth were present, but the three upper incisors were small and obviously on the way to being suppressed. There was a small gap between the front teeth and the cheek teeth, and each molar had four selenodont or crescentic cusps.

The main line of traguloid development leading to the modern Old World chevrotains was in Eurasia, but during middle and late Tertiary times there were some interesting side branches of traguloids that evolved in North America. One of these was the hypertragulid group which retained with some variations the primitive small size and general adaptations of the early traguloid ruminants. *Hypertragulus* was a common animal in the Oligocene faunas of this continent, and it evidently lived in large herds along with oreodonts, early camels, and creodonts.

A particularly interesting line of North American traguloids were the protoceratids, represented by the Oligocene *Protoceras,* the Miocene *Syndyoceras,* and the Pliocene *Synthetoceras.* There was some size growth in these

ruminants so that their last representatives were as large as small deer. But the most spectacular development among the protoceratids was the growth of horns on the skull in the males. In *Protoceras* there were six horns, two on the nose, two above the eyes, and two on the back of the head. In *Syndyoceras* there was a pair of horns above the eyes and two long, diverging horns on the nose. In *Synthetoceras* the horns above the eyes were long and directed backward, and on the nose there was a very long beam or pedicle that diverged like a Y at its tip. This front, Y-shaped horn was longer than the total length of the skull, and it must have been a potent weapon for fighting.

Deer

The deer, the most primitive of the pecorans, were obviously of traguloid ancestry, and it would seem likely that they arose during Oligocene times. *Eumeryx,* of Oligocene age, is an almost ideal progenitor for the deer, and its primitive characters are continued in certain Miocene forms, like *Palaeomeryx* in Europe or *Blastomeryx* in North America. These primitive deer were rather small, and they lacked antlers upon the skull. Usually the upper canine teeth were enlarged as long sabers. The skull was fairly long and low, the back was arched, the tail was very short, the legs and feet were elongated, the two central metapodials were fused into a cannon bone, and four toes were present, of which the lateral ones were much reduced so that only the middle toes functioned. The modern Asiatic musk deer, *Moschus,* is a persistent primitive type, very much like these early Miocene deer.

As the deer evolved there was a strong trend toward increase in size, as is so common in the artiodactyls. Many deer became very large, although some evolutionary lines have remained rather small. Because the deer have been browsers through the extent of their evolutionary history the selenodont cheek teeth have remained low crowned. But the striking specialization during the evolution of the ad-

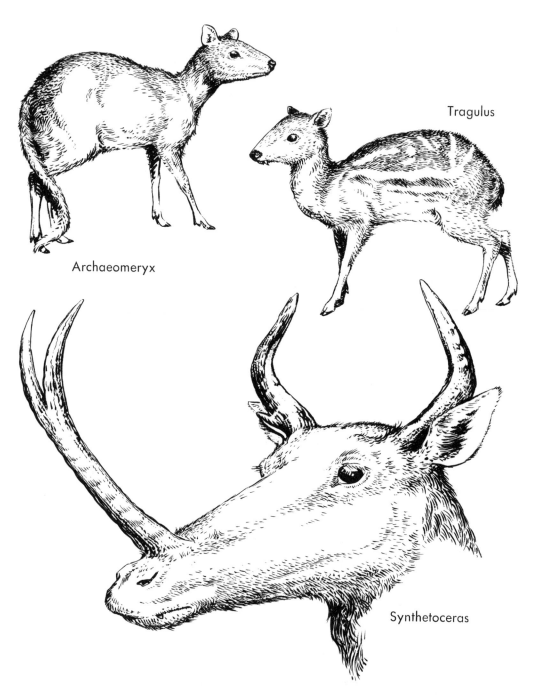

Tragulus

Archaeomeryx

Synthetoceras

Figure 29-10.   Traguloids. *Archaeomeryx* was an Eocene hypertragulid and in most respects a generalized ruminant. *Tragulus* is the modern Oriental chevrotain that has retained many of the characters of the ancestral ruminants. Both these animals about a foot in height. *Synthetoceras* was a Pliocene protoceratid, with a skull about eighteen inches in length. Prepared by Lois M. Darling.

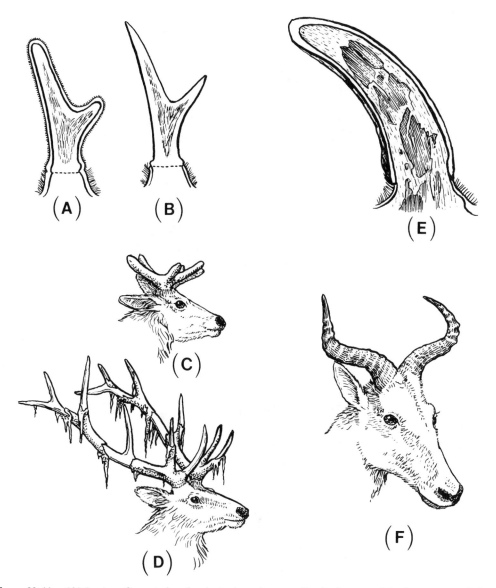

Figure 29-11. **(A)** Section of an artiodactyl antler in the early stage of its development. It is a bony outgrowth from the frontal bone, covered with highly vascularized skin or "velvet." The dotted line shows the area of separation between the antler and the frontal bone. **(B)** A fully-grown antler, after loss of the velvet. **(C)** A stag with antlers in early growth stage, with velvet. **(D)** A stag with fully developed antlers, in which the dead skin is being lost. **(E)** A bovid horn core—an outgrowth of the frontal bone with its permanent horny sheath. **(F)** Head of an African antelope, *Alcelaphus,* showing the rugose horns.

vanced deer has been the development of antlers on the skull of the males. Antlers are outgrowths from the frontal bones, above the eyes. There are two cylindrical bony stumps on the skull that form the bases for the ant-

lers. These are the pedicles, and in life they are covered with skin. From the pedicles new, skin-covered antlers grow each year, increasing in size year after year as the deer reaches the full stature of its maturity. This growth

Muntjac

Wapiti

of bone although remarkably rapid (in some large modern deer attaining a growth rate of more than a centimeter a day) nevertheless takes several months. When the antlers have reached their full size the skin that has covered them dries, and is rubbed off by the deer. The antlers are now hard, bony spikes or branching, many-pointed weapons, with which the male deer fight each other during the mating season. After the mating season the antlers drop off, and new antlers begin to grow to replace them. This remarkable adaptation supplies the male deer with efficient weapons for part of the year, and at this time he is a belligerent animal. But during the time the new antlers are growing, the male deer are shy and timid, and very careful to protect the skin-covered antlers, which are said to be "in velvet."

Antlered deer began to evolve along many adaptive lines during the Miocene epoch. The early antlered deer were small, as were their antlers, which were borne on elongated bony pedicles. These deer, as represented by the Miocene genus *Dicrocerus* of Europe, and by the modern Oriental genus, *Muntiacus,* typically have very small, two-tined antlers. In the late Tertiary genus *Stephanocemas* of Asia, the antlers, although small, were palmate and of rather complex form.

As the primitive deer developed in Eurasia, a group of aberrant deer, the dromomerycines, evolved in North America. They did not survive beyond the Pliocene epoch. In the meantime the more familiar lines of deer that continue at the present time were evolving. With the passage of time, many deer increased in body size, and with this increase there was frequently a proportionately greater increase in antler size. The culmination of these trends was reached in the Pleistocene epoch, and is exemplified by the Irish "elk" *Megaloceros,* a deer found in great numbers in the peat bogs of Ireland, and notable for its tremendous antlers—often spreading eight feet from one side to the other.

When the isthmus arose to connect South America with North America, the deer pushed into the southern continent, as did so many other North American mammals. Deer, however, never invaded Africa below the Atlas Mountains, perhaps, because of the competition from the host of antelopes that had already taken over the forests and plains of that continent.

Today deer of many kinds are widely distributed throughout the continental areas, except for Antarctica, Australia, and (as noted) the middle and southern portions of Africa. They are:

The muntjacs. Asiatic deer with very long pedicles and very small antlers. *Muntiacus* is the characteristic genus.

The cervine deer or stags. Deer that evolved in the Old World, and invaded North America during Pleistocene times. The European stag and the American Wapiti or "elk" (*Cervus*) are superb examples of these noble deer.

The deer of the New World. Deer that evolved in North America and invaded South America during Pleistocene times. The common white-tailed and black-tailed deer (*Odocoileus*) represent this line in North America.

The alcine deer or moose. The true "elk" (*Alces*) of the Old World, which invaded North America in Pleistocene times.

The reindeer and caribou. Arctic deer (*Rangifer*), unusual in that the females have antlers.

The water deer of Asia.

The roebucks (*Capreolus*) of Eurasia.

Giraffes

The giraffes are closely allied to the deer, and it is apparent that these artiodactyls and the deer had a common ancestry. Indeed, the giraffes can be considered as having branched

Figure 29-12.   Two late Cenozoic and Recent deer. The muntjac, a small deer living in Asia, is in many respects similar to the primitive deer of Miocene time. The male has small antlers borne on long pedicles, and there are large canine teeth. The wapiti or stag, a large deer with magnificent antlers in the male, is an Eurasiatic deer that migrated into North America during Pleistocene times. Prepared by Lois M. Darling.

from the deer during Miocene times, after which they diverged widely from their deer cousins.

The primitive giraffes are well exemplified by the modern okapi, a persistent Miocene type of giraffe that was living in the deep forests of the Belgian Congo, unknown to white men, until after the turn of the present century. The okapi is a rather large artiodactyl, standing five or six feet in height at the shoulder. The legs are long, and the front legs are somewhat longer than the hind legs, so that the back slopes from the shoulders to the pelvis. As in other pecorans, the feet are long, the metapodials are fused into a cannon bone, and the middle toes form the functional foot. The skull is elongated, and in the male there are two small outgrowths from the frontal bones, covered with skin. This type of skin-covered "horn" is characteristic of the modern giraffes. The teeth are rather low crowned, and the enamel of the cheek teeth is very much wrinkled, as it is in all the giraffes.

In Miocene and Pliocene times the first giraffes, such as *Palaeotragus,* were strikingly similar to the modern okapi. From them the modern giraffe *(Giraffa)* evolved, by tremendous elongation of the limbs and the neck, and by the growth of large, skin-covered horns above the eyes.

However, there was another branch of the giraffes that developed during late Pliocene and Pleistocene times. This was the group of sivatheres, very large, massive giraffes, with normally proportioned limbs and oxlike bodies. They were characterized by large, strikingly ornate "horns" on the skull, bony outgrowths of such size that it seems possible they may have been covered with hard horn during life. *Sivatherium,* known from the Pleistocene of India, was one of the last of this line of giraffid evolution, a giant pecoran with two huge, flaring bony horns on the back of the skull and two smaller, conical horns in front of them, above the eyes. A tantalizing little bronze figure, made by an ancient Sumerian, several thousand years ago, indicates that sivatheres may still have been living when this early civilization flourished in the Middle East.

The Bovids, Dominant Artiodactyls of the Modern World

The bovids—the pronghorns, sheep, goats, muskoxen, antelopes, and cattle—form a bewildering array of ruminants widely distributed throughout the world. These artiodactyls, the most advanced of the ruminants, arose in Miocene times, but most of their evolutionary development has been confined to Pliocene and subsequent ages. They were pretty clearly of northern origin, but they invaded southern Asia and Africa in late Pliocene and Pleistocene times, and in these continental regions they now live in the greatest abundance and variety. Bovids never reached South America, until they were introduced by man.

These ruminants are characterized by powerful bodies and long limbs for running. The feet are even more progressive than in the deer, and the lateral digits are in an advanced stage of reduction. The cheek teeth are hypsodont or high crowned, often extremely so, and the enamel is generally folded to form rather complex surface patterns when the tooth crown is worn. Thus, the teeth in the bovids are well adapted for grazing. However, the most characteristic feature of the bovids is the presence of horns on the skull, almost always in both males and females. The horns are outgrowths of the frontal bones, covered with a very hard outer casing of horn. Except in the antilocaprids this horny cover is permanent, expanding as the bony pedicle or horn core grows during life. This is certainly a more efficient manner of providing a weapon than the yearly production of antlers in the deer, which necessarily places a great drain on the energy of the animal. The horns of the bovids have been a large factor, together with the habit of ruminating or chewing the cud, in assuring the success of these artiodactyls in the modern world, for the horns are powerful and efficient weapons, and many bovids are quite able to defend themselves against attack from

Giraffe

Sivatherium

Palaeotragus

Figure 29-13. *Palaeotragus* was a Pliocene giraffe that represents the approximate stem from which later giraffes evolved. The modern okapi of the Belgian Congo is a little-changed descendant from a *Palaeotragus*-like ancestor. *Sivatherium* was an ox-like giraffe of Pleistocene age. The recent giraffe of Africa needs no introduction. All to the same scale. Prepared by Lois M. Darling.

Springbok   Stockoceros

Figure 29-14.   Bovids, to the same scale. Here is shown one of the very large Pleistocene bison, which had enormous, sweeping horns. *Stockoceros* was a four-horned prongbuck of Pleistocene age. It was about the same size as the modern pronghorn antelope. The springbok is an African antelope. Prepared by Lois M. Darling.

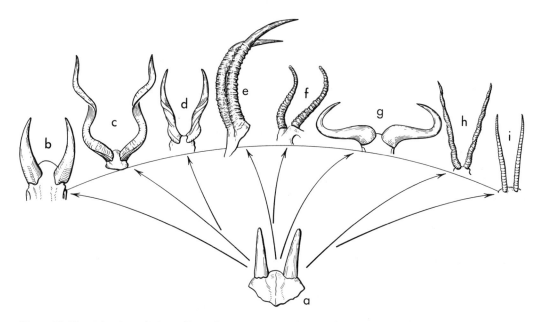

Figure 29-15. Adaptive radiation of horn shapes in some modern antelopes. Not to scale. **(a)** *Eotragus*, a primitive Miocene bovid that represents an ancestral type from which later bovids may have evolved. **(b)** *Boselaphus*, the nilgai. **(c)** *Strepsiceros*, the kudu. **(d)** *Taurotragus*, the eland. **(e)** *Hippotragus*, the sable antelope. **(f)** *Cobus*, the waterbuck. **(g)** *Connochaetes*, the gnu. **(h)** *Antilope*, the blackbuck. **(i)** *Oryx*, the oryx. The nilgai and the blackbuck live in India, the other modern forms in Africa.

predatory carnivores by fighting back. Indeed, some of the bovids, particularly the large cattle, are themselves very aggressive animals, more dangerous than the so-called lords of the jungles, the great cats, the lions, leopards, and tigers.

One group of bovids, the antilocaprines, for many years thought to be contained in a separate family, the Antilocapridae, but recently on the basis of convincing evidence included with the other horned ruminants in the family Bovidae, has through its evolutionary history been limited to North America. The antilocaprines appeared in the Miocene as small, deerlike animals, of which *Merycodus* was characteristic. *Merycodus* had branched horn cores, which look like the antlers of deer, but it is clear that they were never shed during life. Later antilocaprines had flattened or twisted horncores, and lived in considerable variety during Pliocene and Pleistocene times. However, these varied ruminants became extinct except for one genus, *Antilocapra,* the

"antelope" or pronghorn of our western plains, which until the arrival of white men inhabited the West in large herds. As mentioned above the pronghorn is unique among bovids in that the horn sheath is shed and renewed each year.

The remaining ruminants belonging to the family Bovidae, arose and evolved in northern Eurasia during the Pliocene and Pleistocene epochs, in which time they followed numerous and truly confusing lines of adaptive radiation that resulted in the great host of genera and species now living in many parts of the world. It would seem that in late Pliocene times the varied bovids of Eurasia moved south into Africa to populate that continent. A few of them invaded North America, where as species, such as the bison, they multiplied enormously to live in tremendous herds that were numbered in the millions. However, this continent was never the home of a numerous array of genera and species, as are Asia and Africa.

Only within the last few years have some facets of bovid evolution become properly understood. For instance, why is there such a great variety of antelopes now living in Africa? On the face of it, we would think that a good pair of horns would enable a species of antelope to protect itself and spread widely. Yet in Africa there are literally dozens of antelope species with an astonishing variety of horns. There are the straight-horned oryx, the twisted horned elands, the spiral-horned kudus, the curved-horned sable antelopes, the recurved-horned wildebeests, and so on. And many of these antelopes live in the same environment. What are the advantages of a kudu's horns over an eland's horns, or vice versa? Recent studies indicate that the great variety of horn shapes in the antelopes evolved primarily for the purpose of sexual combat. During the mating season the males engage in furious yet relatively harmless contests to establish their individual supremacies within the herd. During their fights they "lock horns" and push, but rarely do they seriously injure each other. The shapes of the horns in the various species make this possible. But, also, these varied horns provide many antelopes with the means of defense against their enemies. Perhaps the modern African antelopes represent numerous populations that have evolved in genetic isolation from each other, and all have been successful.

No attempt will be made to describe the bovids. They can be listed, as follows:

Antilocaprines: The pronghorn and its fossil relatives, North America.

Bovines: The tragelaphines, such as the kudu bongo and eland, Africa; the boselaphines, the nilgai and *Tetracerus* of Asia; the bovines, the Old World cattle and buffalo, the bison of North America and Europe.

Cephalophines: The small duikers, Africa.

Reduncines: The reedbuck, waterbuck, and their relatives, Africa.

Hippotragines: The sable and roan antelopes, oryx, and addax, Africa.

Alcelaphines: Wildebeest, hartebeest, topi, and blesbok, Africa.

Aepycerotines: The impala, Africa.

Antilopines: Blackbuck and gazelle, Asia; springbuck, gerenuk, gazelles, klipspringers, dik-bik, beira, oribi, steinbuck and other related antelopes, Africa.

Peleinines: The rhebuck, Africa.

Caprines: The saigines, such as the saiga, Asia and Europe; the rupicaprines, the goral, serow, and chamois of Eurasia, and the Rocky Mountain "goat" of North America; the ovibovines, the takin of Asia, and the musk ox of North America and Greenland; the caprines, such as the tahr of Asia and the various sheep and goats of Asia, Europe, North Africa, and North America.

A Cave Drawing.

Elephants and Their Kin

THE EARLY MAMMALS OF NORTH AFRICA

The Proboscidea, namely, the moeritheres, deinotheres, barytheres, mastodonts, and elephants, the Desmostylia, an extinct group, the Sirenia or sea cows, the Hyracoidea, which are the conies mentioned in the Bible, and the Embrithopoda, another extinct group, will be discussed in this chapter, thereby bringing to an end this brief history of mammalian evolution. Fossil remains of all the above-named mammals, except for the Desmostylia, are found in the late Eocene and early Oligocene sediments of the Fayum Basin of Egypt, an area that probably is near the centers from which they originated. The Proboscidea, Desmostylia and Sirenia obviously arose from a common ancestry, and for this reason they sometimes have been united within a group above the level of an order, which has been designated as the Tethytheria. The Hyracoidea, although not included within the Tethytheria, have been considered by some authorities also to have been derived from a base in common with the tethytheres. The origin of the Embrithopoda remains a complete mystery.

The proboscideans have been throughout their history large and even gigantic mammals of the forest and the plain. The sea cows have been aquatic herbivores, living in shallow waters along seacoasts, or venturing up rivers that flow into the sea, where they feed upon underwater plants. The desmostylids likewise were water-living herbivores. The conies are small ungulates that look more like rodents than hoofed mammals. The embrithopods were giant, horned ungulates, paralleling in some ways the Oligocene titanotheres of North America or the late Cenozoic rhinoceroses of the Old World.

AN INTRODUCTION TO THE PROBOSCIDEANS

Elephants are familiar and fascinating creatures to all of us. These giant mammals, still living in sadly reduced numbers is Asia and Africa, and widely domesticated by man, seem like permanent fixtures in our world. Yet the modern elephants are actually the last representatives of a dying group, and it is quite possible that even without the intervention of man as a destructive agent they may very well be on their way to extinction within the next few thousand years. Although they are still numerous as individuals, elephants are limited at the present time to two genera, each with a single species, one in Asia and one in Africa. It would hardly be suspected from knowledge of modern elephants alone that their ancestors and collateral relatives inhabited the world in prodigous number and in a bewildering array of genera and species through the middle and later portions of the Cenozoic era. That they did we now know from the fossils.

Remains of these great beasts were among the relics of extinct animals first known to man. They were collected by the ancient Greeks and Romans as curious objects, worthy of preservation; and a leg bone of a fossil elephant was among the treasures of the Tlascalan Indians of Mexico, who were subdued by and then became the allies of stout Cortez. Many of the legends of giants, so popular among men of earlier ages, were based upon the discoveries of fossil elephant bones. The modern science of vertebrate paleontology had its beginnings with the study of these animals. Consequently there has been time for the accumulation of a great mass of data on the past history of the proboscideans.

The fossils show that at various times during the Cenozoic era the proboscideans lived on all the continents of the world, except Australia and Antarctica. During the later phases of the Tertiary period these great beasts evolved along numerous lines of adaptive radiation, some of which continued into Pleistocene times. Consequently the phylogenetic history of the proboscideans is very complex and not at all easy to interpret, especially since there was a great deal of parallel evolution within this order of mammals.

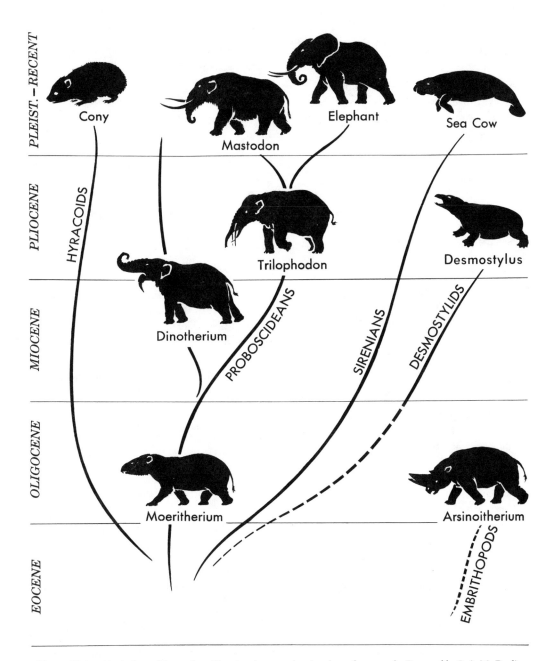

PLEIST. – RECENT

PLIOCENE

MIOCENE

OLIGOCENE

EOCENE

Cony

Mastodon

Elephant

Sea Cow

HYRACOIDS

Trilophodon

Desmostylus

Dinotherium

PROBOSCIDEANS

SIRENIANS

DESMOSTYLIDS

Moeritherium

Arsinoitherium

EMBRITHOPODS

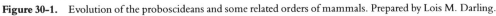

Figure 30-1. Evolution of the proboscideans and some related orders of mammals. Prepared by Lois M. Darling.

THE MOERITHERES

The first proboscideans known from the fossil record were the moeritheres, named from the typical genus, *Moeritherium*, from Egypt. Recent discoveries and studies have shown that the earliest moeritheres lived in southern Asia during early and middle Eocene times. *Anthracobune*, first described in 1940, and the more recently described *Pilgrimella* and *Lammidhania* were for many years considered as being anthracothere artiodactyls, but it is now generally recognized that they are in fact very primitive moeritheres. Furthermore, the recent discovery of a skull of early Eocene age in Algeria, would seem to indicate the presence there of an ancestral proboscidean.

Perhaps our most satisfactory knowledge of an ancestral proboscidean is to be had from *Moeritherium* itself, from the Upper Eocene and Lower Oligocene beds of Egypt. Although *Moeritherium* probably is not on the line of direct ancestry to later proboscideans, it serves very nicely as a morphological archetype, to afford an impression of what a very early proboscidean was like. It was a heavily built animal about the size of a pig, with a long body, and with stout legs that terminated in broad, spreading feet, having flat hoofs on the end of the toes. The tail was short. All in all the body of *Moeritherium* was generalized and about what might be expected in a medium-sized, heavy-footed ungulate of Eocene age.

The skull, however, was specialized in a number of interesting ways. It was elongated with the eye set far forward in front of the most anterior premolars, thus making the cranial region of the skull very much lengthened. Because of the forward position of the eye the zygomatic arch or cheek bone was also very long. The back surface of the skull, the occiput, was broad and forwardly sloped, to give a large area of attachment for strong neck muscles. The lower jaw was deep, and its back portion, the large ascending ramus, extended up so that the articulation between the jaw and the skull was placed high above the level of the teeth.

The second upper and lower incisors of *Moeritherium* were enlarged and were set in a transverse line across the front of the skull and the lower jaw. The first incisors were very small teeth, crowded between these rather tusklike second incisors; and in the skull the third incisor and the canine were small teeth, separated by a gap from the large incisors in front. The third incisor and the canine had been completely suppressed in the lower jaw. The cheek teeth were separated by gaps from the anterior teeth, and the molars, both above and below, were doubly cross-crested, each crest being formed by two large cusps placed side by side.

The external nostrils were situated at the front of the skull, and it is obvious that although *Moeritherium* may have had a thick upper lip, it did not have a trunk or proboscis.

These were the basic proboscidean characters, and from such a base the proboscideans evolved along several lines of adaptive radiation during middle and late Cenozoic times. Although the evolution of these mammals was remarkably varied, it was marked by certain dominant trends running through the wide range of proboscidean forms. These trends were:

1. Increase in size. Almost all the proboscideans became giants.
2. Lengthening of the limb bones and the development of short, broad feet. This has been a common evolutionary trend among very large mammals.
3. Growth of the skull to extraordinarily large size.
4. The shortening of the neck. Since the skull and its associated structures became large and heavy, the neck was reduced in length to shorten the lever between the body and the head.
5. Elongation of the lower jaw. In many of the later proboscideans there was a secondary shortening of the lower jaw, but lengthening of the jaw was an early, primary trend.
6. Growth of a trunk. Elongation of the upper lip and the nose probably went along with elongation of the lower jaw. Subsequently the nose was further elongated to form a very mobile trunk or proboscis.
7. Hypertrophy of the second incisors to form tusks, used for defense and for fighting.
8. Limitation and specialization of the cheek teeth in various ways, as adaptations for chewing and grinding plant food.

THE LINES OF PROBOSCIDEAN EVOLUTION

Excluding the barytheres (a peculiar and little-known group, to be discussed subsequently) there were two principal lines of proboscidean development after the moerithere stage, namely, the deinotheres and the euelephantoids. The deinotheres were quite distinct from the beginning, and they evolved along a very narrow path of adaptations through most of Cenozoic times, becoming extinct during the Pleistocene epoch. The euelephantoids, on the other hand, flowered into many diverging branches during middle and late Cenozoic times, to produce a great concourse of large, trunked, tusk-bearing giants that spread to almost all corners of the earth. Indeed the diversity among the euelephantoids was so great that we might question the propriety of classifying all of them within a single large group. However, the various euelephantoids have more in common with each other than any of them have with the moeritheres or with the deinotheres, and therefore such an arrangement is quite logical.

Divergence within the euelephantoids was so marked during their evolutionary history that for practical purposes it is useful to think of them as being composed of four groups, rather than as a single entity. These groups, which may be given superfamily rank, are the long-jawed mastodonts or gomphotheres (very generally known as the trilophodonts), the short-jawed mastodonts, the stegodonts, and finally the elephants and their relatives. The two groups of mastodonts constitute parallel series, evolving side by side during Tertiary times, some extending into the Pleistocene epoch. The elephants represent a group that arose at a late date from the long-jawed mastodont stem, to evolve with great rapidity and profusion during Pliocene and Pleistocene times.

With these general considerations of proboscidean evolution in mind, let us now turn to a more detailed discussion of the several lines of adapative radiation in these large and interesting mammals.

THE DEINOTHERES

The first deinotheres appear in Miocene sediments, fully specialized and showing no intermediate antecedents to connect them with their presumed moerithere ancestors. There is no record anywhere as to the Oligocene evolution of the deinotheres, but it must have been proceeding at a rapid rate to bridge the gap between a moerithere and the first *Deinotherium*.

The history of the deinotheres from Miocene times until their extinction during the Pleistocene epoch was so stereotyped that all these proboscideans may be placed within the single genus *Deinotherium*. It is obvious that there was little morphological progress in the deinotheres after their first appearance, and the changes that took place during the long lapse from early Miocene into Pleistocene times were those of details, mainly increase in size. Here is an interesting example of rapid initial evolutionary development (as yet unknown to us) followed by a long period of evolutionary stability at a high level of specialization.

The first deinotheres were modestly proportioned giants, but in late Cenozoic times they became some of the largest of the proboscideans, standing ten feet or more in height at the shoulder. These were long-legged proboscideans, comparable to the modern elephants in this respect. In the skull and dentition, however, the deinotheres were quite unlike any of the other proboscideans. The skull was rather flat, not high as usual among the advanced proboscideans, and it was quite tuskless. In the lower jaws there were two large tusks that curved down from the front of the jaws and then back toward the body. They formed a sort of huge hook on the front of the jaws. It is difficult to say just how such tusks might have been used; very likely they were employed for digging into the ground and pulling up roots or plants. (Incidentally one of the pioneers of paleontology thought that the deinotheres lived in rivers, and that they used the strange tusks to anchor themselves to the bank at night, where they slumbered in watery repose!) The cheek teeth of the deinotheres, like those of most mammals, were arranged in long rows in the skull and the lower jaw (a rather unproboscidean character as we shall see), and most of the teeth had two sharp cross-crests constituting each crown.

There was evidently a well-developed trunk, as in other proboscideans.

Deinotheres lived in Eurasia and Africa, but never entered the New World.

Even though they may seem strange to us, the deinotheres were nevertheless well-adapted mammals, for theirs was a long history. Indeed the persistence of the single genus *Deinotherium* through almost three geological epochs is an outstanding example of unusual mammalian stability in a changing world. The extinction of the deinotheres during Pleistocene times was probably part of the pattern of proboscidean extinction that so characterized the later phases of this last division of the Cenozoic era.

THE BARYTHERES

Barytherium inhabited Egypt during late Eocene times. This animal, imperfectly known from the lower jaw and some limb bones, had a general proboscidean appearance, and for this reason it is tentatively placed in the Proboscidea. It is so poorly known at the present time as a single genus, based upon incomplete fossils and without antecedents or descendants, that it is of little paleontological significance.

THE MASTODONTS

We now go to early Tertiary times in Egypt. Not long after *Moeritherium* lived in what is now the valley of the Nile, some proboscideans appeared that showed definite advances

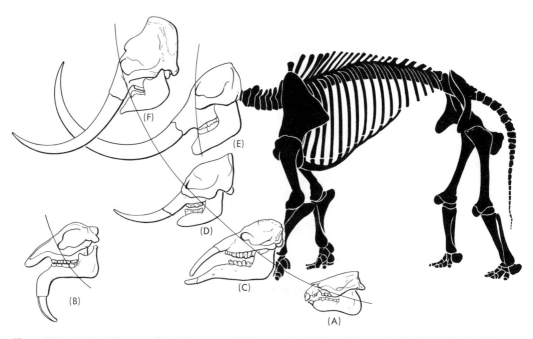

Figure 30–2.   Proboscideans, to about the same scale. The mastodon skeleton is about nine feet high. **(A)** *Moeritherium,* a moeritheriid of late Eocene and early Oligocene age. **(B)** *Deinotherium,* a deinotheriid, of middle and late Cenozoic age. **(C)** *Serridentinus,* a Miocene mastodont. **(D)** *Stegomastodon,* a Pleistocene mastodont. **(E)** *Mastodon americanus,* the Pleistocene American mastodon. **(F)** *Parelephas,* a Pleistocene mammoth. The lines indicate broad relationships, from moeritheres through early mastodonts to late specialized mastodonts. Elephants were derived from advanced mastodonts. The dinotheres were isolated from other proboscideans throughout their history.

beyond the ancestral moeritheres. They were the first of the mastodonts, the lower Oligocene genera *Palaeomastodon* and *Phiomia*. *Phiomia* was considerably larger than any of the moeritheres, standing perhaps as much as seven or eight feet tall at the shoulder. It had an elephantlike skeleton, with comparatively long legs. The skull was greatly enlarged by the expansion of sinus cavities above and in front of the braincase, so that the back portion of the skull had increased in height. The nasal bones were retracted far back on this high, swollen skull, an indication that *Phiomia* was provided with a well-formed trunk. And in front there were two tusks, projecting forward and down, tusks that had been derived from the enlarged second incisors of *Moeritherium*. The lower jaw was very long, and in front it carried two tusks that protruded horizontally. The cheek teeth were low crowned, and each molar was provided with three transverse pairs of bluntly conical cusps.

All in all both *Phiomia* and *Palaeomastodon* probably looked something like a medium-sized elephant with a long lower jaw that was provided with straight, horizontally directed tusks. From ancestors of which *Phiomia* and its close relative *Palaeomastodon* are approximations the numerous euelephantoids evolved.

There is a strange gap of several million years comprising much of the Oligocene epoch without any known record of mastodont evolution. It will be remembered that the same was true for the deinotheres. Where were the proboscideans during this phase of geologic history? This is an intriguing question, and some day we may discover the answer to it.

Gomphotherium, more commonly known as *Trilophodon*, was the central type of what may be called the bunodont mastodonts, those forms in which the cheek teeth were composed of three or more pairs of heavy cones arranged along the axis of each tooth. It lived during Miocene and early Pliocene times. This mastodont was in effect a large edition of *Phiomia* with certain refinements. The lower jaws were much elongated, and provided

with two tusks. The first two molars had three pairs of transversely arranged cones which became, when worn, three low cross-crests. The third molar was elongated by the development of a heel behind the last two of the paired cones. There was evidently a long and flexible trunk.

From a *Gomphoterium* stem the bunodont mastodonts evolved in various directions. In some of the bunodont mastodonts accessory cusps were added to the sides of the main molar cones, giving to the tooth a complexly sinuous enamel pattern when it was worn. Naturally this increased the effectiveness of the teeth as grinding mills. This trend is seen in the Miocene-Pliocene genus *Serridentinus*, closely allied to the trilophodonts, and reaches its culmination in the upper Pliocene *Synconolophus* and the lower Pleistocene *Stegomastodon*. In these last two mastodonts the lower tusks were suppressed so that they disappeared, and the lower jaw was shortened. The upper tusks became large and curved strongly upward. Here we see a parallelism among some of the bunodont mastodonts to other lines of advanced proboscideans. During Pleistocene times a branch of these mastodonts invaded South America, and became widely distributed on that continent as the genus *Cuvieronius*.

In another lateral branch of bunodont mastodonts the front of the lower jaws and the tusks were strongly downturned. These were the Miocene and Pliocene rhynchotheres, of which *Rhynchotherium* was characteristic. Some of the most remarkable of the bunodont mastodonts were the so-called shovel tuskers of Pliocene age, *Amebelodon* of North America and *Platybelodon* of Asia. In these mastodonts the lower tusks became very broad instead of having the primitive round cross-section, so that they literally formed huge scoops or shovels on the front of the jaws. Presumably these shovels were used for digging up plants, perhaps from the bottom of shallow waters.

While the bunodont mastodonts were evolving there was a second and independent evolutionary line of what may be called the

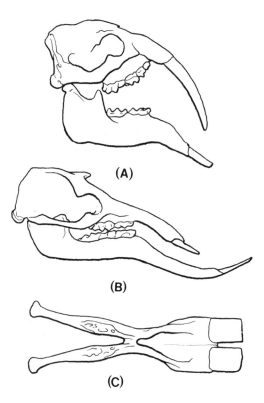

(A)

(B)

(C)

Figure 30-3. Skull and jaws of gomphothere probosci-deans, showing the elongation of the lower jaws with tusks, characteristic of these mastodonts. **(A)** *Rhyncho-therium*, from the Miocene of North America; **(B)** *Platybelodon*, from the Miocene of Asia; **(C)** the lower jaw of *Platybelodon* in crown view, showing the broad, shovel-like shape of the front of the mandible and tusks. All about one-fifteenth natural size.

zygodont mastodonts, those forms in which the cheek teeth were composed of strong cross-crests, with three or more crests to each tooth. From the outset of their history they had virtually tuskless lower jaws. These mast-odonts, perhaps arising from a *Palaeomastodon* type of ancestor, evolved as the Miocene *Miomastodon*, the Pliocene *Pliomastodon* and its relatives, and the Pleistocene *Mastodon* (which according to the rules of zoological nomen-clature should be called *Mammut*, but almost never is). The American mastodon, *Mastodon americanus*, is one of the best known of fossil vertebrates, for its remains have been found in great abundance throughout most of North America. This was a large proboscidean, not as tall as a large modern elephant, but heavily built. It had sharply cross-crested molars, as did all members of this line of mastodont evo-lution, and the upper tusks were large and strongly curved. The American mastodon persisted until the very end of Pleistocene times, and skeletons that show virtually no al-teration of the bones have been found in swamp deposits. From soft materials pre-served with some of the skeletons we know that this mastodont was covered with long, reddish-brown hair, and that it browsed upon the leaves of trees. There seems little doubt now that early man in America was contem-poraneous with the mastodon, and methods of dating by Carbon 14 give conclusive proof that mastodons were living on this continent as recently as about eight thousand years ago. Sometime after the entry of early man into North America, but several thousand years before the coming of white men, the mastodon became extinct. Why, we do not know.

An evolutionary line derived from some of the zygodont mastodonts was that of the stegolophodonts, named from the late Ceno-zoic genus *Stegolophodon*. In *Stegolophodon* the lower jaw was shortened, the upper tusks were large, and the molar teeth were cross-crested, the crests being formed by a sort of breaking up of the original large mastodont cones into conelets, arranged in transverse rows across the tooth. Correlated with this development of cross-crests was the lengthening of the molar teeth and the multiplication in the number of crests, to four in the intermediate molars and as many as six in the third molars. These events took place in Eurasia during late Mio-cene and early Pliocene times.

From the stegolophodonts it was but a short evolutionary step to *Stegodon*. The stegodonts, appearing in the Old World in late Pliocene times and continuing into the Pleistocene ep-och, were large, very long-legged animals, in which the skull was deep, the upper tusks very long and curved, the lower jaw short and tuskless, and the molar teeth greatly elongated, with numerous low cross-crests on each molar crown. Indeed there were as many as twelve or thirteen crests in the third molars of some of the advanced stegodonts.

The elongation of the cheek teeth in *Stegodon* introduced a problem with regard to the arrangement of the teeth within the skull and the lower jaws, and the matter of tooth succession. In many of the earlier mastodonts the teeth were arranged in longitudinal rows, with all the premolars and all the molars in place at once. As proboscidean teeth became larger, two evolutionary alternatives were possible for the accommodation of the teeth. Either the skull and the lower jaw could become very long, to take care of a series of long teeth, one behind the other, or the tooth succession could be altered so that only a few teeth were in place at one time. The stegodonts and the elephants developed in the second way. In these highly specialized proboscideans the cheek teeth come into use one or two at a time. The skull and the jaw are greatly deepened, and the teeth are formed behind and above the tooth sockets in the skull, and behind and somewhat below the sockets in the lower jaw. Then the teeth push forward into place; as the back teeth push ahead, the teeth in front of them are shoved forward in the jaws, where they finally break away at the front and disappear. This specialization reaches an extreme in the Pleistocene mammoths and the modern elephants.

It had long been thought that the stegodonts were intermediate between some of the mastodonts and the true elephants. A more recent view considers the stegolophodonts and stegodonts, descended from zygodont mastodonts, as having evolved independently of and parallel to the elephants. If this is so it constitutes a remarkable example of evolutionary parallelism.

THE ELEPHANTS

If one accepts the stegodonts as having been parallel rather than ancestral to the elephants, the stem of elephant evolution must be sought elsewhere. This stem would seem to be provided by the bunodont mastodont, *Stegotetrabelodon*, from the late Miocene and Pliocene of Africa. Although the lower jaws of this mastodont are provided with long tusks, the cheek teeth are cross-crested, each crest being formed of a transverse series of conelets. From such teeth the teeth of early elephants may have been derived by a deepening and a compression of the crests from front to back, until in the elephants they took the form of tall, parallel plates, rather than V-shaped ridges. Thus the tall grinding teeth of the most advanced proboscideans came into being.

This process of transformation of the cheek teeth from low-crowned into very tall teeth was remarkably fast in the geologic sense. The first true elephants of lower Pleistocene times had comparatively low teeth, and before the end of this geologic epoch, spanning a time lapse of perhaps two million years, the change had taken place. Elephants and men are mammals that show an appreciable amount of evolutionary development during Pleistocene times.

The Pleistocene or the great Ice Age was the age of mammoths (which is the name commonly applied to extinct elephants) on all the continents except South America, Australia, and Antarctica. These giant mammals wandered far and wide across Eurasia, Africa, and North America, being immigrants into this continent from the Old World. There were various species of mammoths; one group was related to the modern Asiatic elephant, *Elephas*, another to the modern African elephant, *Loxodonta*.

One of the largest of the mammoths was the great imperial mammoth that lived in North America, and attained shoulder heights of fourteen feet or more. Perhaps the best known of the mammoths was the woolly mammoth, living in northern Eurasia and North America. This mammoth was well known to ancient Stone Age man in Europe, and he drew pictures of it on the walls of caves.

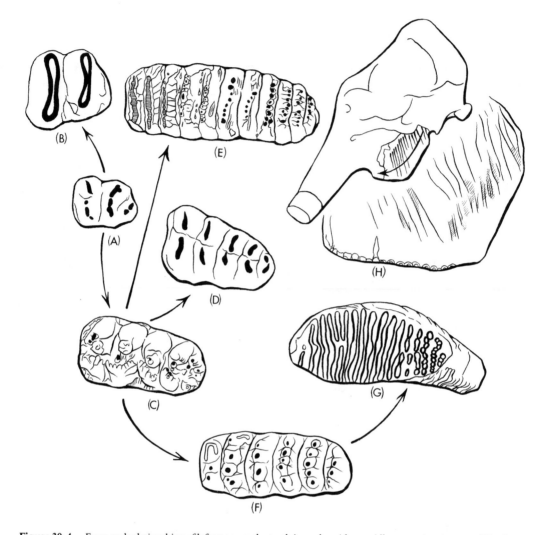

Figure 30-4.   Form and relationships of left upper molar teeth in proboscideans. All crown views, except (H), about one fourth natural size. **(A)** *Moeritherium,* an Oligocene moerithere. **(B)** *Deinotherium,* a Miocene-Pleistocene deinothere. **(C)** *Serridentinus,* a Miocene-Pliocene mastodont. **(D)** *Mastodon,* a Pleistocene mastodont. **(E)** *Stegodon,* a Pleistocene stegodont. **(F)** *Stegotetrabelodon,* a Pliocene proboscidean intermediate between the mastodonts and the elephants. **(G)** *Parelephas,* a Pleistocene mammoth. **(H)** Side view of a large elephant molar, with the skull shown above it. The direction of eruption of an upper molar in the skull is indicated by the arrow. In the ancestral Moeritheres (A), the crowns of the cheek teeth consisted of low cones arranged transversely to form incipient crests. In the deinotheres the cheek teeth became sharply cross-crested, and changed little throughout the evolutionary history of this group. In the mastodonts the tooth crowns were composed either of cones or of crests. From certain mastodonts the Pliocene and Pleistocene stegodonts arose, with long, cross-crested teeth. The stegotetrabelodonts, also derived from mastodonts, led to the Pleistocene and Recent elephants, in which the cheek teeth are very tall and consist of transverse plates. These teeth are so large that there is room in each jaw for only one or parts of two teeth at a time. Consequently the teeth erupt one after another, the last molar coming into place when the animal has reached a fully adult stage of growth.

In the last two centuries several carcasses of the woolly mammoth have been found frozen in the ice, in Siberia and Alaska. From this evidence we know that the woolly mammoth was covered by a dense coat of hair, enabling it to live in an arctic climate.

Man has lived with mammoths and elephants during the course of his evolutionary history. To the earliest men living in Eurasia and Africa the giant mammoths must have been truly fearsome beasts, against which there was little protection. But as man evolved and became a tool and weapon maker, he also became a hunter of mammoths. Numerous discoveries in Europe show that Stone Age man pursued the mammoths, prevailing against these giant beasts by the use of guile and well-conceived strategy. Mammoths were frequently trapped in deep pits, where they could be stoned to death, or killed with heavy deadfalls. And when man came from Asia into the New World he found mammoths on this continent, to be hunted and killed. There is definite evidence to show that early men in America hunted mammoths, perhaps as recently as eight or ten thousand years ago.

Various species of mammoths lived until the very end of the Pleistocene epoch. Then they became extinct, and only two proboscideans, the African and Asiatic elephants, survived into modern times. Primitive men in our present day world are still elephant hunters, but with the rise of Old World civilizations man became the master of the elephants instead of their foe. For several thousand years elephants have been caught and domesticated, and they have played no little part in the rise and spread of culture in Asia and the Middle East. These great mammals are remarkably docile, once they have been kept in captivity for a while, and because of their long life and their great strength they have been very useful servants of man. Elephants are now being displaced by the gasoline engine, and their use as work animals has been greatly restricted. However, modern man is a lover of elephants, and it is probable that he will do his best to see that these noble mammals do not disappear from the face of the earth.

Why did the mammoths become extinct? This question, like so many of the questions having to do with problems of extinction, is extremely difficult to answer. In fact, it is probable that we shall never know the real reason for the disappearance of mammoths a few thousand years ago, after their successful reign through Pleistocene time. Very likely the extinction of the mammoths was the result of complex causes. Men may have had something to do with it, but we can hardly believe that primitive men were of prime importance in bringing an end to these numerous, widely distributed, and gigantic mammals. It seems possible that factors of which we have little inkling at the present time brought about the disappearance of the mammoths, and that these and other factors may be operating at the present time to cause the eventual extinction of the two remaining species of elephants.

THE SEA COWS

Although they are classified as ungulates of a sort, the sirenians or sea cows have become completely modified for an aquatic mode of life. Adaptations for life in the water took place at an early date among these mammals, because the earliest-known sea cows, of Eocene age, were already highly specialized sirenians. These early sea cows are well represented in the upper Eocene beds of Egypt, but they are also found in beds of similar age in Europe, and as far distant as the West Indies. It seems very probable, therefore, that the early sea cows were widely distributed through the world, as might be expected of animals able to swim along the shores of the continents.

Modern sea cows are large mammals, with strange, blunt-nosed heads, torpedo-shaped bodies, flipperlike front limbs, and a broad, horizontal tail fin. The skin is naked and tough. The hind limbs are suppressed, and the pelvis is reduced to a rodlike bone. The ribs are very massive and heavy, making a sort of

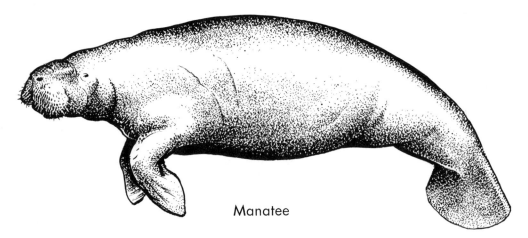

Manatee

Figure 30-5. A modern sea cow. A large manatee is ten feet or more in length. Prepared by Lois M. Darling.

ballast for the body in these mammals. The skull is rather long and low, and in its back portion it resembles to some degree the skull in the very primitive proboscideans, the ancestors of the elephants. The front of the skull forms a narrow rostrum that in the modern sea cows is sharply down-turned. The cheek teeth may be doubly cross-crested or they may have bunodont cusps, like the teeth in some of the Tertiary proboscideans. As the anterior teeth become worn they are replaced by the posterior teeth pushing forward. This manner of tooth replacement also characterizes the proboscideans, again a feature linking the two orders. Sea cows are of course adept swimmers, and as mentioned above they frequent rivers that flow into the sea, where they feed upon aquatic vegetation.

The earliest sea cows were slightly more primitive than the modern forms. For instance, in the Eocene genus *Protosiren* the front of the skull or rostrum was not turned down, as it was in later sirenians like *Halitherium* of the Miocene or the recent dugong. In *Eotheroides*, another Eocene form, the pelvis was still a recognizable pelvis, not a rod of bone as in the later sea cows. But these are differences of degree, not of kind, and the first sirenians were completely aquatic mammals.

As they evolved the sea cows followed two

lines of development, probably from Eocene times on. The manatees evolved on both sides of the Atlantic basin, inhabiting the shores of Africa and America. The dugongs evolved widely along the world's coasts. The Steller's sea cow was a gigantic sirenian, weighing as much as four tons, that lived in the Bering Sea. It was exterminated about two centuries ago, only a few years after it had been discovered and scientifically described in 1741 by Georg Steller, the naturalist on the Bering expedition.

THE DESMOSTYLIDS

Within recent years some very interesting mammals of middle and late Tertiary age have been discovered along both sides of the Pacific Ocean. They are the desmostylids, found in the Pacific Northwest, California and Japan.

The earliest known desmostylids are represented by *Behemotops* found in marine Oligocene deposits of the Pacific Northwest. This interesting mammal, presently known from a well-preserved lower jaw about forty centimeters in length, and fragments of a hind limb, shows features that link it with early proboscideans such as *Anthracobune* and *Moeritherium*. There are large canine tusks, similar to those of later desmostylids, and the molar teeth, each consisting of four rounded, bunodont cusps, are remarkably similar to those of ancestral proboscideans. The fragments of limb bones indicate that *Behemotops* had strong legs. From such an ancestry the later desmostylids evolved.

Desmostylus had a long, mastodontlike skull, with tusks. The nasal opening was retracted, and it seems likely that this mammal had a short trunk. The cheek teeth were peculiar; they consisted of vertical columns, arranged in pairs along the length of each tooth. Skeletons of this and another genus, *Paleoparadoxia*, recently discovered in Japan and in California, are surprising to say the least, for they indicate an animal with a large body, very heavy limbs, and broad, elephantlike feet.

Desmostylus is found in sediments of marine origin. One is led to believe, therefore, that this mammal was a sort of "marine hippopotamus"

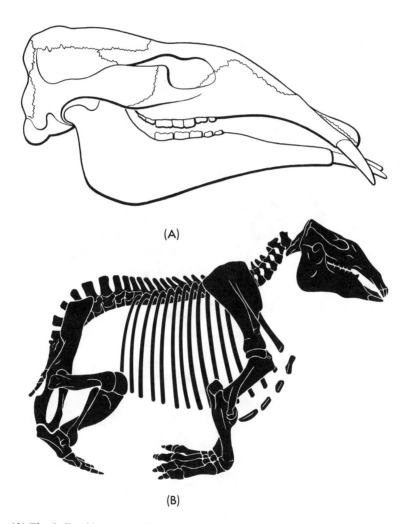

(A)

(B)

Figure 30-6. **(A)** The skull and lower jaw of Desmostylus, found in late Tertiary marine deposits along the Pacific coasts of North America and Asia. About one-tenth natural size. **(B)** The skeleton of *Palaeoparadoxia*, a late Tertiary desmostylian, some eight feet in length. Note the resemblance of the *Desmostylus* skull to the skulls in primitive sirenians and mastodonts. Note also the heavily constructed skeleton of *Palaeoparadoxia*.

wading and swimming in the shallow waters along the seacoasts. But the comparison with hippopotamuses is one of function only. In its anatomy *Desmostylus* shows resemblances to the sea cows on the one hand, and to the early proboscideans on the other. Perhaps this was a truly intermediate mammal. But the evidence of *Behemotops* suggests that the relationships of the desmostylids are to be found particularly with the early proboscideans.

THE CONIES

The conies of Africa and the Middle East (not to be confused with certain lagomorphs known as conies) are small rabbit-like or rodentlike mammals that live among the rocks of steep hillsides, or in trees. Among ancient peoples, and even among the earlier of modern naturalists, these mammals were thought to be rabbits of some sort, but with the growth of modern systematic zoology the position of the conies has been clarified; these are small ungulates that constitute the Hyracoidea, a separate order of mammals.

In several respects the conies, or dassies, or hyraxes, have paralleled the lagomorphs, or perhaps more closely certain rodents. This is apparent in their small size and in their habits.

Arsinoitherium

Saghatherium

Figure 30-7. Two Oligocene mammals of northern Africa, drawn to the same scale. *Arsinoitherium*, as large as a rhinoceros, is the only known genus belonging to the order Embrithopoda. *Saghatherium* was a large hyracoid or cony. Prepared by Lois M. Darling.

Conies are herbivorous. In the skull there is on each side a single rootless incisor tooth, very much like the large, chisel-shaped incisor teeth of rodents. This tooth is opposed by two incisors in the lower jaw. A gap separates the incisor teeth from the cheek teeth, which are high crowned, with the molars showing a pattern somewhat like that seen in the rhinoceroses. The feet are completely distinctive, there being four toes on the front feet, three on the hind, terminating in small, hooflike nails. A single pad makes up the bottom of each foot.

Various genera and species of conies now live in Africa and adjacent regions along the Mediterranean Sea. The first members of the Hyracoidea appear in the Oligocene sediments of Egypt, and the history of these mammals can be traced in a limited way in the Mediterranean region through Tertiary times. There was a certain amount of evolutionary radiation of the Tertiary hyraxes, and some of them like *Megalohyrax* of the Oligocene epoch became as large as pigs. However, the larger hyraxes failed to survive, perhaps because of competition from other ungulates, and only the small, rabbitlike conies, represented by the modern *Procavia*, carried the line into recent times.

EMBRITHOPODA

Arsinoitherium, of Oligocene age, is the single genus of the Embrithopoda. This was a large ungulate, as big as a rhinoceros, with a heavy skeleton, strong, elephantlike limbs, and broad, spreading feet. The skull was large, and the teeth formed a continuous series from incisors to last molars. The cheek teeth were tall crowned, which is surprising in an ungulate as early as *Arsinoitherium*, and each molar crown was strongly cross-crested. The most striking character of *Arsinoitherium* was a pair of massive, bony horns, side by side, the confluent bases of which occupied the top of the skull from the nostrils back to the middle of the braincase.

Arsinoitherium is truly a paleontological mystery. There are no known fossils that might represent its ancestors, and it apparently left no descendants. This animal stands alone in time and its zoological position; it appears suddenly in the lower Oligocene sediments of Egypt, and that is the end of it. It is the solitary representative of an order of mammals about which we may learn more at some future date.

Cenozoic Landscape.

The Age of Mammals

CENOZOIC CONTINENTS
AND CLIMATES

With the transition from the age of Reptiles into the Age of Mammals, the world entered the modern phase of its geologic history. This was the time when peninsular India, having broken away from its African connection, was drifting rapidly toward mainland Asia—to the Cenozoic collision that buckled the earth's crust, thus forming the ancestral Himalayan range. This was the time when South America, ever drifting to the west and completely free from its African connection, became an island continent, except for a very early Cenozoic isthmian connection to North America, a connection that was reestablished at the close of the Pliocene epoch. This was the time (specifically during the Eocene epoch) when Australia was severed from Antarctica, to drift to the northeast with its cargo of marsupials, which had previously reached Australia, perhaps from South America by an Antarctic bridge. This was the time (again during the Eocene epoch) when North America and western Europe were rather briefly joined by a North Atlantic connection. And this was the time when the Bering land bridge became established, to be intermittently a corridor for the exchange of mammals between North America and Asia during Cenozoic time.

The establishment of modern continental connections was, in addition to plate tectonic movements, to a considerable degree the result of uplifts that began in late Cretaceous times. Continental areas, which had been low and partially inundated by shallow seas during middle and late Mesozoic times, were lifted to new heights during late Cretaceous and early Cenozoic history. There were broad recessions of the shallow seas, and, what is particularly important, the modern mountain systems were born. This was the time of the initial uplift of the Alps and of the Himalaya chains in the Old World, of the Rocky Mountain and of the Andean chains in the New World. It was the beginning of a long period of mountain-making that is actively continuing at the present day.

The delineation of modern continents, the uplift of great land masses, and the rise of mountain systems were accompanied by the beginning of worldwide climatic changes, of the utmost importance to the evolution of life during the last seventy million years. In middle and late Mesozoic times when the continental blocks were closely connected much of the world was tropical and subtropical. Uniform temperatures, with but slight seasonal changes, ranged from the equator to the northern and southern regions of Laurasia and Gondwana, respectively, so that tropical plants and dinosaurs lived from northern Eurasia and Alaska to the tips of the southern continents and to Australia. As the continents were increasingly separated and uplifted and the new mountain systems began to grow, there were gradual alterations of world environments in the direction of increased variety and differences. Climatic zones became established, and as time went on they became ever more sharply defined from each other. The alternation of seasons became more marked, especially in the higher latitudes, so that cold winters followed hot summers, year in and year out. These changes took place gradually during Cenozoic times, to culminate in the extreme climatic conditions of the great Ice Age, in one phase of which we are now living.

Of course such profound changes in world climates brought about changes in the plant life of the globe. Even before the close of the Cretaceous period the modern deciduous trees had made their appearance, to lend much botanical variety to landscapes. Forests, which in previous ages had consisted of ancient tree ferns and various conifers, were freshened by the variegated shapes and leaf patterns of oaks, willows, sassafras, and other trees quite familiar to us. And with the progression of the Cenozoic era flowering plants and grasses evolved, thereby introducing new habitats for the development of animal life. Green savannas and broad prairies spread across large continental areas, to provide the setting in which the many hoofed mammals of Cenozoic times lived, in a long succession of complex faunas.

Continental drift, uplifts and the rise of new mountain systems, the diversification of climates, and the evolution and spread of the flowering plants were all factors that affected the evolution of animal life on the earth. These events together, probably with the interaction of other factors of which we have no knowledge, resulted in the extinction of the dominant reptiles, including the great dinosaurs, at the close of the Cretaceous period and cleared the way for the rise and the profuse radiation of the mammals. To the student of vertebrate evolution, the change from reptilian dominance in the Cretaceous period to mammalian dominance in the Tertiary period was one of the great events in the history of life on the earth.

The changes briefly outlined above affected continents and life on the continents. Parallel with these transformations that took place during the Mesozoic to Cenozoic transition there were changes in the waters as well. It follows that, as tropical conditions became restricted to the equatorial belt, there would be changes in the temperatures of the oceans. The warm seas, which we know lapped the shores of northern and southern lands during Mesozoic times, gradually contracted to the middle region of the earth, and cold seas spread through the latitudes toward the poles. Perhaps these changes had not set in at the close of the Cretaceous period, but it is significant that during the transition the numerous and widely distributed marine reptiles of late Mesozoic times disappeared from the waters of the earth, just as the dinosaurs became extinct on land. With the advent of the Tertiary period there were no more plesiosaurs, ichythyosaurs, or mosasaurs. The gaps created by their disappearance were filled by new animals, especially the whales.

Even before the close of Cretaceous history the bony fishes had enjoyed a long period of diverse adaptive radiation, during which time they became the most numerous and varied of the aquatic vertebrates. Their development continued into Tertiary times, and indeed has carried on to the present day. The bony fishes are supreme in the waters, and as has been pointed out earlier in this book, they far outnumber all other vertebrates in variety, in numbers of genera and species, and in numbers of individuals. We live in the Age of Bony Fishes, as well as the Age of Mammals.

THE DEVELOPMENT OF CENOZOIC FAUNAS

Our record of the succession of Cenozoic mammals is much more complete than the comparable records for late Paleozoic and Mesozoic amphibians and reptiles. There are several reasons why this should be so. Perhaps the most important one is that Cenozoic continental sediments have suffered less destruction and on the whole are more completely preserved than the continental deposits of Paleozoic and Mesozoic age; therefore a larger proportion of the actual fossil record of Cenozoic mammals has been preserved than of the earlier land-living vertebrates. For instance, the sequence of mammalian faunas ranging from the lower Paleocene to the end of the Pleistocene in North America gives a reasonably continuous story of the changing assemblages of mammals through time on this continent. Compare this succession of faunas with the record of Jurassic land-living vertebrates, a record so incomplete that most of our knowledge about Jurassic dinosaurs and other land-living reptiles is gained from faunas of late Jurassic age.

Another factor that contributes to our superior knowledge of Cenozoic mammal faunas, as compared with earlier tetrapod faunas, is the nature of the fossils themselves. Because of the complexity of their teeth, and because teeth are hard and more frequently preserved than other parts, fossil mammals may be studied from the teeth alone (although of course every effort is always made to base studies of

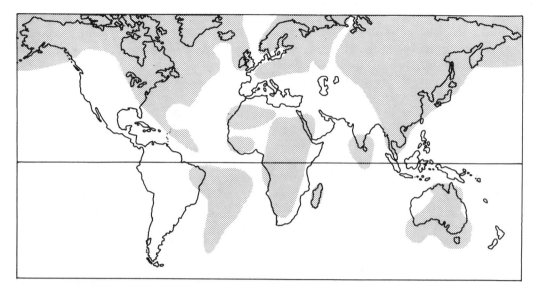

Figure 31-1. The postulated relationships of continents during Mid-Cenozoic time (shaded) according to the concept of Plate Tectonics and Continental Drift, and the distributions of land-living tetrapods, contrasted with the positions of modern continents (outlines).

mammals on as complete materials as can be obtained). With reptiles, on the other hand, it is frequently impossible to make any significant conclusions or to attempt comprehensive faunal studies without more complete material—at least partial skulls or fragmentary skeletons.

Again, the fossils of Cenozoic mammals are generally much more abundant than the fossils of Paleozoic and Mesozoic amphibians and reptiles.

In broad terms, the succession of Cenozoic mammalian faunas may be grouped into four major events.

1. In the first place, there was the initial faunal radiation of Paleocene and Eocene times. This was the period of archaic faunas, and of the dominance of mammalian groups that failed to survive into later times.

2. Second, there was a replacement (except in South America and Australia) of the archaic faunas of the Paleocene and Eocene epochs by the ancestors of the modern mammals. This was a phenomenon of late Eocene and Oligocene times.[1]

3. During later Cenozoic times there was a modernization of faunas, a process that extended over many

millions of years to produce the advanced, specialized mammals of the modern world.

4. Finally there was the establishment of present-day faunas, a process taking place in Pleistocene and "post-Pleistocene" times. The bases for our contemporary faunas were set by the emergence of faunal assemblages during the Pleistocene period. But the final event that gave a contemporary complexion to the mammalian faunas of the world, making them distinct from the strictly Pleistocene faunas, was a wave of extinction that took place within the last few thousand years. It brought an end to many of the larger Pleistocene mammals, so that our modern faunas are impoverished relics of the rich Pleistocene faunas.

The initial radiation of archaic mammals was the distinctive zoological phenomenon of early Tertiary times. Insectivores and early marsupials had already lived for millions of years in association with the dinosaurs. Now the niches occupied by the dinosaurs were va-

---

[1]Antarctica is omitted in the discussions of Cenozoic tetrapod faunas because at this date few Cenozoic tetrapods have been discovered there. In recent years, as has been described in an earlier chapter, some marsupial fossils, closely related to certain early Cenozoic South American marsupials, have been found on the Antarctic peninsula. This may indicate that marsupials reached Australia from South America by way of an Antarctic connection.

cated, and various mammals, derived from insectivore or marsupial ancestors, quickly grew up to occupy these vacancies. Archaic hoofed mammals—condylarths, uintatheres, and the like—took over the role of large plant-eaters, and the varied creodonts were the predators that hunted them. In South America the notoungulates, litopterns, and astrapotheres became the herbivores, and the function of predatory carnivores was taken over by the marsupials. In Australia, which evidently became an island continent before the advent there of placental mammals, the marsupials became supreme, occupying ecological niches filled by placental mammals in other lands. In the other continents, however, the archaic ungulates and the creodonts were replaced during late Eocene and Oligocene times by the perissodactyls, the artiodactyls, and the proboscideans, and by the carnivores, whereas the faunas were enriched by the development of the rodents, the primates, and the bats. (Various aquatic mammals, such as the whales and sea cows, having returned to the water from terrestrial ancestors, are elements of a separate story.) Thus the bases for modern faunal groups were established.

As time passed during the latter portion of the Cenozoic era, the hoofed mammals and the carnivores, the rodents, primates, and other lesser groups became increasingly specialized toward their modern state. Three-toed horses gradually evolved into single-toed horses; ancient deer disappeared as they were replaced by their descendants, the modern deer; primitive dogs grew into modern wolves, foxes, and the like. The details have already been set forth in several chapters preceding this one.

So it was that with the advent of the Pleistocene epoch the mammalian faunas of the world were essentially modern in their aspect, but much richer than they now are. They contained numerous mammoths and mastodonts, woolly rhinoceroses, giant ground sloths and glyptodonts, varied giraffes, and so on. If these mammals had continued into modern times this would indeed be an interesting world for explorers and for visitors to zoological parks. But they didn't. And the domination of the land, that for ages had been shared by many large mammals, was taken over by a single species, man.

INTERCONTINENTAL MIGRATIONS IN CENOZOIC TIMES

As mammals evolved during Cenozoic times they frequently wandered back and forth from one continent to another. These intercontinental migrations, and in many cases the lack of such migrations, were of great importance in determining the composition of faunas in the several continental regions. For instance, various factors, particularly those of ecological relationships, prevented many mammals from migrating freely from one continental region to another, and these sedentary animals gave to the fauna of a particular region its distinctive characters. On the other hand, some mammals, especially the larger ungulates and carnivores, roamed widely, and their migrations led to faunal resemblances over wide expanses of land surface.

One of the striking features of Cenozoic mammalian faunas was the general unity of the animals living in the great circumpolar land mass of Europe, northern Asia and North America. During the Eocene epoch there was a northern connection between western Europe and eastern North America, allowing an interchange of primitive mammals. Through much of Cenozoic times there was a land bridge in the Bering region, across which mammals migrated east and west. Therefore, it is not surprising that Tertiary and Pleistocene faunas of northern Eurasia and North America show many resemblances, a condition that is reflected in the modern faunas of this region. The north circumpolar land mass is recognized, on the basis of the distribution of modern mammals, as a major zoogeographical region, the great Holarctic Region, and it is probable that the faunal relationships characteristic of this region extended back through time into the Tertiary period.

Primates appeared at an early date in the faunas of the Holarctic Region, but after a brief sojourn in this area they disappeared, to evolve in other parts of the world. However, many of the major groups of mammals did evolve through Cenozoic times in the Holarctic Region, particularly the creodonts, most of the carnivores, the proboscideans, the perissodactyls, and the artiodactyls. The wide distribution of most of these mammals gave to the Holarctic faunas their community of appearance.

At the present time the fauna of southeastern Asia is distinct from that of northern Asia, and is thereby designated as the Oriental fauna, characteristic of the Oriental Region of zoogeography. In the modern Oriental Region primates are prominent elements in the local faunas. Carnivores are well represented, and include viverrids and hyenas, which are now largely absent from the Holarctic Region. Likewise the Oriental Region is typified by elephants, tapirs, rhinoceroses, and many artiodactyls. The differences whereby the modern Oriental fauna is distinguished from the Holarctic fauna are in large part the result of mountain uplifts, of the raising of the Himalaya chain, and of the mountains that separate northern and southern China, to form high barriers that prevented late Cenozoic animals from wandering north and south in Asia.

The Himalayan and related uplifts did not constitute the formidable barriers to migrations in the early and middle parts of the era that they do at the present time. Consequently there was a considerable flow of mammals in and out of the Oriental Region during much of Cenozoic history. Horses erupted from North America and made their way into India and China in great numbers. Bears, which possibly arose in India, made the long trek into North America. The pandas, which we think of as Oriental mammals, wandered as far west as England. And many other similar cases of migrations into and out of the Orient might be cited.

Africa is still another major zoogeographic region in the modern system of classification. This, the Ethiopian Region, includes the continent south of the Atlas Mountains and the Sahara. During much of Cenozoic time the Mediterranean Sea was contracted as compared with its present extent, so that there were avenues for the migration of mammals north and south between Africa and Europe. Also mammals pushed back and forth between Africa and the Oriental Region. Indeed, much of the modern African fauna of artiodactyls represents an influx into that continent of animals that lived in southern Europe and the Orient during late Pliocene and Pleistocene times. In the Ethiopian Region primates were prominent during Cenozoic times, and it was here that some of the very advanced manlike apes arose. Creodonts were present in early Tertiary times, and most of the carnivores were prominent in later faunas. However, bears and deer never reached Africa. Proboscideans, perissodactyls, and artiodactyls were important in the development of African faunas, as they were in the Orient. The distinct characters of modern African faunas is the result in part of animals being present in the Ethiopian Region that have become extinct elsewhere, frequently within the last few thousand years. For instance, spotted hyenas, elephants, hippopotamuses, various antelopes, and other animals, now characteristically African, lived in Europe or Asia during very late Pleistocene times.

Two zoogeographic regions of continental extent were isolated from the rest of the world during much of the Cenozoic era. They are the Neotropical Region, consisting of South America and part of Central America, and the Australian Region, consisting of Australia, New Guinea, and certain other islands. The Neotropical Region, as we have seen, was connected to North America at the beginning of the Tertiary period, after which it was completely isolated until the end of Tertiary times. The Australian Region has been isolated ever since Eocene time, when it broke away from Antarctica.

This isolation contributed to the very distinct faunas of these two regions. During much of Cenozoic history the Neotropical fauna consisted to a large degree of ungulates unique to that region, carnivorous marsupials, and edentates. Primates appeared in the Neotropical faunas during Tertiary times; just how it is difficult to say. Then at the close of the Tertiary period there was a great influx of Holarctic mammals into this region, to give the Neotropical fauna its modern cast—that of a mixture of ancient survivors and recent immigrants.

The isolation of Australia was complete, and Cenozoic history in that region was a period of marsupial radiation. The only placentals in the Australian Region before the coming of man were bats and rodents. Since his arrival, man has introduced various placental mammals into Australia, much to the detriment of the native marsupials.

THE DISTRIBUTION OF CENOZOIC VERTEBRATE FAUNAS

Most of our knowledge of the earliest Cenozoic faunas comes from the Paleocene of North America, the only region, in fact, where the continental Paleocene is reasonably well represented by fossil mammals. In New Mexico are the Puerco and Torrejon faunas, the classic mammalian assemblages of early and middle Paleocene age, respectively; in Utah is the Dragon fauna, intermediate in age between the New Mexican assemblages. In Wyoming and Colorado are the Clark Fork and Tiffany faunas, representing the upper Paleocene. Thus a sequence through the Paleocene can be established by the study of these faunas, and some idea as to the course of evolution during this phase of earth history may be obtained.

The Paleocene is poorly represented in other parts of the world. Mammals of late Paleocene age are found in the Cernay deposits of France and the Thanet sands of England, in the Gashato formation of Mongolia, and in the Rio Chico beds of Patagonia, and in the Itaborai deposits of Brazil.

When we reach the Eocene, the record is much more abundant. Numerous Eocene faunas are now known from North America and Europe, and in Asia and South America certain portions of the Eocene are well represented by fossil faunas. In western North America the sequence runs from the so-called Wasatch faunas of early Eocene age through the Bridger faunas of middle Eocene age to the Uinta and Duchesne River faunas of the late Eocene. The Green River formation of middle Eocene age contains an abundant and very important fish fauna. The parallel series in Europe ranges from the London Clay of England and several lower Eocene faunas found in France, through the middle Eocene calcaire grossier and other contemporaneous faunas of France, to the gypsum beds of Montmartre and the lower portion of the Quercy phosphorites, of late Eocene age. In Italy there is an interesting fish fauna of middle Eocene age, the Monte Bolca fauna.

Only the upper portion of the Eocene record is known from Asia, but here are several faunas of great importance. They are the Irdin Manha and Shara Murun faunas discovered and collected by the Central Asiatic Expeditions of the American Museum of Natural History, and the Pondaung fauna of Burma. In northern Africa are the upper Eocene Birket-el-Qurun beds and Qasr-el-Sagha beds of the Fayum district of Egypt that have yielded the most primitive known proboscideans and sirenians.

In South America the early stages in the history of the unique faunas of that continent are found in sediments of early and middle Eocene age, namely, the Casa Mayor beds below and the Musters beds above.

One of the most picturesque and best-known Oligocene series in the world is found in the famous White River badlands of South Dakota. Here the entire Oligocene sequence

CORRELATION OF CENOZOIC VERTEBRATE-BEARING SEDIMENTS

		South America	North America	Europe	Asia	Africa
Pleistocene		Pampean	Rancho la Brea, Hay Springs, Broadwater, Blanco, San Pedro	McKittrick, Tehama	Rexroad, Hagerman	Caves, Terraces, Perrier, Val d'Arno (Villa-franchian), Cromer	Caves, Terraces, Irrawaddy, Pinjor, Tatrot, Nihowan	Caves, Terraces, Olduvai, Omo, Vaal River
		Chapadmalal				
Pliocene	U	Monte Hermosa	Hemphill	Ash Hollow, Rattlesnake	Red Crag, Rousillon, Montpellier		Bon Hanifa
	M	Tunuya	Thousand Creek	Santa Fe	Alachua	Perpignan	Dhok Pathan	
	L	Huayqueria; Entre Rios, Catamarca	Clarendon	Ricardo; Burge, Valentine	Goliad	Concud, Vallés-Penedés, Mt. Leberon, Eppelsheim, Pikermi, Samos	Nagri; Honan, Shanshi; Maragha	Wadi Natrûn; Maghreb, Mallal
Miocene	U	La Ventana	Barstow	Pawnee Creek, Mascall	Sebastopol, Grive St. Alban	Chinji; Tung Gur	Ft. Ternan
	M	Chasico, Rio Frias	Heming-ford; Snake Creek	Sheep Creek, Marsland	Hawthorn	Vallé-Penedés, Sansan, Simorre, Grive St. Alban, Steinheim	Kamlial	Moghara
	L	Santa Cruz	Arikaree	Harrison, Monroe Creek, Gering	Oakville, Thomas Farm, John Day	Vallés-Penedés, St. Gerand-le-Puy, Orléans	Bugti; Loh	Namib, Rusinga, Lake Rudolph

Correlation chart of Paleocene, Eocene, and overlying mammal-bearing formations

Epoch		South America	North America	Western Europe	Asia	Africa
	U	Colhué Huapi	Brulé; Whitney	La Rochette, Mainz	Hsanda Gol, Kazakstan	Gebel Qatrani
	M		Orella	Quercy; Aveyron, Weinheim, Flonheim	Houldjin	Qasr-el-Sagha
	L	Deseado	Chadron; Pipestone Springs, Cypress Hills	Ronzon, Hampstead	Ardyn Obo, Ulan Gochu	Birk et-el-Qurun
Eocene	U	Divisadero Largo	Duchesne River; Sespe; Uinta; Washakie	Montmartre gypsum, Isle of Wight	Shara Murun, Ulan Shireh; Irdin Manha; Pondaung	
Eocene	M	Musters	Bridger; Green River	Calcaire grossier, Issel, Monte Bolca, Gieselthal		
Eocene	L	Casa Mayor	Wasatch; Huerfano; San José; Amalgre	Soissons, London Clay		
Paleocene	U	Rio Chico; Itaborai	Clark Fork; Tiffany; Fort Union; Silver Coulee, Sentinel Butte, Paskapoo, Polecat Bench	Thanet, Walbeck, Cernay	Gashato	
Paleocene	M		Torrejon; Dragon			
Paleocene	L		Puerco			

is represented by the Chadron formation or "titanothere beds" of early Oligocene age, and the Brulé formation or "oreodon beds" of middle and late Oligocene age. The lower portion of the Brulé, the Orella member, represents the middle Oligocene; the upper portion, the Whitney member, represents the upper Oligocene in the White River area.

In France the Quercy phosphorites continue from the upper Eocene into the Oligocene, the lower portion of which is especially well represented by the Ronzon fauna. In Asia the lower Oligocene is represented by the Ardyn Obo fauna, the middle Oligocene by the Houldjin fauna, and the upper Oligocene by the Hsanda Gol fauna, all of which are in Mongolia. The early Tertiary sediments of North Africa continue into the lower portion of the Oligocene period, where in the Fayûm the first long-jawed mastodons are found. South America at present has a record of lower and upper Oligocene continental sediments, with a middle Oligocene break—perhaps an imperfection of the geologic record. In Patagonia the lower Oligocene sediments are the Deseado beds, and the upper Oligocene sediments are the Colhué Huapi beds. In these beds are found fossils that show the progressive development of the South American mammals along their particular lines of evolutionary development, so different from what is seen in faunas on other continents.

The Miocene and Pliocene of North America are abundantly represented by numerous rich faunas, many of which have been discovered during the last two or three decades and are as yet not completely studied or described. They give a hint as to the abundance of life on this continent during middle and late Tertiary times, when mammals were beginning to assume, through the long processes of evolution, their modern form. The lower Miocene faunas worthy of special notice are those of the Gering, Monroe Creek, and Harrison formations, belonging to the Arikaree group, all found in western Nebraska and South Dakota, of the John Day formation of Oregon, and of

the Thomas Farm formation in Florida. Of middle Miocene age are the Hemingford, Sheep Creek, and Marsland faunas of Nebraska, and the Hawthorn formation of Florida; the upper Miocene is represented by the Pawnee Creek fauna of Colorado, the Mascall fauna of Oregon, and the Barstow fauna of California.

Europe is likewise rich in faunas of Miocene and Pliocene age. Here are found the St. Gerand-le-Puy fauna of lower Miocene relationships, the Sansan, Simorre, St. Gaudens, Steinheim, and Oeningen faunas of middle Miocene age, and the upper Miocene Sebastopol, and Grive St. Alban faunas. In Baluchistan are the lower Miocene Bugti beds, in which are found the fossils of the giant rhinoceros *Baluchitherium*. The Loh formation represents the lower Miocene in Mongolia, and the Tung Gur formation is the upper Miocene horizon in this part of Asia. In India are the Kamlial and Chinji beds, the initial sediments of the Siwalik series, which forms a most important and thick sequence of mammal-bearing sediments along the lower flanks of the Himalaya uplift.

One of the most famous mammalian faunas of the Miocene is the Santa Cruz assemblage of South America. This fauna, representing a high point in the development of the mammals in South America, can be regarded as of early Miocene age.

The history of mammalian evolution in North America is carried into the Pliocene by the Clarendon fauna of Texas, the Santa Fe faunas of New Mexico, and the Valentine and Burge faunas of Nebraska. Of middle to late Pliocene age are the Thousand Creek, Rattlesnake, Hemphill, and Ash Hollow faunas of various western states, whereas the Benson, Hagerman, and possibly the Blanco faunas bring the Pliocene to a close in western North America. (The Blanco and related faunas, regarded by many students as of late Pliocene affinities, may be more intimately related to Pleistocene history than to the close of the Pliocene sequence.) In Florida the middle Pliocene is represented by the fauna of the Alachua clays.

The lower Pliocene of Eurasia is the time of the Pontian fauna, when the horse *Hipparion*, having entered the Old World from North America, spread widely through the eastern part of the earth. Well-known Pontian faunas are those of Concud in Spain, Mt. Leberon in France, Eppelsheim in Germany, Pikermi in Greece, and the island of Samos in the eastern Mediterranean. The same faunal complex is continued in the Maragha beds of Persia and in the Nagri formation of the Siwalik series of India. The middle and upper Pliocene in France is continued by the Perpignan below and the Montpellier and Rousillon faunas above. In India the Nagri and the Dhok Pathan faunas, continuing the Siwalik sequence above the Chinji stage, represent part or perhaps all of the Pliocene above the Pontian.

The Pliocene is present in South America as a series of fossil-bearing formations known as the Mesopotamian beds of early Pliocene age, the Monte Hermosa beds of middle Pliocene age, and the Chapadmalal beds of late Pliocene age.

Finally we come to a consideration of the Pleistocene mammalian faunas. They usually can be identified as of Pleistocene age, but often it is not easy to determine to which portion of the Pleistocene epoch they should be assigned. The Pleistocene was a relatively short period of time, and for the most part the evolution of mammals was not very great during this phase of geologic history. Moreover, the deposition of Pleistocene sediments, frequently as river terraces or river gravels, makes their age determination difficult. Finally, some Pleistocene faunas are known from cave deposits, and many of these are almost impossible to date within the Pleistocene.

The Pleistocene epoch opened with the appearance of the Villafranchian fauna of Europe, a mammalian assemblage containing the horse *Equus* (an immigrant from North America), cattle of the *Bos* type, and mammoths of the *Elephas* type, these last two of Old World

origin. The spread of these diagnostic mammals may be accepted as indicating the advent of the Pleistocene in most parts of the world, except South America and Australia. Other mammals of North American origin invaded South America at the opening of the Pleistocene epoch; Australia, of course, remained isolated as it had been during Tertiary times.

A few of the particularly significant Pleistocene faunas might be mentioned at this place. In addition to the lower Pleistocene Villafranchian fauna, there are found in Europe the Perrier fauna of France, the Crag faunas of the North Sea coast of England, and various cave and river terrace faunas. In Asia the Siwalik series is continued by the Tatrot and Pinjor faunas, of early and middle Pleistocene age, and the upper Irrawaddy fauna of Burma is generally correlative with the Pinjor assemblage. Various cave faunas from western and southern China are of middle Pleistocene age. Africa is also characterized by cave faunas, some of which have in recent years yielded important fossils of primitive manlike primates. The Olduvai faunas of Africa are middle and upper Pleistocene assemblages that can be related to contemporaneous faunas in Asia. In North America are the lower Pleistocene faunas of Broadwater, Nebraska, perhaps the Blanco and Hagerman faunas of Texas and Idaho, the Rexroad fauna of Kansas and the San Pedro fauna of Arizona. Of later age are, the Hay Springs fauna of Nebraska, various cave and swamp deposits containing fossil mammals, and the famous tar pits of Rancho la Brea, California. The abundant Pampean fauna of South America is the last in the sequence of mammalian faunas on that continent, before the establishment of modern faunas.

THE CORRELATION OF FAUNAS

The problem of correlation can be difficult, especially from one continent to another. It is not always clear as to whether comparable faunas are exactly comparable in age, and many correlations must be accepted as tentative and approximate rather than as absolute. However, if these considerations are kept in mind, the various faunas of the Cenozoic can be correlated rather satisfactorily.

Europe was a series of islands during much of Cenozoic times, so that marine deposits containing the shells of invertebrates and continental deposits containing fossil mammals are frequently contiguous or even interfingered with one another. Therefore, it has been possible to build up a series of European age names, based on the marine succession, and to tie the continental sediments into this sequence.

Because of the unity of the Holarctic Region during the Cenozoic era, the North American faunas can be related upon the basis of many resemblances to the standard European succession. But since the North American sequence of mammalian faunas is unusually complete, American paleontologists have in recent years set up an age sequence for this continent, based upon the outstanding mammalian faunas. The Asiatic and African faunas are fitted into the Cenozoic sequence upon the basis of faunal resemblances, and there is good reason to believe that these faunas have been correlated with a fair degree of accuracy.

However, the problem of the South American faunas is separate and difficult, because these faunas evolved in isolation from the rest of the world and therefore cannot be directly related to the well-known faunas of North America and Eurasia. Correlations of the South American faunas consequently must be based upon varied lines of evidence, geologic as well as paleontologic, and at best many of them must be regarded as of a tentative nature.

The general aspect of the correlation of Cenozoic mammal faunas, as discussed in the foregoing pages, is set forth by the preceding correlation chart. Perhaps this will help to make clear the relationships of the numerous faunas on which our knowledge of the Age of Mammals has been erected.

A Classification of the Chordates

The classification presented here is intended to assist the reader in maintaining his or her orientation among the orders and lesser categories of the vertebrates. For this reason it is not a comprehensive classification, but is instead a synoptic outline in which the larger relationships are emphasized. With this purpose in mind, the classification is not carried down below the grade of families, and in many orders it is not carried down that far. It is felt that this unequal emphasis will provide the system that is of the greatest practical use to the reader of this book. In a further attempt to emphasize the larger relationships that are of particular importance in a book of this kind, genera have been included only sparingly in the classification. However, all genera mentioned in the text and shown in the illustrations are included in the index.

Several textbooks listed in the bibliography of this book provide comprehensive classifications. For vertebrates in general, the books by Romer (1966), Carroll (1988), and Benton (1988) were consulted in developing the classification below. For fishes, the text by Moy-Thomas and Miles (1971) and recommendations by Dr. David K. Elliott (Northern Arizona University, Flagstaff) also were used. Several sources were consulted for the classification of reptiles, including Romer (1956), Kemp (1982), and Pickering and Molnar (1984). The main reference for bird classification was Cracraft's (1988) recent synthesis, and for mammals Simpson's (1945) classification and the books by Halstead (1978), Vaughan (1986), and Savage and Long (1986) were major resources.

PHYLUM CHORDATA:

Chordates or animals with a notochord.

Subphylum Cephalochordata: *Amphioxus (Branchiostoma)*.
Subphylum Urochordata: sea squirts or tunicates.
Subphylum Vertebrata: backboned animals or vertebrates.

Class Agnatha: jawless vertebrates; lampreys, hagfishes, and ★ostracoderms.
  Subclass Pteraspidomorphi (Diplorhina): agnathans without a dorsal nostril.
      Order Heterostraci (Pteraspida): ★ostracoderms with head shield of large plates.
      Order Thelodontida (Coelolepida): ★unarmored ostracoderms, denticles covering body.
  Subclass Cephalaspidomorpha (Monorhina): agnathans with a dorsal nostril between the eyes.
      Order Cephalaspida (Osteostraci): ★armored ostracoderms with a flattened head shield.
      Order Galeaspida: ★a parallel branch of cephalaspidomorphs.
      Order Anaspida: ★small, armored, deep-bodied ostracoderms.
      Order Petromyzontida: lampreys.
  Subclass uncertain:
      Order Myxinida: hagfishes.

Class Placodermi: ★early jawed fishes, mostly heavily armored.
      Order Ptyctodontida: ★small, armored placoderms.
      Order Pseudopetalichthyida: ★placoderms with sharklike fin supports.
      Order Rhenanida: ★skatelike placoderms.
      Order Acanthothoraci: ★very primitive placoderms.
      Order Petalichthyida: ★armored forms related to arthrodires.
      Order Phyllolepida: ★flattened, heavily plated placoderms.

★Extinct.

Order Arthrodira: *armored forms with jointed necks.
Order Antiarchi: *small placoderms with small, movable pectoral appendages.

Class Chondrichthyes: the broad category of cartilaginous fishes, including sharks.

Subclass Elasmobranchii: sharks.

Order Cladoselachida: *ancestral sharks.

Order Eugeneodontida (Edestida): *Paleozoic sharks with symphyseal tooth whorls.

Order Ctenacanthida: *ancestors of modern sharks.

Superfamily Ctenacanthoidea: *basal ctenacanth sharks.

Superfamily Hybodontoidea: *dominant, early Mesozoic predaceous sharks.

Order Xenacanthida (Pleurocanthida): *early freshwater sharks.

Order Galeomorpha: dominant predaceous sharks of the present.

Superfamily Heterodontoidea: sharks with heavy crushing teeth for mollusc-feeding.

Superfamily Orectoloboidea: nurse sharks.

Superfamily Lamnoidea: modern sharks such as the Great White.

Superfamily Carcharhinoidea: tiger sharks, requiem sharks, bull sharks.

Superfamily Hexanchoidea: six-gilled sharks.

Order Squalomorpha: dogfishes and related sharks.

Order Batoidea: skates and rays.

Subclass Holocephali: sharklike fishes with crushing tooth plates.

Order Chimaerida: chimaeras or ratfishes. (plus *"bradyodonts" and several small orders)

Subclass uncertain

Order Iniopterygida: *bizarre chondrichthyans with winglike pectoral fins.

Order Petalodontida: *Carboniferous-Permian, skatelike chondrichthyans.

Class Acanthodii: *spiny fishes.

Order Climatiformes: *primitive acanthodians.

Order Ischnacanthiformes: *specialized forms with reduced spines.

Order Acanthodiformes: *the latest acanthodians.

Class Osteichthyes: bony fishes.

Subclass Actinopterygii: ray-finned bony fishes.

Infraclass Chondrostei: primitive ray-finned fishes.

Order Palaeonisciformes: *ancestral ray-finned fishes.

Suborder Palaeoniscoidei: *basal Devonian chondrosteans.

Order Redfieldiiformes: *Triassic freshwater chondrosteans.

Order Perleidiformes: *progressive Triassic chondrosteans.

Order Saurichthyiformes: *elongate, pikelike chondrosteans.

Order Polypteriformes: surviving chondrosteans living in Africa; bichir.

Order Acipenseriformes: sturgeons and paddlefishes and their extinct relatives. (plus several small orders)

Infraclass Neopterygii: intermediate and advanced ray-finned fishes.

(Holosteans: intermediate ray-finned fishes; not a natural group.)

Order Lepisosteiformes: modern garfishes and their relatives.

Order Semionotiformes: *early holosteans.

Order Pycnodontiformes: *deep-bodied holosteans.

*Extinct.

Order Macrosemiiformes: *a distinct group of holosteans.

Order Amiiformes: modern bowfin and its relatives.

Order Pachycormiformes: *isolated Mesozoic holosteans.

Order Aspidorhynchiformes: *heavily scaled, elongated holosteans.

Division Teleostei: advanced ray-finned fishes.

Order Pholidophoriformes: *fishes intermediate in structure between holosteans and teleosts.

Order Leptolepiiformes: *generalized teleosts.

Order Ichthyodectiformes: *large, Jurassic-Cretaceous predaceous teleosts.

Subdivision Osteoglossomorpha: primitive Cretaceous teleosts and their descendants.

Order Osteoglossiformes: primitive, tropical, freshwater fishes.

Subdivision Elopomorpha: primitive, varied teleosts.

Order Elopiformes: ancestors of the tarpons.

Order Anguilliformes: eels.

Order Notacanthiformes: certain, deep-sea fishes.

Subdivision Clupeomorpha: persistently primitive teleosts.

Order Ellimmichthyiformes: *Cretaceous-early Cenozoic clupeomorphs.

Order Clupeiformes: herrings and their relatives.

Subdivision Euteleostei: Cretaceous-Recent teleosts.

Order Salmoniformes: salmon and trout.

Superorder Ostariophysi: a majority of modern freshwater fishes.

Order Gonorhynchiformes: milkfishes and their relatives.

Order Characiformes: tetras and piranhas.

Order Cypriniformes: characins, minnows, carps, suckers, loaches.

Order Siluriformes: catfishes.

Superorder Stenopterygii: specialized, deep-sea teleosts.

Order Stomiiformes: deep-sea hatchetfishes and their relatives.

Superorder Scopelomorpha: deep-sea fishes.

Order Aulopiformes: varied deep-sea teleosts.

Order Myctophiformes: small, deep-sea fishes.

Superorder Paracanthopterygii: advanced teleosts, paralleling the acanthopterygians.

Order Percopsiformes: pirate perch and freshwater relatives.

Order Batrachoidiformes: toadfishes.

Order Gobiesociformes: clingfishes.

Order Lophiiformes: anglerfishes.

Order Gadiformes: cod and haddock.

Order Ophidiiformes: brotulas, cusk-eels, pearlfishes.

Superorder Acanthopterygii: spiny teleosts, a majority of modern forms.

Order Atheriniformes: flying fishes.

Order Cyprinodontiformes: killifishes.

Order Beryciformes: primitive acanthopterygians, squirrel fishes.

Order Zeiformes: John Dory and other tropical teleosts.

Order Lampriformes: moon fishes.

Order Gasterosteiformes: sticklebacks, sea horses.

Order Scorpaeniformes: sculpins, sea robins.

Order Perciformes: the majority of spiny teleosts.

Order Pleuronectiformes: flatfishes.

Order Tetraodontiformes: plectognath fishes.

(plus several small orders and superorders)

*Extinct.

430

Subclass Sarcopterygii: lobe-finned bony fishes
    Order Crossopterygii: progressive, air-breathing fishes.
        Suborder Rhipidistia: *ancestors of the tetrapods.
           Infraorder Osteolepiformes: *the central line of rhipidistians.
           Infraorder Porolepiformes (Holoptychiformes): *a side branch.
        Suborder Onychodontiformes: *Devonian crossopterygians lacking internal nostrils.
        Suborder Coelacanthiformes (Actinistia): predominantly marine crossopterygians, including the surviving *Latimeria*.
    Order: Dipnoi: lungfishes.

Class Amphibia: amphibians
Subclass Labyrinthodontia: *labyrinthodonts.
    Order Ichthyostegalia: **Ichthyostega* and its relatives.
    Order Loxommatida: *labyrinthodonts with key-hole shaped eye sockets.
    Order Temnospondyli: *labyrinthodonts in which the vertebral intercentrum is elaborated.
        Superfamily Colosteoidea: *primitive temnospondyls with elongate bodies and small limbs.
        Superfamily Trimerorhachoidea: *small to medium sized forms.
        Superfamily Edopoidea: *temnospondyls that trended to terrestrial habits.
        Superfamily Eryopoidea: *the most terrestrial temnospondyls.
        Superfamily Rhinesuchoidea: *aquatic and semiaquatic forms of Permian age.
        Superfamily Capitosauroidea: *fully aquatic temnospondyls with flattened bodies.
        Superfamily Rhytidosteoidea: *broad-headed Triassic temnospondyls.
        Superfamily Trematosauroidea: *brackish water or marine forms, some with long snouts.
        Superfamily Brachyopoidea: *short-faced temnospondyls with traplike jaws.
        Superfamily Metoposauroidea: *large, big-headed temnospondyls of the Late Triassic.
        Superfamily Plagiosauroidea: *temnospondyls with extremely short and wide skulls.
    Order Anthracosauria: *labyrinthodonts in which the vertebral pleurocentrum is elaborated.
        Suborder Embolomeri: *very long-bodied aquatic anthracosaurs
        Suborder Gephyrostegida: *terrestrial anthracosaurs with relatively large limbs.
        Suborder Seymouriamorpha: *the most terrestrial of anthracosaurs.
Subclass Lepospondyli: *small to medium sized amphibians with spool-shaped vertebral centra.
    Order Aistopoda: *limbless snakelike lepospondyls.
    Order Nectridea: *aquatic newt-like lepospondyls.
    Order Microsauria: *the most varied lepospondyls; aquatic to lizardlike in habits.
    Order Lysoropha: *long-bodied lepospondyls.
    Order Adelogyrinida: *an isolated group of lepospondyls.
Subclass Lissamphibia: modern amphibians.

*Extinct.

Order Gymnophiona (Apoda): modern legless amphibians or caecilians.
Order Caudata (Urodela): newts and salamanders.
Order Proanura: *protofrogs.
Order Anura: frogs and toads.

Class Amphibia or Reptilia

Family Diadectidae: *Diadectes* and related genera.

Class Reptilia: reptiles; scaled or armored tetrapods reproducing by an amniote egg.
  Subclass Anapsida: reptiles with solid skull roof; no temporal openings.
    Order Captorhinida: *the earliest group of reptiles.
      Suborder Captorhinomorpha: *generally small, carnivorous captorhinidans.
      Suborder Procolophonia: *small, specialized captorhinidans.
      Suborder Pareiasauria: *large captorhinidans of Permian age.
      Suborder Millerosauria: *lizardlike successors of captorhinomorphs.
    Order Mesosauria: *ancient aquatic reptiles.
    Order Chelonia: turtles and tortoises.
      Suborder Proganochelyida: *ancestral turtles.
      Suborder Pleurodira: side-neck turtles.
      Suborder Cryptodira: vertical-neck turtles.
  Subclass Diapsida: the majority of reptiles; primitively with both upper and lower temporal openings, separated by the postorbital-squamosal bones, on each side
      Order Araeoscelida: *ancestral diapsids of Carboniferous-Permian age.
      Order Choristodera: *champsosaurs; Cretaceous-Eocene aquatic diapsids.
    Infraclass Lepidosauromorpha: lizards, snakes, and their relatives.
      Order Eosuchia (Younginiformes): *ancestral lepidosauromorphs.
    Superorder Lepidosauria: lizards, snakes, and sphenodontans.
      Order Spheodontida: the modern tuatara and its ancestors.
      Order Squamata: lizards and snakes.
        Suborder Lacertilia: the broad group of lizards.
        Suborder Serpentes: snakes.
    Infraclass Archosauromorpha: advanced diapsids trending to adaptations for rapid and efficient locomotion.
      Order Rhynchosauria: *small to medium sized, terrestrial archosauromorphs with jaws forming a beak.
      Order Thalattosauria: *early marine archosauromorphs.
      Order Trilophosauria: *Triassic archosauromorphs of lizard-like habits.
      Order Protorosauria: *primitive land-living archosauromorphs.
    Superorder Archosauria: advanced archosauromorphs; the ruling reptiles.
      Order Thecodontia: *ancestral archosaurians.
        Suborder Proterosuchia: *early primitive thecodonts.
        Suborder Rauisuchia: *large, predaceous thecodonts.
        Suborder Ornithosuchia: *small to medium sized, theropodlike thecodonts.
        Suborder Aetosauria: *heavily, armored thecodonts.
        Suborder Phytosauria (Parasuchia): *aquatic, crocodilelike thecodonts.
      Order Crocodylia: crocodilians.
        Suborder Protosuchia: *ancestral crocodilians.
        Suborder Mesosuchia: *Mesozoic crocodilians.
        Suborder Eusuchia: modern crocodilians; gavials, crocodiles, alligators.
      Order Pterosauria: *flying reptiles.

*Extinct.

Suborder Rhamphorhynchoidea: *primitive pterosaurs.
Suborder Pterodactyloidea: *advanced pterosaurs or pterodactyls.
Superorder Dinosauria: *dinosaurs.
Order Saurischia: *lizard-hipped or saurischian dinosaurs.
Suborder Staurikosauria: *ancestral saurischian dinosaurs.
Suborder Theropoda: *carnivorous dinosaurs.
Infraorder Coelurosauria: *small to medium sized theropods.
Infraorder Deinonychosauria: *theropods with specialized feet.
Infraorder Carnosauria: *large and gigantic theropods.
Suborder Sauropodomorpha: *herbivorous saurischian dinosaurs.
Infraorder Prosauropoda: *Triassic-Jurassic, medium to large sized sauropdomorphs.
Infraorder Sauropoda: *brontosaurs; large to gigantic sauropodomorphs.
Order Ornithischia: *bird-hipped or ornithischian dinosaurs.
Suborder Ornithopoda: *duck-billed dinosaurs and their relatives.
Family Fabrosauridae: *small, ancestral ornithopods.
Family Heterodontosauridae: *early primitive ornithopods with differentiated teeth.
Family Hypsilophodontidae: conservative ornithopods of fabrosaurid ancestry.
Family Iguanodontidae: *large Jurassic and Cretaceous ornithopods.
Family Hadrosauridae: *Cretaceous duck-billed dinosaurs.

Suborder Pachycephalosauria: *dome-headed dinosaurs.
Suborder uncertain
Family Scelidosauridae: *primitive armored/plated dinosaurs.
Suborder Stegosauria: *plated dinosaurs.
Suborder Ankylosauria: *armored dinosaurs.
Suborder Ceratopsia: *horned dinosaurs.
Infraclass Euryapsida: *marine diapsids which have lost the lower temporal opening.
Superorder Sauropterygia: *nothosaurs and plesiosaurs.
Order Nothosauria: *small to medium sized, primitive sauropterygians.
Order Plesiosauria: *large advanced sauropterygians.
Superfamily Plesiosauroidea: *long-necked plesiosaurs.
Superfamily Pliosauroidea: *short-necked plesiosaurs.
Superorder Placodontia: *mollusk-eating euryapsids.
Superorder Ichthyosauria: *ocean-living reptiles of fishlike form.
Subclass Synapsida: *mammallike reptiles, with only a lower temporal opening on each side of the skull.
Order Pelycosauria: *early synapsids.
Suborder Ophiacodontia: *primitive pelycosaurs.
Suborder Sphenacodontia: *carnivorous pelycosaurs.
Suborder Edaphosauria: *herbivorous pelycosaurs.
Order Therapsida: *advanced synapsids.
Suborder Eotitanosuchia: *primitive therapsids.
Suborder Dinocephalia: *large, massive herbivores.
Suborder Dicynodontia: *beaked therapsids, often with tusks.
Suborder Theriodontia: *advanced carnivorous therapsids.
Infraorder Gorgonopsia: *primitive theriodonts.
Infraorder Therocephalia: *early advanced theriodonts.
Infraorder Bauriamorpha: *specialized theriodonts.

*Extinct.

Infraorder Cynodontia: *late advanced theriodonts.
Infraorder Tritylodontia: *highly adapted theriodonts.
Infraorder Ictidosauria: *theriodonts close to mammals.

Class Aves: birds

Subclass Archaeornithes: *primitive Jurassic toothed birds.
Order Archaeopterygiformes: *Archaeopteryx.
Subclass Ornithurae: advanced Cretaceous and Cenozoic birds.
Infraclass Hesperornithae: *Cretaceous flightless, loonlike toothed birds.
Order Hesperornithiformes: *Hesperornis and its relatives.
Infraclass Carinatae: birds with strong adaptations for flight such as a keeled sternum.
Division Ichthyornithes: *Cretaceous toothed flying birds.
Order Ichthyornithiformes: *Ichthyornis and its relatives.
Division Neornithes: advanced toothless birds.
Superorder Palaeognathae: birds with a primitive type of palate.
Order Tinamiformes: tinamous.
Order Struthioniformes: ostriches.
Order Rheiformes: rheas.
Order Casuariiformes: cassowaries, emus.
Order Aepyornithiformes: *elephant birds.
Order Dinornithiformes: *moas.
Order Apterygiformes: kiwis.
Superorder Neognathae: birds with an advanced type of palate.
Order Galliformes: grouse, quail, turkeys, pheasants.
Order Anseriformes: ducks, geese, swans.
Order Podicipediformes: grebes.
Order Diatrymiformes: *Diatryma and related genera.
Order Gaviiformes: loons.
Order Sphenisciformes: penguins.
Order Pelecaniformes: pelicans, frigate birds.
Order Procellariiformes: albatrosses, petrels.
Order Gruiformes: cranes, rails, limpkins, *Phorusrhacos
Order Charadriiformes: shore birds, gulls, auks.
Order Columbiformes: pigeons, doves, the dodo.
Order Psittaciformes: lories, parrots, macaws.
Order Coliiformes: colies.
Order Ciconiiformes: herons, storks.
Order Cuculiformes: cuckoos, roadrunners.
Order Falconiformes: vultures, hawks, falcons, eagles.
Order Strigiformes: owls.
Order Caprimulgiformes: goatsuckers.
Order Apodiformes: swifts, hummingbirds.
Order Coraciiformes: kingfishes, rollers, hoopoes, hornbills.
Order Piciformes: barbets, toucans, woodpeckers.
Order Passeriformes: perching birds; flycatchers, ovenbirds, lyrebirds, songbirds.

Class Mammalia: mammals; tetrapods with hair and mammary glands with which they suckle their young, and with a dentary-squamosal jaw joint.

Subclass Eotheria: *very primitive Triassic and Jurassic mammals.
Order Docodontia: *docodonts.

*Extinct.

Order Triconodontia: *triconodonts.
Subclass Prototheria: egg-laying mammals.
   Order Monotremata: the modern platypus and echidna.
Subclass Allotheria: *a long line of early mammals.
   Order Multituberculata: *early rodentlike mammals with multicuspid teeth.
     Suborder Plagiaulacoidea: *primitive multituberculates.
     Suborder Taeniolabidoidea: *large, persisting multituberculates.
     Suborder Ptilodontoidea: *small forms with shearing teeth.
Subclass Theria: the majority of mammals; therians.
  Infraclass Pantotheria: *the first therians.
   Order Eupantotheria: *ancestors of marsupials and placentals.
   Order Symmetrodonta: *symmetrodonts.
  Infraclass Metatheria: pouched mammals.
   Order Marsupialia: marsupials.
    Suborder Didelphoidea: opossums.
    Suborder Caenolestoidea: "mouse" opossums.
    Suborder Dasyuroidea: marsupial "rats", carnivores, *thylacine.
    Suborder Parameloidea: bandicoots.
    Suborder Diprotodonta: kangaroos, wallabies, phalangers, koalas,
      wombats.
  Infraclass Eutheria: placental mammals.
   Order Insectivora: insectivores, the most primitive placentals.
    Suborder Proteutheria: *primitive insectivores.
    Suborder Soricomorpha: shrews.
    Suborder Erinaceomorpha: hedge hogs.
   Order Macroscelida: elephant shrews.
   Order Scandentia: tree shrews.
   Order Chiroptera: bats.
    Suborder Megachiroptera: fruit-eating bats.
    Suborder Microchiroptera: insect and fish eating bats.
   Order Dermoptera: flying "lemurs" and their relatives.
   Order Taeniodonta: *taeniodonts.
   Order Tillodontia: *tillodonts.
   Order Edentata: edentates.
    Suborder Cingulata: armadillos and *glyptodonts.
    Suborder Pilosa: ant eaters, tree sloths, *ground sloths.
   Order Pholidota: pangolins.
   Order Primates: primates; lemurs, tarsiers, monkeys, apes, and humans.
    Suborder Plesiadapiformes: *ancestral primates.
    Suborder Strepsirhini: lemurs.
    Suborder Haplorhini: tarsiers.
    Suborder Platyrhini: New World monkeys.
    Suborder Catarrhini: Old World monkeys, apes, and humans.
   Order Rodentia: rodents.
    Suborder Sciurognathi: sciurognaths.
     Infraorder Protrogomorpha: *primitive rodents.
     Infraorder Sciuromorpha: squirrels, gophers.
     Infraorder Castorimorpha: beavers and their relatives.
     Infraorder Myomorpha: mice, rats, and their relatives.
     Infraorder Theridomorpha: *early Cenozoic European rodents.
    Suborder Hystricognathi: hystricognaths.

*Extinct.

Infraorder Hystricomorpha: Old World porcupines.
Infraorder Phiomorpha: cane rats, gundis, "mole rats".
Infraorder Caviomorpha: South American rodents.
Order Lagomorpha: rabbits and hares.
Order Acreodi: *large, terrestrial predators, ancestral to cetaceans.
Order Cetacea: whales, porpoises, dolphins.
  Suborder Archaeoceti: *ancestral whales.
  Suborder Odontoceti: porpoises, dolphins, toothed whales.
  Suborder Mysticeti: whalebone (baleen) whales.
Order Creodonta: *ancient carnivorous placental mammals.
  Suborder Oxyaenoidea: *early creodonts.
  Suborder Hyaenodontia: *varied and persisting creodonts.
Order Carnivora: modern carnivorous placental mammals.
  Suborder Fissipedia: land-living carnivores.
    Superfamily Miacoidea: *ancestral fissipeds.
      Family Miacidae: *miacids.
    Superfamily Canoidea: dogs, bears, pandas, raccoons, mustelids.
      Family Canidae: dogs, wolves, foxes.
      Family Ursidae: bears.
      Family Ailuridae: pandas.
      Family Procyonidae: raccoons, coatis, ring-tails.
      Family Mustelidae: weasels.
    Superfamily Feloidea: cats, hyaenas.
      Family Viverridae: Old World civets.
      Family Hyaenidae: hyaenas.
      Family Felidae: cats.
  Suborder Pinnipedia: marine carnivores.
      Family Phocidae: seals.
      Family Otariidae: sea lions.
      Family Odobenidae: walruses.
Order Condylarthra: *ancestral hoofed mammals or ungulates.
Order Tubulidentata: aardvarks.
Order Pantodonta: *large early Tertiary ungulates.
Order Dinocerata: *gigantic horned ungulates.
Order Notoungulata: *the most varied of the South American ungulates.
  Suborder Notioprogonia: *primitive notoungulates.
  Suborder Toxodonta: *large specialized ungulates.
  Suborder Typotheria: *small, rabbitlike notoungulates.
  Suborder Hegetotheria: *small notoungulates.
Order Litopterna: *camellike and horselike South American ungulates.
Order Astrapotheria: *large, possibly amphibious South American ungulates.
Order Trigonostylopia: *early South American ungulates.
Order Pyrotheria: *very large South Americann ungulates.
Order Perissodactyla: odd-toed hoofed mammals.
  Suborder Hippomorpha: horses, *brontotheres or titanotheres.
    Superfamily Equoidea: horses, zebras, asses.
    Superfamily Brontotheroidea: *brontotheres or titanotheres.
  Suborder Ancylopoda: *chalicotheres or clawed perissodactlys.
  Suborder Ceratomorpha: rhinoceroses, tapirs.
    Superfamily Tapiroidea: tapirs.

*Extinct.

Superfamily Rhinoceratoidea: rhinoceroses.
Order Artiodactyla: even-toed hoofed mammals.
  Suborder Paleodonta: *early artiodactyls.
    Superfamily Dichobunoidea: *ancestral artiodactyls.
    Superfamily Entelodontoidea: *entelodonts.
  Suborder Suina: piglike artiodactyls.
    Superfamily Suoidea: pigs, peccaries.
    Superfamily Anthracotherioidea: *anthracotheres.
    Superfamily Hippopotamoidea: hippopotamuses.
    Superfamily Anoplotherioidea: *anoplotheres or three-toed artiodactyls.
  Suborder Ancodonta: *ancodonts.
    Superfamily Cainotherioidea: *cainotheres.
    Superfamily Merycoidodontoidea: *oreodonts.
  Suborder Tylopoda: camels and llamas.
  Suborder Ruminantia: advanced artiodactyls.
    Superfamily Traguloidea: tragulids and their relatives.
    Superfamily Cervoidea: deer.
    Superfamily Giraffoidea: okapis, giraffes.
    Superfamily Bovoidea: antelopes, cattle.
Order Proboscidea: elephants and their relatives.
  Suborder Moeritherioidea: *ancestral proboscideans.
  Suborder Deinotherioidea: * dinotheres.
  Suborder Euelephantoidea: elephants and *mastodonts, *mammoths, *stegodonts.
    Superfamily Gomphotherioidea: *long-jawed mastodonts and their descendants.
    Superfamily Mastodontoidea: *crested-toothed mastodonts.
    Superfamily Stegodontoidea: *stegodonts.
    Superfamily Elephantoidea: elephants, *mammoths.
  Suborder Barytherioidea: *barytheres.
Order Sirenia: sea cows, dugongs.
Order Desmostylia: *desmostylians; large marine waders.
Order Hyracoidea: hyraxes, conies.
Order Embrithopoda: *Arsinoitherium, a gigantic rhinoceroslike mammal.

*Extinct.

References

This bibliography is made up of a selected list of references that have a bearing on the subject of vertebrate evolution. The literature on this subject is vast, so that no attempt is made here to present a comprehensive list of publications. The bibliographies of vertebrate paleontology, included at the end of this reference list, cite the published studies on fossil vertebrates since the beginning of the science. The bibliographies in Alfred S. Romer's *Vertebrate Paleontology* and Robert L. Carroll's *Vertebrate Paleontology and Evolution* are excellent selections of significant works on the various groups of fossil vertebrates. The works listed below are those readily available, for the most part, to readers who do not have access to highly specialized libraries of natural history. Some books may be available in more recent editions than here listed.

EVOLUTION

Since much is being written on this subject at the present time, only a few references of particular importance to the reader interested in the evolution of the vertebrates are listed here. The books by Carter, Grant, Simpson (1959) and Stanley, are excellent discussions of the subject in its general aspects. The important books by Huxley, Mayr, and Simpson (1953), are more extended than the general textbooks.

Carter, G.S. 1957. *A Hundred Years of Evolution*. New York, The Macmillan Co. x + 206 pp.
——. 1967. *Structure and Habit in Vertebrate Evolution*. Seattle, University of Washington Press. xiv + 520 pp.
Futuyma, Douglas J. 1986. *Evolutionary Biology*, second edition. Sunderland, Sinauer Associates, Inc. 600 pp.
Grant, Verne. 1985. *The Evolutionary Process*. New York, Columbia University Press. xii + 499 pp.
Gregory, William K. 1951. *Evolution Emerging: A Survey of Changing Patterns from Primeval Life to Man*. New York, The Macmillan Co. Vol. 1, xxvi + 736 pp. (text); Vol. 2, viii + 1013 pp. (figures and plates).
Hotton, Nicholas, III. 1968. *The Evidence of Evolution*. New York, American Heritage Publishing Co., Inc. 160 pp.
Huxley, Julian S. 1942. *Evolution, the Modern Synthesis*. London, George Allen and Unwin, Ltd. 645 pp.
Mayr, E. 1963. *Animal Species and Evolution*. Cambridge, Harvard University Press. vii + 797 pp.
Moore, Ruth (and the editors of Life), 1962. *Evolution*. Life Nature Library. New York, Time Incorporated. 192 pp.
Pollard, J.W. 1984. *Evolutionary Theory*. New York, John Wiley & Sons. xxii + 271 pp.
Simpson, George Gaylord, 1953. *The Major Features of Evolution*. New York, Columbia University Press, 434 pp.
——, 1959. *The Meaning of Evolution*. New York, The New American Library, Mentor Book, 192 pp.
——, 1983. *Fossils and the History of Life*. New York, Scientific American Books, Inc. 239 pp.
Sober, Elliott, 1984. *Conceptual Issues in Evolutionary Biology*. Cambridge, Massachusetts Institute of Technology Press, xiv + 725 pp.
Stanley, Steven M., 1979. *Macroevolution*. San Francisco, W.H. Freeman and Company. xi + 332 pp.

VERTEBRATE ZOOLOGY

McNeill, Hildebrand, Wake (Hyman), and Young are all standard textbooks of vertebrate zoology. The book by Romer and Parsons, is of particular importance because it is more than the usual type of comparative anatomy text; it includes extensive consideration of the anatomy in fossil vertebrates. The book by Goodrich also gives considerable attention to the fossils as they bear on vertebrate anatomy. Young's book is an excellent text that deals with many aspects of the structure, physiology, and evolution of vertebrates, fossil and recent. The books by Simpson and Beck, and Simpson, Stebbins and Nybakken, are outstanding texts on biology and its many aspects. The two books (Blue, and Green versions) written by a large group of carefully selected authorities for the Biological Sciences Curriculum Study, under the auspices of the National Science Foundation, deserve special mention. Each volume is in itself complete.

Alexander, R. McNeill. 1981. *The Chordates*, second edition. London, Cambridge University Press. vi + 480 pp.

American Institute of Biological Sciences: Biological Sciences Curriculum Study.

———. 1987. *Biological Sciences: An Ecological Approach*, sixth edition (BSCS Green Version). Dubuque, Iowa, Kendall/Hunt Publishing Co.

———. 1990. *Biological Science: A Molecular Approach*, sixth edition (BSCS Blue Version). Lexington, Mass., D.C. Heath and Co.

Bone, Q. 1979. *The Origin of Chordates*. Burlington, Carolina Biological Supply Company. 16 pp.

Goodrich, Edwin S. 1958. *Studies on the Structure and Development of Vertebrates*, 2 vols. New York, Dover Publications, Inc. Vol. 1, pp. i-1xix + 1-485; Vol. 2, pp. 486-837.

Grassé, Pierre-P. (and collaborators). 1954. *Traité de Zoologie*. Tome XII (Comparative Anatomy). Paris, Masson et Cie. 1145 pp.

Gregory, William K. 1929 (1965). *Our Face from Fish to Man*. New York, Capricorn Books Edition. xl + 295 pp.

Hildebrand, Milton. 1988. *Analysis of Vertebrate Structure*, second edition. New York, John Wiley & Sons. xv + 710 pp.

Hildebrand, Milton, Bramble, Dennis M., Liem, Karel F., and Wake, David B. (editors) 1985. *Functional Vertebrate Morphology*. Cambridge, Harvard University Press. 430 pp.

McFarland, William N. 1985. *Vertebrate Life*, second edition. New York, Macmillan Publishing Co., Inc.

Radinsky, Leonard B. 1987. *The Evolution of Vertebrate Design*. Chicago, University of Chicago Press. 188 pp.

Rogers, Elizabeth. 1986. *Looking at Vertebrates*. New York, Longman Inc. xi + 195 pp.

Romer, Alfred S. and Parsons, Thomas S. 1986. *The Vertebrate Body*, sixth edition. Orlando, Florida, Saunders College Publishing. 688 pp.

Scientific American (readings). 1974. *Vertebrate Structures and Functions*. San Francisco, W.H. Freeman and Company. 440 pp.

Simpson, George Gaylord, and W.S. Beck. 1965. *Life: An Introduction to Biology*. Second edition. New York, Harcourt, Brace and World, Inc. xviii + 869 pp.

———, ———, Stebbins, Robert C. and Nybakken, James W. 1972. *General Zoology*, fifth edition. New York, McGraw-Hill Book Co. ix + 897 pp.

Wake, Marvalee et al. 1979. *Hyman's Comparative Vertebrate Anatomy*, third edition. Chicago, University of Chicago Press.

Young, J.Z. 1950. *The Life of the Vertebrates*. London, Oxford University Press, 767 pp.

GENERAL WORKS ON VERTEBRATE PALEONTOLOGY

Vertebrate Paleontology, by Romer and *Vertebrate Paleontology and Evolution,* by Carrol, are the outstanding books of this subject at the present time. They give full and well-balanced treatments that contain virtually all the pertinent information necessary in a general account of the fossil vertebrates. *The Vertebrate Story* by Romer is a much more popular account. The volumes edited by Orlov and by Piveteau are massive compendia of the subject. The books by Olson and by Stahl are particularly valuable in that they present detailed discussions dealing with problems of vertebrate evolution.

Benton, Michael J. (editor). 1988. *The Phylogeny and Classification of the Tetrapods.* (2 vols). Oxford, Clarendon Press. x + 377 pp.

de Beaumont, Gerard. 1973. *Guide des Vertébrés Fossiles.* Neuchatel (Switzerland), 479 pp.

Carroll, Robert L. 1988. *Vertebrate Paleontology and Evolution.* New York, W.H. Freeman and Company. xiv + 698 pp.

Halstead, L.B. 1968. *The Pattern of Vertebrate Evolution.* San Francisco, W.H. Freeman and Company. xii + 209 pp.

Jacobs, Louis L. (editor). 1980. *Aspects of Vertebrate History.* Flagstaff, Museum of Northern Arizona Press. xv + 407 pp.

Jarvik, Erik. 1980. *Basic Structure and Evolution of Vertebrates* (vols. 1 and 2). London, Academic Press. xvi + 575 pp. and xiii + 337 pp.

Jefferies, R.P.S. 1986. *The Ancestry of the Vertebrates.* London, Cambridge University Press. viii + 376 pp.

Joysey, K.A. and Kemp T.S. (editors). 1972. *Studies in Vertebrate Evolution.* New York, Winchester Press. 284 pp.

Kuhn-Schnyder, Emil and Rieber, Hans. 1984. *Paläozoologie.* New York, Georg Thieme Verlag Stuttgart. x + 390 pp.

Olson, Everett C. 1971. *Vertebrate Paleozoology.* New York, John Wiley and Sons, Inc. xv + 839 pp.

Orlov, J.A. (editor). 1959-1964. *Fundamentals of Paleontology* (in Russian). Three volumes are devoted to the vertebrates. Moscow.

Patterson, Colin and Greenwood, P.H. (editors) 1967. *Fossil Vertebrates.* London, Academic Press Inc. v + 260 pp.

Peyer, Bernhard. 1950. *Geschichte der Tierwelt.* Zürich, Büchergilde Gutenburg. 228 pp.

Piveteau, Jean (editor). 1952-1966. *Traité de Paléontologie.* 7 vols., Vols. 4-7 (Vertebrates). Paris, Masson et Cie.

Rich, P.V., ven Tets, G.F., and Knight, F. 1985. *Kadimakara, Extinct Vertebrates of Australia.* Victoria, Pioneer Design Studio Pty. Ltd. 284 pp.

Romer, Alfred S. 1966. *Vertebrate Paleontology*, third edition. Chicago, University of Chicago Press. ix + 468 pp.
——. 1968. *Notes and Comments on Vertebrate Paleontology*. Chicago, The University of Chicago Press. viii + 304 pp.
——. 1971. *The Vertebrate Story*. Chicago, The University of Chicago Press. 437 pp.
Spinar, Zdeněk and Burian, Zdeněk. 1984. *Paleontologie Obratlovcu*. Praha (Prague), Academia. 859 pp.
Stahl, Barbara J. 1974. *Vertebrate history: Problems in Evolution*. New York, McGraw-Hill Book Co. ix + 594 pp.

FISHES

The book by Moy-Thomas and Miles is an excellent treatment of the early fishes. The *Traité de Paléontologie* volumes, edited by Piveteau, provide a comprehensive survey of fossil fishes.

Alexander, R. McNeill. 1974. *Functional Design of Fishes*. London, Hutchinson University Library. 160 pp.
Bemis, William E., Burggren, Warren W., and Kemp, Norman E. (editors). 1987. *The Biology and Evolution of Lungfishes*. New York, Alan R. Liss, Inc. viii + 383 pp.
Budker, Paul. 1971. *The Life of Sharks*. New York, Columbia University Press. xviii + 222 pp.
Gosline, William A. 1971. *Functional Morphology and Classification of Teleostean Fishes*. Honolulu, The University Press of Hawaii. ix + 208 pp.
Lagler, Karl F., Bardach, John E., and Miller, Robert R. 1977. *Ichthyology*, second edition. New York, John Wiley & Sons, Inc. 506 pp.
Migdalski, Edward C. and Fichter, George S. 1976. *The Fresh and Salt Water Fishes of the World*. New York, Alfred A. Knopf. 316 pp.
Moy-Thomas J.A. and Miles, R.S. 1971. *Paleozoic Fishes*. New York, Chemical Publishing Co. 149 pp.
Obruchev, D.V. (editor). 1967. *Fundamentals of Paleontology*. (Orlov, Y.A. chief editor.) Vol. XI. Agnatha, Pisces. Jerusalem, S. Monson, x + 825 pp. (English translation of the 1964 Russian publication).
Ommanney, F.D. (and the editors of Life). 1963. *The Fishes*. Life Nature Library. New York, Time Incorporated. 192 pp.
Piveteau, Jean (editor). 1964. *Traité de Paléontologie*. Tome IV, Vol. 1, *Vertebres (generalities)*. *Agnathes*. Paris, Masson et Cie. 387 pp.
——. 1966. Ibid. Tome IV, Vol. 3, *Actinopterygiens*. *Crossopterygiens*. *Dipneustes*. Paris, Masson et Cie. 442 pp.

AMPHIBIANS AND REPTILES

The book by Romer on the osteology of the reptiles is a detailed work that gives particular attention to the fossil forms. It contains, in addition, a very complete classification of the reptiles. Volume V of the Traité de Paléontologie is a modern, comprehensive survey of fossil amphibians, reptiles, and birds. This is also true of the Russian volume, edited by Rozhdestvensky and Tatarinov, in the series edited by Orlov. In recent years there has been a spate of books about dinosaurs. The book by Desmond presents one side of a vigorously argued thesis. The work edited by Thomas and Olson contains the varying views of several authors on this subject.

Bakker, Robert T. 1986. *The Dinosaur Heresies*. New York, William Morrow and Co. 481 pp.
Bellairs, A. d'A. 1969. *The Life of Reptiles*. London, Weidenfeld and Nicolson. Vol. 1, pp. i-xi, 1-282. Vol. 2, pp. 283–590.
Carr, Archie (and the editors of Life). 1963. *The Reptiles*. Life Nature Library. New York, Time Incorporated. 192 pp.
Charig, 1979. *A New Look at the Dinosaurs*. London, British Museum of Natural History. 160 pp.
Colbert, Edwin H. 1961. *Dinosaurs—Their Discovery and Their World*. New York, E.P. Dutton and Co., Inc. 300 pp.
——. 1966. *The Age of Reptiles*. New York, W.W. Norton and Co., Inc. The Norton Library, N374. 228 pp.
——. 1983. *Dinosaurs, An Illustrated History*. Maplewood, New Jersey, Hammond Inc., 224 pp.
——. 1984. *The Great Dinosaur Hunters and Their Discoveries*. New York, Dover Publications, Inc. xii + 283 pp.
Czerkas, Sylvia J. and Olson, Everett C. (editors) 1987. *Dinosaurs Past and Present* (vols. 1 and 2). Seattle, University of Washington Press. xvi + 161 pp. and xiii + 149 pp.
Desmond, Adrian J. 1976. *The Hot-Blooded Dinosaurs. A Revolution in Paleontology*. New York, The Dial Press/James Wade. 238 pp.
Dong, Zhiming. 1988. *Dinosaurs from China*. Beijing, China Ocean Press. 114 pp.
Duellmman, William E. and Trueb, Linda. 1986. *Biology of Amphibians*. New York, McGraw-Hill Book Co. xvii + 670 pp.
Ferguson, Mark W.J. (editor) 1984. *The Structure, Development, and Evolution of Reptiles*. London, Academic Press. xxii + 697 pp.
Gans, Carl (general editor). 1969 to present. *Biology of the Reptilia* (series). New York, Academic Press.

Hotton, Nicholas III, MacLean, Paul D., Roth, Jan J., and Roth, E. Carol (editors). 1986. *The Ecology and Biology of Mammal-like Reptiles*. Washington D.C., Smithsonian Institution Press. x + 326 pp.

Kemp, T.S. 1982. *Mammal-Like Reptiles and the Origin of Mammals*. London, Academic Press. xiv + 363 pp.

McGowan, Christopher. 1983. *The Successful Dragons*. Toronto, Samuel Stevens. 263 pp.

McLoughlin, John C. 1979. *Archosauria: A New Look at the Old Dinosaurs*. New York, Viking Press. viii + 117 pp.

———. 1980. *Synapsida: A New Look into the Origin of Mammals*. New York, Viking Press. xii + 148 pp.

Norman, David. 1985. *The Illustrated Encyclopedia of Dinosaurs*. New York, Crescent Books. 208 pp.

Padian, Kevin (editor). 1986. *The Beginning of the Age of Dinosaurs*. Cambridge, Cambridge University Press. xii + 378 pp.

Panchen, A.L. (editor) 1980. *The Terrestrial Environment and the Origin of Land Vertebrates*. London, Academic Press. xii + 633 pp.

Paul, Gregory S. 1988. *Predatory Dinosaurs of the World*. New York, Simon and Schuster. 464 pp.

Piveteau, Jean (editor). 1955. *Traité de Paléontologie*. Tome V. *Amphibiens, Reptiles, Oiseaux*. Paris, Masson et Cie. 1113 pp.

Porter, Kenneth R. 1972. *Herpetology*. Philadelphia, W.B. Saunders Co. xi + 524 pp.

Romer, Alfred S. 1947. *Review of the Labyrinthodontia*. Cambridge, Mass., Bull. Mus. Comp. Zool. Vol. 99, No. 1. 368 pp.

———. 1956. *Osteology of the Reptiles*. Chicago, University of Chicago Press. xxi + 772 pp.

Rozhdestvensky, A.K., and Tatarinov, L.P. (editors). 1964. *Fundamentals of Paleontology* (Orlov, Y.A. chief editor): *Amphibians, Reptiles, and Birds*. Moscow. 722 pp. (in Russian).

Russell, Dale A. 1977. *A Vanished World: The Dinosaurs of Western Canada*. National Museums of Canada. Natural History Series No. 4. 142 pp.

Schmalhausen, I.I. 1968. *The Origin of Terrestrial Vertebrates*. New York, Academic Press. xviii + 314 pp.

Sun, Ailing, 1989. *Before Dinosaurs*. Beijing, China Ocean Press. 109 pp.

Thomas, D.K. and Olson, Everett C. (editors). 1980. *A Cold Look at the Warm-Blooded Dinosaurs*. American Association for the Advancement of Science Symposium Series. xxx + 514 pp.

Wilford, John Noble. 1985. *The Riddle of the Dinosaur*. New York, Alfred A. Knopf. viii + 304 pp.

BIRDS

The book by Heilmann is a standard work. The books by the Darlings and by Peterson are excellent texts on modern birds.

de Beer, G. 1954. *Archaeopteryx lithographica*. London, British Museum (Natural History). xi + 64 pp.

Dalton, Stephen. 1977. *The Miracle of Flight*. New York, McGraw-Hill Book Co. 168 pp.

Darling, Lois and Louis Darling. 1962. *Bird*. Boston, Houghton Mifflin Company. xiii + 261 pp.

Feduccia, Alan. 1980. *The Age of Birds*. Cambridge, Harvard University Press. 196 pp.

Gilbert, B. Miles, Martin, Larry D., and Savage, Howard G. 1985. *Avian Osteology*. Laramie, Wyoming, Modern Printing Co. i + 252 pp.

Hecht, Max K., Ostrom, John H., Viohl, Gunter, and Wellnhofer, Peter (editors). 1985. *The Beginnings of Birds*. Willibaldsburg, Freunde des Jura-Museums Eichstatt. 382 pp.

Heilmann, G. 1927. *The Origin of the Birds*. New York, D. Appleton and Co. 210 pp.

Olson, Storrs L. (editor) 1976. *Collected Papers in Avian Paleontology Honoring the 90th Birthday of Alexander Wetmore*. Washington, D.C., Smithsonian Contributions to Paleobiology No. 27. xxvi + 211 pp.

Padian, Kevin (editor). 1986. *The Origins of Birds and the Evolution of Flight*. San Francisco, California Academy of Sciences. viii + 98 pp.

Peterson, R.T. (and the editors of Life). 1963. *The Birds*. Life Nature Library. New York, Time Incorporated. 192 pp.

MAMMALS

There are numerous good modern books in English on recent mammals. Vaughan's book is an invaluable modern work on mammalogy. Volume XVII (in two parts) of the monumental *Traité de Zoologie* is an authoritative, comprehensive text on fossil and on recent mammals. Volume VI and VII of the *Traité de Paléontologie* presents a comprehensive survey of fossil mammals. Simpson's classification of mammals is an outstanding monograph that carries the classification of all mammals, fossil and recent, down to genera. It contains clear and authoritative discussions of mammalian relationships by the leading modern student of fossil mammals. Osborn's *Age of Mammals* and Scott's *History of Land Mammals in the Western Hemisphere* are standard works, widely used. Kurtén's book is an up to date treatment of the subject. Gregory's *Orders of Mammals* is a basic work on mammalian classification. *Climate and Evolution*, by Matthew, is a classic work, devoted to the geographic history of the mammals. Young's book is a parallel to his volume on the vertebrates. Many books and papers are concerned with various mammalian orders, families, or lesser groups, but here only a few outstanding works on primate evolution—a subject of particular interest to most of us—will be listed.

Carrington, Richard (and the editors of Life). 1963. *The Mammals*. Life Nature Library. New York, Time Incorporated. 192 pp.

Clapman, Frances M. (editor). 1976. *Our Human Ancestors*. New York, Warwick Press. 160 pp.

Eimerl, Sarel, and Irven DeVore (and the editors of Life). 1965. *The Primates*. Life Nature Library. New York, Time Incorporated. 199 pp.

Grassé, Pierre-P. (and collaborators) 1955. *Traité de Zoologie*, Tome XVII, *Mammifères*. Paris, Masson et Cie. Pts. I and II, 2300 pp.

Gregory, William K. 1910. *The Orders of Mammals*. New York, Bull. Amer. Mus. Nat. Hist., Vol. 27. 524 pp.

Gromova, V.I. (editor). 1962. *Fundamentals of Paleontology* (Orlov, Y.A. chief editor): *Mammals*. Moscow. (in Russian).

Halstead, L.B. 1978. *The Evolution of the Mammals*. London, Eurobook Limited. 116 pp.

Howell, F. Clark (and the editors of Life). 1965. *Early Man*. Life Nature Library. New York, Time Incorporated. 200 pp.

Keast, Allen, Erk, Frank C., and Glass, Bentley. 1972. *Evolution, Mammals, and Southern Continents*. Albany, State University of New York Press. 543 pp.

Kurtén, Bjorn. 1968. *Pleistocene Mammals of Europe*. Chicago, Aldine Publishing Company. viii + 317 pp.

——. 1971. *The Age of Mammals*. New York, Columbia University Press. 250 pp.

——. 1972. *The Ice Age*. New York, G.P. Putnam's Sons. 178 pp.

——. 1980. *Pleistocene Mammals of North America*. New York, Columbia University Press. xvii + 442 pp.

Lillegraven, Jason A. Kielan-Jaworowska, Zofia, and Clemens, William A. 1979. *Mesozoic Mammals: The First Two-Thirds of Mammalian History*. Berkeley, University of California Press. x + 311 pp.

Matthew, William D. 1939. *Climate and Evolution*, second edition, revised. The New York Academy of Science, Special Publications. Vol. 1, 223 pp.

Merriam, Nick. 1989. *Early Humans*. New York, Alfred A. Knopf. 63 pp.

Osborn, Henry F. 1910. *The Age of Mammals in Europe, Asia, and North America*. New York, The Macmillan Co. 635 pp.

Piveteau, Jean (editor). 1957. *Traité de Paléontologie*. Tome VII. *Primates. Paleontologie humaine*. Paris, Masson et Cie. 675 pp.

——. 1958. Ibid. Tome VI, Vol. 2, *Mammifères, Evolution*. Paris, Masson et Cie. 962 pp.

——. 1961. Ibid. Tome VI, Vol. I, *Mammifères, Origine Reptilienne. Evolution*. Paris, Masson et Cie. 1135 pp.

Ranzi, Carlo 1982. *Seventy Million Years of Man*. New York, Greenwich House. 121 pp.

Savage, R.J.G. and Long, M.R. 1986. *Mammal Evolution: An Illustrated Guide*. New York, Facts on File Publications. 259 pp.

Scott, William B. 1937. *A History of Land Mammals in the Western Hemisphere*. New York, The Macmillan Co. 786 pp.

Simpson, George Gaylord. 1945. *The Principles of Classification and a Classification of Mammals*. New York, Bull. Amer. Mus. Nat. Hist., Vol. 85. 350 pp.

Sutcliffe, Antony J. 1986. *On the Track of Ice Age Mammals*. Dorchester, Dorset Press. 224 pp.

Vaughan, Terry A. 1986. *Mammalogy*, third edition. Philadelphia, W.B. Saunders Co. vii + 576 pp.

Wood, Bernard. 1976. *The Evolution of Early Man*. London, Eurobook Limited. 124 pp.

Woodburne, Michael O. 1987. *Cenozoic Mammals of North America*. Berkeley, University of California Press. xv + 336 pp.

Young, J.Z. 1957. *The Life of Mammals*. Oxford, Oxford University Press. 820 pp.

GEOLOGY

A few modern books on historical geology are listed here.

Cloud, Preston. 1988. *Oasis in Space: Earth History from the Beginning*. New York, W.W. Norton & Company. xviii + 508 pp.

Greenwood, P.H. (general editor). 1981. *The Evolving Earth*. London, Cambridge University Press. vii + 263 pp.

Kay, Marshall, and Colbert, Edwin H. 1965. *Stratigraphy and Life History*. New York, John Wiley and Sons, Inc. 736 pp.

Stanley, Steven M. 1986. *Earth and Life Through Time*. New York, W.H. Freeman and Co. xi + 690 pp.

PALEONTOLOGY AND PLATE TECTONICS

Colbert, Edwin H. 1973. *Wandering Lands and Animals: The Story of Continental Drift and Animal Populations* (With a Foreword by Laurence M. Gould). Dover Publications, Inc. xxv + 323 pp.

McKerrow, W.S. and Scotese, C.R. (editors). 1989. *Atlas of Palaeozoic Basemaps*. In: *Palaeozoic Palaeogeography and Biogeography*. Geological Society of London Special Paper.

Scotese, Christopher (coordinator). 1987. *Atlas of Mesozoic and Cenozoic Plate Tectonic Reconstructions*. Technical Report No. 90, Paleoceanographic Mapping Project, Institute of Geophysics, University of Texas.

Smith, A.G. and Hurley, A.M., and Briden, J.C. 1981 *Phanerozoic Paleocontinental World Maps*. Cambridge, Cambridge University Press. 102 pp.

Sullivan, Walter. 1974. *Continents in Motion: The New Earth Debate*. New York, McGraw-Hill Book Company. xiv + 399 pp.

Wegener, Alfred. 1966. *The Origin of Continents and Oceans*. Translated from the fourth edition by John Biram. New York, Dover Publications, Inc. ix + 246 pp.

Wilson, J. Tuzo and others. 1972. *Continents Adrift*. (Readings from the Scientific American.) San Francisco, W.H. Freeman and Company. 172 pp.

PAST CLIMATES

Brooks, C.E.P. 1970. *Climate Through the Ages*, second revised edition. New York, Dover Publications, Inc. 395 pp.

Nairn, A.E.M. (editor). 1961. *Descriptive Palaeoclimatology*. New York, Interscience Publishers. xi + 380 pp.

FOSSIL PLANTS

Andrews, Henry, N. 1947. *Ancient Plants and the World They Lived In*. Ithaca, Comstock Publishing Co. 279 pp.

Banks, H.P. 1970. *Evolution and Plants of the Past*. Belmont, Wadsworth Publishing Company. x + 170 pp.

Stewart, Wilson N. 1983. *Paleobotany and the Evolution of Plants*. Cambridge, Cambridge University Press. x + 405 pp.

Thomas, Barry. 1981. *The Evolution of Plants and Flowers*. London, Eurobook Limited. 116 pp.

White, Mary E. 1988. *The Greening of Gondwana: The 400 Million Year Story of Australia's Plants*. French Forest, N.S.W., Reed Books, Pty., Ltd. 256 pp.

EXTINCTIONS

The several episodes of major vertebrate extinctions are receiving much attention at the present time. A few books on this subject are listed.

Berggren, W.A. and Van Couvering, John A. (editors). 1984. *Catastrophies and Earth History: The New Uniformitarianism*. Princeton, Princeton University Press. ix + 464 pp.

Elliott, David K. (editor) 1986. *Dynamics of Extinction*. New York, John Wiley & Sons. x + 294 pp.

Goldsmith, Donald. 1985. *Nemesis: The Death Star and Other Theories of Mass Extinction*. New York, Walker and Company. ix + 166 pp.

Martin, Paul S. and Klein, Richard G. (editors). 1984. *Quaternary Extinctions: A Prehistoric Revolution*. Tucson, the University of Arizona Press. x + 892 pp.

PICTURE BOOKS

The series of large format books, with text by Augusta and by Spinar, and plates (usually in color) by Burian, deserve special mention. The texts are authoritative, and the many beautiful restorations by Burian present accurate and imaginative scenes of past life. Perhaps no other sources will give the reader the vivid glimpses into a vanished world that are available in these volumes.

Augusta, Josef, and Burian, Zdeněk. 1960. *Prehistoric Animals*. London, Spring Books. 47 pp. 60 plates with accompanying text.

——. 1960. *Prehistoric Man*. London, Paul Hamlyn. 45 pp. 52 plates with accompanying text.

——. 1961. *Prehistoric Reptiles and Birds*. London, Paul Hamlyn. 104 pp. 39 plates.

——. 1962. *A Book of Mammoths*. London, Paul Hamlyn. 50 pp. 19 plates.

——. 1964. *Prehistoric Sea Monsters*. London, Paul Hamlyn. 64 pp. plates.

——. 1966. *The Age of Monsters*. London, Paul Hamlyn. 79 pp. 23 plates.

Cox, Barry. 1970. *Prehistoric Animals*. New York, Grosset and Dunlap. 159 pp.

Peters, David. 1986. *Giants of Land, Sea and Air, Past & Present*. New York, Alfred A. Knopf. 73 pp.

——. 1989. *A Gallery of Dinosaurs & Other Early Reptiles*. New York, Alfred A. Knopf. 64 pp.

Spinar, Zdeněk V. 1972. *Life before Man*. Illustrated by Zdeněk Burian. London, Thames and Hudson. 228 pp.

Wolf, Josef. 1978. *The Dawn of Man*. Illustrated by Zdeněk Burian. New York, Harry N. Abrams, Inc. 231 pp.

BIBLIOGRAPHIES

As mentioned above, the selected bibliographies in Romer's and Carroll's vertebrate paleontology texts are good sources for outstanding works on fossil vertebrates. The Hay bibliographies list all papers on fossil vertebrates of North America up to 1928, and the Camp and Gregory bibliographies do this on a worldwide basis, from 1928 to the present time. The bibliography by Romer, Edinger, and Wright on fossil vertebrates other than the ones from North America, covers the period up to 1928.

Bacskai, Judith, Laurie J. Bryant, Joseph T. Gregory, George V. Shkurkin, and Melissa C. Winans. 1983. *Bibliography of Fossil Vertebrates, 1973–1977*, Volume I. The American Geological Institute. xii + 557 pp.

——— 1983. *Bibliography of Fossil Vertebrates, 1973–1977*, Volume II. The American Geological Institute. viii + 559–1101 pp.

Camp, Charles L. and V.L. Vanderhoof. 1940. *Bibliography of Fossil Vertebrates, 1928–1933*. Geological Society of America, Special Paper No. 27, 503 pp.

Camp, C.L., D.N. Taylor, and S.P. Welles. 1942. *Bibliography of Fossil Vertebrates. 1934–1938*. Geological Society of America, Special Paper No. 42, 663 pp.

Camp, C.L., S.P. Welles, and M. Green. 1949. *Bibliography of Fossil Vertebrates, 1939–1943*. Geological Society of America, Memoir 37, 371 pp.

———. 1953. *Bibliography of Fossil Vertebrates. 1944–1948*. Geological Society of America, Memoir 57, 465 pp.

Camp, C.L., and H.J. Allison. 1961. *Bibliography of Fossil Vertebrates, 1949–1953*. Geological Society of America, Memoir 84, xxxvii + 532 pp.

Camp, C.L., H.J. Allison, and R.H. Nichols, 1964. *Bibliography of Fossil Vertebrates. 1954–1958*. Geological Society of America, Memoir 92, xxv + 647 pp.

Camp, C.L., H.J. Allison, R.H. Nichols, and H. McGinnis. 1968. *Bibliography of Fossil Vertebrates, 1959–1963*. Geological Society of America, Memoir 117. xlii + 644 pp.

Camp, C.L., R.H. Nichols, B. Brajnikov, E. Fulton, and J.A. Bacskai. 1972. *Bibliography of Fossil Vertebrates, 1964–1968*. Geological Society of America, Memoir 134. lxi + 1117 pp.

Gregory, J.T., J.A. Bacskai, B. Brajnikov, and K. Munthe. 1973. *Bibliography of Fossil Vertebrates, 1969–1972*. The Geological Society of America, Inc., Memoir 141. xlvi + 733 pp.

Gregory, Joseph T., Judith A. Bacskai and George V. Shkurkin. 1981. *Bibliography of Fossil Vertebrates, 1978*. The American Geological Institute. xi + 380 pp.

Gregory, Joseph T., Judith A. Bacskai, George V. Shkurkin, and Laurie J. Bryant. 1981. *Bibliography of Fossil Vertebrates, 1979*. The American Geological Institute. xii + 442 pp.

———, 1983. *Bibliography of Fossil Vertebrates, 1980*. The American Geological Institute. viii + 471 pp.

Gregory, Joseph T., Judith A. Bacskai, George V. Shkurkin, and Melissa C. Winans. 1984. *Bibliography of Fossil Vertebrates. 1981*. The Society of Vertebrate Paleontology. x + 564 pp.

———. 1985. *Bibliography of Fossil Vertebrates, 1982*. The Society of Vertebrate Paleontology. x + 614 pp.

———. 1986. *Bibliography of Fossil Vertebrates, 1983*. The Society of Vertebrate Paleontology. viii + 628 pp.

Gregory, Joseph T., Judith A. Bacskai, George V. Shkurkin, Melissa C. Winans, and Laurie J. Bryant. 1987. *Bibliography of Fossil Vertebrates, 1984*. The Society of Vertebrate Paleontology. viii + 607 pp.

Gregory, Joseph T., Judith A. Bacskai, George V. Shkurkin, Melissa C. Winans, and Bonnie H. Rauscher. 1988. *Bibliography of Fossil Vertebrates, 1985*. The Society of Vertebrate Paleontology. viii + 568 pp.

———, 1989. *Bibliography of Fossil Vertebrates, 1986*. The Society of Vertebrate Paleontology. ix + 509 pp.

Hay, O.P. 1902. *Bibliography and Catalogue of the Fossil Vertebrata of North America*. Bull. U.S. Geol. Surv., No. 179, 868 pp.

———. 1929. *Second Bibliography and Catalogue of the Fossil Vertebrata of North America*. Carnegie Inst. of Washington, Publication No. 390, Vol. I, 916 pp.

———. 1930. *Second Bibliography and Catalogue of the Fossil Vertebrata of North America*. Carnegie Inst. of Washington, Publication No. 390, Vol. 2, 1074 pp.

Romer, Alfred S., N.E. Wright, T. Edinger, and R. van Frank. 1962. *Bibliography of Fossil Vertebrates Exclusive of North America, 1509–1927*. Geological Society of America, Memoir 87, Vol. 1, A–K, pp. 1–1xxxix, 1–772., Vol. 2, L–Z, pp. 773–1544.

Key to Abbreviations of Bone Names Used in Illustrations

BONES OF THE SKULL AND THE LOWER JAW

a	articular		op	opercular (fish)
an	angular		op	opisthotic (tetrapod)
as	alisphenoid		opo	
ba	branchial arch		p	parietal
bo	basioccipital		pap	paroccipital process
bpt	basipterygoid		pd	predentary
bs	basisphenoid		pf	postfrontal
			pof	
c	coronoid		pl	palatine
			pm	premaxilla
d	dentary		pn	postnarial
			po	postorbital
ec	ectopterygoid		pop	preopercular
eo	exoccipital		pos	postsplenial
ep	epipterygoid		pp	postparietal
esl	lateral extrascapular		pq	palatoquadrate
esm	medial extrascapular		pr	postrostral
exn	external nares		pra	prearticular
			prf	prefrontal
f	frontal		pro	prootic
			ps	parasphenoid
g	gular		pt	pterygoid
hm	hyomandibular		q	quadrate
			qj	quadratojugal
in	internarial			
inn	internal nares		sa	surangular
it	intertemporal		san	
			sm	septomaxilla
j★	jugal		sop	subopercular
ju			sor	supraorbital
			sp	splenial
l	lacrymal		sq	squamosal
			st	supratemporal
m	maxilla		sta	stapes
md	mandible			
			t	tabular
n	nasal			
na	nares		v	vomer
			vac	vacuity

★In spite of every effort to be consistent, a few of these abbreviations got out of hand. The reader's kind indulgence is requested.

448

BONES OF THE POSTCRANIAL SKELETON

ba   branchial arch

ce   centrale

ef   entepicondylar foramen

f   fibula
fi   fibulare

ic   intercentrum
il   ilium
in   intermedium
is   ischium

na   neural arch

p   pubis
pc   pleurocentrum

r   radius
ra   radiale

sp   spine

t   tibia
ti   tibiale

u   ulna
ul   ulnare

Sources and Credits
For Illustrations

All of the illustrations, with the exception of Figures 4–8, 8–7, 12–4, and 12–7, are new, and were drawn specifically for this book.

The chapter headings, the restorations, and the phylogenetic diagrams were made by Mrs. Lois Darling under the supervision of the authors. The figures in this category, in addition to the chapter headings, are as follows: Figures 2–1, 2–7, 3–6, 3–7, 4–3, 4–5, 4–6, 4–9, 5–4, 7–1, 7–9, 7–15, 8–6, 9–3, 9–9, 11–2, 12–3, 12–5, 12–11, 12–12, 12–13, 14–7, 15–2, 15–6, 15–7, 17–1, 17–5, 18–8, 19–3, 19–5, 20–4, 21–3, 21–5, 22–1, 22–3, 22–8, 23–3, 23–4, 24–3, 25–3, 25–5, 26–4, 26–5, 26–6, 27–1, 27–2, 28–5, 28–10, 28–11, 28–12, 29–5, 29–7, 29–8, 29–9, 29–10, 29–12, 29–13, 29–14, 30–1, 30–5, 30–7.

Figures 1–1, 1–6, 4–10, 10–1, 10–2, 11–6, 16–1, 16–2, 16–3, and 16–4 were drawn from original layouts made by the senior author.

The remaining illustrations, with the exception of Figure 4–8, were done under the supervision of the authors, Many of these figures were prepared by the drafting department of John Wiley and Sons, Inc. Figures 9–6, 9–10, 13–1, 21–1, and 23–2 were drawn by Miss Pamela Lungé. Figures 2–2, 2–6, 3–5, 4–1, 4–2, 4–4, 5–3, 5–6, 7–3, 7–4, 7–10, 7–11, 7–13, 7–16, 7–17, 8–4, 9–1, 9–4, 10–1, 11–1, 11–3, 11–4, 12–10, 13–2, 13–4, 14–3, 14–4, 14–5, 14–6, 15–1, 15–4, 15–5, 16–1, 16–2, 16–3, 17–3, 18–1, 18–2, 18–4, 18–5, 18–9, 19–2, 19–4, 20–1, 20–2, 21–2, 21–6, 22–2, 22–4, 22–6, 22–7, 24–1, 24–2, 25–2, 25–6, 26–1, 26–2, 28–3, 28–7, 29–2, 29–6, 29–11, 30–3, and 31–1 were drawn by Miss Louise Waller.

Many of them were of necessity based in part upon information from published works. The sources for these figures are listed below. In this listing, the names of authors, followed by dates, indicate source materials taken from scientific papers or monographs. Sources from copyrighted publications are indicated in each case in the usual fashion.

Fig. 1–2. Adapted from *Animals Without Backbones* by R. Buchsbaum, copyright 1945, by permission of The University of Chicago Press.

Fig. 1–3. From *Wonderful Life* by Stephen Jay Gould, 1989, W.W. Norton & Company.

Fig. 1–4. (A) From Vertebrate Life by William N. MacFarland et al., 1979, MacMillan Publishing Co., Inc. (B) and (C) From *Vertebrate Hard Tissues* by L.B. Halstead, 1974, Wykeham Publications, Ltd.

Fig. 1–5. (A), (B), (C), (D), and (E) From *The Vertebrate Body*, fourth edition, by A.S. Romer, 1970, W.B. Saunders Company. (F) and (G) from *Vertebrate Hard Tissues* by L.B. Halstead, 1974, Wykeham Publications, Ltd.

Fig. 2–2. Illustration by Louise Waller, based on information from Dr. David K. Elliott.

Fig. 2–3. After Alexander Ritchie, 1968.

Fig. 2–4. Adapted from *Man and the Vertebrates*, by A.S. Romer, copyright 1941, by permission of The University of Chicago Press.

Fig. 2–5. After E.A. Stensiö, 1932.

Fig. 2–6. (A) After *Paleozoic Fishes*, second edition, by J.A. Moy-Thomas and R.S. Miles, 1971, W.B. Saunders Company. (B) After *Chordate Structure and Function* by Allyn J. Waterman et al., 1971, The MacMillan Company.

Fig. 3–1. (A) After D.M.S. Watson, 1937. (B) After T.S. Westoll.

Fig. 3–2. After D.M.S. Watson, 1937.

Fig. 3–3. From models in the American Museum of Natural History, New York.

Fig. 3–4. After A. Heintz, 1931.

Fig. 3–5. (A) and (B) After Stensio, 1969.

Fig. 4–1. Illustration by Louise Waller, adapted from *Vertebrate History: Problems in Evolution* by Barbara J. Stahl, 1974, MacGraw-Hill Book Company.

Fig. 4–2. Illustrations by Louise Waller, adapted from *Evolution of Chordate Structure* by Hobart M. Smith, 1960, Holt, Rinehart, and Winston, Inc.

Fig. 4–4. After *Paleozoic Fishes*, second edition, by J.A. Moy-Thomas and R.S. Miles, W.B. Saunders Company.

Fig. 4–7. (A) After H. Aldinger, 1937. (B) After D. Rayner, 1948.

Fig. 4–8. From an exhibit in The American Museum of Natural History; courtesy of Bobb Schaeffer.

Fig. 5–1. (A) After D.M.S. Watson, 1921. (B) From a model by E. Jarvik.

Fig. 5–2. (A) From a model in the American Museum of Natural History. (B) From various sources. (C) After *The Vertebrate Story* by A.S. Romer, 1959, University of Chicago Press.

Fig. 5–3. After W.K. Gregory and H. Raven, 1941.

Fig. 5–5. (A) and (C) Adapted from *Vertebrate Paleontology*, by A.S. Romer, copyright 1945, by permission of The University of Chicago Press. (B) and (D) After E. Jarvik, 1952.

Fig. 5–6. Illustration by Louise Waller, adapted from *Evolution Emerging*, by W.K. Gregory, copyright 1951. By permission of the American Museum of Natural History.

Fig. 7–2. After E. Jarvik, 1952, 1955.

Fig. 7–3. Illustration by Louise Waller, adapted from various sources.

Fig. 7–4. After A.L. Pachen, 1972. The skull and skeleton of *Eogyrinus attheyi*. Watson (Amphibia: Labyrinthodontia). *Phil. Trans. Roy. Soc. Lond.* B, 263:279–326.

Fig. 7–5. After T.E. White, 1939.

Fig. 7–6. Adapted from *Evolution Emerging*, by W.K. Gregory, copyright 1951, by permission of the American Museum of Natural History.

Fig. 7–7. From a skull in the American Museum of Natural History, New York.

Fig. 7–8. From a skeleton in the American Museum of Natural History, New York.

Fig. 7–10. After H.J. Sawin, 1945, Amphibians from the Dockum Triassic of Howard County, Texas. *Univ. Texas Publ.* No. 4401:361–399.

Fig. 7–11. Illustration by Louise Waller, based on T. Nilsson, 1946, A new find of *Gerrothorax rhaeticus* Nilsson, a plagiosaurid from the Rhaetic of Scania. *Acta Univ. Lund.* 42:1–42.

Fig. 7–12. (A) After E. Jarvik, 1952. (B), (C), and (D) After A.S. Romer, 1947.

Fig. 7–13. (A) After R.L. Carroll and P. Gaskin, 1978. The Order Microsauria. *Mem. Am. Phil. Soc.* 126:1–211.

Fig. 7–14. (A) After S.W. Williston, 1909. (B) After E.C. Olson, 1951. (C) Adapted from *Evolution Emerging*, by W.K. Gregory, copyright 1951, by permission of the American Museum of Natural History.

Fig. 7–16. Adapted from R. Estes and O.A. Reig, 1973, The early fossil record of frogs: A review of the evidence. In: *Evolutionary Biology of the Anurans*, J. Vial, ed., pp. 11–63. University of Missouri Press, Columbia.

Fig. 7–17. (A) From I.I. Schmalhausen, 1968. (B) From McFarland et al., 1985.

Fig. 8–1. (A) Adapted from *Man and the Vertebrates*, by A.S. Romer, copyright 1941, by permission of the University of Chicago Press. (B) After A.S. Romer and L.I. Price, 1939.

Fig. 8–2. After R.L. Carroll, 1964.

Fig. 8–3. Adapted from *Vertebrate Paleontology*, by A.S. Romer, copyrighted 1945, by permission of The University of Chicago Press.

Fig. 8–4. Adapted from *Evolution Emerging*, by W.K. Gregory, copyright 1951, by permission of the American Museum of Natural History.

Fig. 8–5. (A) Adapted from *Vertebrate Paleontology*, by A.S. Romer, copyright 1945, by permission of The University of Chicago Press. (B) After A.S. Romer and L.I. Price, 1940. (C) After C.W. Andrews, 1910–1913. (D) After R.F. Ewer, 1965.

Fig. 8–7. After E.H. Colbert, 1941.

Fig. 9–1. (A) After R. Carroll, 1972. (B) After R. Reisz, 1972. Pelycosaurian reptiles from the Middle Pennsylvanian of North America. *Bull. Mus. Comp. Zool. Harv.* 144:27–62. Reprinted with permission of the author.

Fig. 9–2. (A,B,C,E,F) After A.S. Romer and L.I. Price, 1940. (D) After E.C. Case, 1907.

Fig. 9–4. (A) After D. Sigogneau and P.K. Chudinov, 1972. Reflections on some Russian eotheriodonts (Reptilia, Synapsida, Therapsida). *Paleovertebrata* 5:79–109. (B) Adapted from *Traite de Paleontologie*, Tome IV, by J. Piveteau (ed.), copyright 1961, by permission of the Masson S.A., Paris. (C) After P.K. Chudinov, 1965. New facts about the fauna of the Upper Permian of the U.S.S.R. *J. Geol.* 73:117–130. By permission of the University of Chicago Press.

Fig. 9–5. After R. Broom, 1912.

Fig. 9–6. Adapted from Robert Broom and M.A. Cluver, 1941.

Fig. 9–7. From E.H. Colbert, 1948.

Fig. 9–8. (A,C) Adapted from *Vertebrate Paleontology*, by A.S. Romer, copyright 1945, by permission of the University of Chicago Press. (B,E,D) After W.K. Gregory and C.L. Camp, 1918.

Fig. 9–10. After Farish A. Jenkins, Jr. 1971.

Fig. 9–11. From a cast in The American Museum of Natural History.

Fig. 11–1. (A) Adapted from R. Ewer, 1965. (B) Adapted from B. Krebs, in Schweiz. Paleont. Abhandlung, Vol. 81, 1965, by permission, Birkhauser Verlag AG, Basel, Switzerland.

Fig. 11–3. After C.L. Camp, 1930.

Fig. 11–4. Adapted from R. Wild, 1973.

Fig. 11–5. After C. Gow, 1975.

Fig. 11–6. After F. von Huene, 1956.

Fig. 12–1. Adapted from *The Dinosaur Book*, by E.H. Colbert, 1945.

Fig. 12–2. (A,B,E) After C.W. Gilmore, 1914, 1920, 1925). (C) After T. Maryanska and H. Osmolska, 1974. Copyright, 1974, by permission of Zaklad Paleobiologü. (D) After R.S. Lull and N. Wright, 1942. (E) After R.S. Lull, 1933.

Fig. 12–4. From E.H. Colbert, 1989.

Fig. 12–6. (A) After C.W. Gilmore, 1936. (B) After J.H. Ostrom, 1969.

Fig. 12–7. From R.A. Thulborn, 1972.

Fig. 12–8. After A.W. Crompton and A. Charig, 1962.

Fig. 12–9. After B. Brown, 1914.
Fig. 12–10. After J.H. Ostrom, 1961.

Fig. 13–1. (A) After E.H. Colbert, 1970. (B) and (C) From Carroll, 1988.
Fig. 13–2. Illustration by Louise Waller, based on information from Dr. Carl Gans.
Fig. 13–3. (A) After H.G. Seeley, 1901. (B) After G.F. Eaton, 1910.
Fig. 13–4. Illustration by Louise Waller, based on various sources.

Fig. 14–1. After G. Steinmann and L. Doderlein, 1890.
Fig. 14–2. From the *Origins of Birds*, by Gerhard Heilmann. Copyright 1927, D. Appleton and Company. Redrawn and adapted by permission of the publishers, Appleton-Century-Crofts, Inc.
Fig. 14–3. (A) After F.A. Lucas, 1901, *Animals of the Past*. McClure Phillips, New York. (B) After O.C. Marsh, 1980, *Odontornithes: A Monograph on the Extinct Toothed Birds of North America*. U.S. Government Printing House, Washington, D.C., 201 pp.
Fig. 14–4. (A) After W.D. Matthew and W. Granger, 1917, The skeleton of Diatryma, a gigantic bird from the Lower Eocene of Wyoming. *Bull. Am. Mus. Nat. Hist.* 37:307–326. (B) After C.W. Andrews, 1901, On the extinct birds of Patagonia. I. The skull and skeleton of *Phororhacos inflatus* Ameghino. *Trans. Zool. Soc. Lond.* 15:55–86.
Fig. 14–5. After Andrews, 1901.
Fig. 14–6. Illustration by Louise Waller, adapted from a 1980 drawing by the National Geographic Society, Washington, D.C.

Fig. 15–1. (A) After P.J. Currie, 1981. *Hovasaurus boulei*, an aquatic eosuchian from the Upper Permian of Madagascar. *Palaeont. Afr.* 24:99–168. (B) After E. Kuhn-Schnyder, 1974. Die Transfauna der Tessiner Kalkalpen. *Neujahrsblatt Naturf. Ges. Zuerich*. 176:1–119.
Fig. 15–3. (A) From B. Peyer, 1950, *Geschichte der Tierwelt*. Buchergilde Gutenbert, Zurich. (B) After F. Broili, 1912.
Fig. 15–4. From E. Kuhn-Schnyder, 1963, I Sauri del Monte San Giorgio. *Comunicazioni dell'Institutio di Paleontologia dell'Universita di Zurgio* 20:811–854.
Fig. 15–5. After E. Fraas, 1902, Die Meer-Crocodilier (Thallatosuchia) des oberen Jura unter specieller Beruecksichtigung von *Dracosaurus* und *Geosaurus*. *Palaeontographica* 49:1–72.

Fig. 17–2. (A) After E.H. Colbert and C.C. Mook, 1951. (B) After E. Fraas, 1902. (C,E) Adapted from *Osteology of the Reptiles*, by S.W. Williston, copyright 1925, by permission of Harvard University Press. (D) After C.W. Andrews, 1910–1913. (F) After L.I. Price, 1945.
Fig. 17–3. After A.S. Romer, 1966.
Fig. 17–4. (A) Adapted from *Osteology of the Reptiles*, by S.W. Williston, copyright 1925, by permission of Harvard University Press. (B) After H. Gadow, 1909.

Fig. 18–1. After A.S. Romer, 1970.
Fig. 18–2. Adapted from: (A,B) E.H. Colbert, 1948; (C,D) A.W. Crompton, 1958. (E,F) W.K. Gregory in *Evolution Emerging*, copyright 1951, by permission of the American Museum of Natural History, and other sources.
Fig. 18–3. From various sources, including adaptation of a figure by permission from *American Mammals*, by W.J. Hamilton, Jr., copyright 1939, McGraw-Hill Book Company, Inc.
Fig. 18–4. From K.A. Kermack, Frances Mussett, and H.W. Rigney, 1973.
Fig. 18–5. Adapted from F.A. Jenkins, Jr., and R. Parrington, 1976.
Fig. 18–6. (A) After H.G. Seeley, 1895. (B–K) After G.G. Simpson, 1961.
Fig. 18–7. After W. Granger and G.G. Simpson, 1929.
Fig. 18–9. Original drawings from various sources.

Fig. 19–1. (A) After W.J. Sinclair, 1906. (B,C,D) Adapted from *Evolution Emerging*, by W.K. Gregory, copyright 1951, by permission of the American Museum of Natural History, (E,G) Adapted from *Organic Evolution*, by R.S. Lull, copyright, 1940, by permission of The Macmillan Company, New York. (F) Adapted from *Textbook of Paleontology*, by K.A. von Zittel, copyright 1925, by permission of Macmillan and Company, Ltd., London.
Fig. 19–2. Adapted from *Evolution Emerging*, by W.K. Gregory, copyright 1951, by permission of the American Museum of Natural History.
Fig. 19–4. (A) After M. Woodburne in *Science*, Vol. 218, 1982, p. 285. Copyright, 1982, by the AAAS, by permission of *Science* and the author. (B) After G.G. Simpson, 1948.

Fig. 20–1. (A) (B) Adapted from *Paleoltologie obratlovcu*, by Z.V. Spinar, copyright 1984, by permission.
Fig. 20–2. (A,B) After G.G.Simpson, 1961. (C-E) Adapted from *Evolution Emerging*, by W.K. Gregory, 1951, by permission of the American Museum of Natural History.
Fig. 20–3. After E.H. Colbert, 1939.

Fig. 21–1. Adapted from Z. Kielan-Jaworowska, 1978.
Fig. 21–2. Adapted from *Mammal Evolution*, by R.J.G. Savage and M.R. Long, copyright 1986, by courtesy of the Natural History Museum.
Fig. 21–4. (A,B) After R. Lydekker, 1894. (C,D) After C. Stock, 1925.
Fig. 21–6. (A) After C.L. Gazin, 1953. (B) After J. Wortman, 1897.

Fig. 22–2. (A,B) Adapted from *Evolution Emerging*, by W.K. Gregory, 1951. By permission of The American Museum of Natural History. (C) After L. Russell.

Fig. 22–4. After de Blainville.

Fig. 22–5. (A,B) After W.K. Gregory, 1921. (C) After A. Gaudry, 1862. (D,E) Adapted from *Evolution Emerging*, by W.K. Gregory, copyright 1951. By permission of The American Museum of Natural History.

Fig. 22–6. Adapted from *The Antecedents of Man*, by W.E. LeGros Clark, copyright 1984. By permission of Edinburgh University Press.

Fig. 22–7. All adapted from *Evolution Emerging*, by W.K. Gregory, copyright 1951. By permission of The American Museum of Natural History.

Fig. 23–1. (A) After W.D. Matthew, 1910. (C) After W.B. Scott, 1905. (D) After O.A. Peterson, 1905. (A–D) Adapted from *Vertebrate Paleontology*, by A.S. Romer, copyright 1945, by permission of the University of Chicago Press.

Fig. 23–2. After A.E. Wood, 1962.

Fig. 24–1. Adapted from P. Gingerich in *Science*, Vol. 220, p. 404. Copyright 1983 by the AAAS. By permission of *Science* and the author.

Fig. 24–2. (A) Adapted from H.F. Osborn, 1909. (B,C) Adapted from R. Kellogg, 1928. (E) Adapted from *Mammalogy*, Third Edition, by Terry A. Vaughan, copyright 1986 by Saunders College Publishing, a division of Holt, Rinehart and Winston, Inc. By permission of the publisher.

Fig. 25–1. Adapted from J. Wortman, 1901–1902, and W.D. Matthew, 1909.

Fig. 25–2. Adapted from W.D. Matthew, 1909.

Fig. 25–4. Adapted from (A) W.D. Matthew, 1935; (B) W.B. Scott and G. Jepsen, 1936; (C) W.D. Matthew, 1930; (D) J.C. Merrian and C. Stock, 1925; (E) S.H. Reynolds, 1911; (F) O. Zdansky, 1911; (G) W.D. Matthew, 1910.

Fig. 25–6. (C) and (D) Adapted from *Mammalogy*, Third Edition, by Terry A. Vaughan, copyright 1986 by Saunders College Publishing, a division of Holt, Rinehart and Winston, Inc. By permission of the publisher.

Fig. 26–1. Adapted from an original photograph by R.E. Sloan. By permission of R.E. Sloan.

Fig. 26–2. (A) Adapted from W.D. Matthew, 1937. (B) Adapted from *Evolution Emerging*, by W.K. Gregory, copyright 1951. By permission of the American Museum of Natural History.

Fig. 26–3. After H.F. Osborn, 1898.

Fig. 27–3. Adapted from *A History of Land Mammals in the Western Hemisphere*, by W.B. Scott, 1937.

Fig. 27–4. (A) After H. Burmeister, 1866. (B) After W.J. Sinclair, 1909. (C,E) After W.B. Scott, 1912, 1928. (D) After E.S. Riggs, 1935. (F) After F.B. Loomis, 1914.

Fig. 28–1. After L. Radinsky, 1966.

Fig. 28–2. (A–D) Adapted from *Evolution Emerging*, by W.K. Gregory, 1951. Copyright 1951, by permission of the American Museum of Natural History. (E) After W.J. Holland and O.A. Peterson, 1913.

Fig. 28–3. Adapted from *Evolution Emerging*, by W.K. Gregory, 1951. Copyright 1951, by permission of the American Museum of Natural History.

Fig. 28–4. After W.B. Scott, 1941. (B) After Cope.

Fig. 28–6. (A) Adapted from *Evolution Emerging*, by W.K. Gregory, 1951. Copyright 1951, by permission of The American Museum of Natural History. (C,E) After H.F. Osborn, 1918. (B,D,F) After W.D. Matthew, 1927.

Fig. 28–7. (A) Adapted from *Evolution Emerging*, by W.K. Gregory, 1951. Copyright 1951, by permission of The American Museum of Natural History.

Fig. 28–8. (A) After J. Wortman, 1896. (B) After Flower and Lydekker, 1891. (C) After E.M. Schlaikjer, 1937. (D) After W.D. Matthew, 1927. (F,G) After H.F. Osborn, 1929, 1898.

Fig. 28–9. (A,B,D) Adapted from *Evolution Emerging*, by W.K. Gregory, 1951. Copyright 1951, by permission of the American Museum of Natural History. (E,F) After H.F. Osborn, 1929, 1898. (C) After E.M. Schlaikjer, 1937.

Fig. 29–1. (A) After W.B. Scott 1940. (B) Adapted from *Evolution Emerging*, by W.K. Gregory, copyright 1951, by permission of the American Museum of Natural History. (C) Adapted from *Vertebrate Paleontology*, by A.S. Romer, copyright 1945; by permission of the University of Chicago Press. (D) After O.A. Peterson, 1904. (E) Adapted from *A History of Land Mammals in the Western Hemisphere*, by W.B. Scott, 1837. (F) After W.D. Matthew, 1904.

Fig. 29–2. Adapted from W.J. Sinclair, 1914.

Fig. 29–3. (A,F,G.H,J) After F.B. Loomis, 1925. (B) Adapted from *Textbook of Paleontology*, by K.A. von Zittel, copyright 1925, by permission of Macmillan and Company, Ltd., London. (C) After W.B. Scott, 1940. (D,I) After E.H. Colbert, 1935, 1941. (E) After O.A. Peterson, 1909. (K) After G. Pilgrim, 1937.

Fig. 29–4. (A) After W.B. Scott, 1940. (B) After S.H. Reynolds, 1922. (C) After H.S. Pearson, 1923. (D) After W.J. Sinclair, 1914. (E) After J. Leidy, 1869. (F) After O.A. Peterson, 1904. (G) After W.D. Matthew, 1908. (H) After R.A. Stirton, 1931. (I) After A. Gaudry, 1867.

Fig. 29–6. Adapted from K. Rose in *Science*, vol 216, p. 621. Copyright 1982, by A.A.A.S. By permission of *Science* and the author.

Fig. 29–11. Adapted from *The Evolution of the Mammals*, by B. Halstead, 1978.
Fig. 29–15. Adapted from *Evolution Emerging*, by W.K. Gregory, copyright 1951, by permission of The American Museum of Natural History.

Fig. 30–2. Adapted from H.F. Osborn, 1936, 1942.
Fig. 30–3. (A) After an original drawing by M. Colbert. (B,C) After H.F. Osborn, 1932.
Fig. 30–4. After V. Maglio, 1973. Others after H.F. Osborn, 1936, 1942.
Fig. 30–6. (A) Adapted from *Evolution Emerging*, by W.K. Gregory, copyright 1951, by permission of The American Museum of Natural History. (B) After C. Repenning, 1965.

Index